WILDLIFE HABITAT CONSERVATION

Wildlife Management and Conservation

Paul R. Krausman, Series Editor

WILDLIFE
Habitat Conservation

Concepts, Challenges, and Solutions

EDITED BY
MICHAEL L. MORRISON
& HEATHER A. MATHEWSON

Published in Association with *THE WILDLIFE SOCIETY*

JOHNS HOPKINS UNIVERSITY PRESS | BALTIMORE

Johns Hopkins University Press
2715 North Charles Street
Baltimore, Maryland 21218-4363
www.press.jhu.edu

Library of Congress Cataloging-in-Publication Data

Wildlife habitat conservation : concepts, challenges, and
solutions / edited by Michael L. Morrison and Heather A.
Mathewson.
    pages cm. — (Wildlife management and conservation)
    Includes bibliographical references and index.
    ISBN 978-1-4214-1610-6 (hardcover : alk. paper) — ISBN
978-1-4214-1611-3 (electronic) — ISBN 1-4214-1610-7
(hardcover : alk. paper) — ISBN 1-4214-1611-5 (electronic)
1. Habitat conservation. I. Morrison, Michael L.
II. Mathewson, Heather A. (Heather Alexis), 1974–
    QH75.W529 2015
    333.95'4—dc23   2014020806

A catalog record for this book is available from the British
Library.

*Special discounts are available for bulk purchases of this book.
For more information, please contact Special Sales at 410-516-
6936 or specialsales@press.jhu.edu.*

Johns Hopkins University Press uses environmentally
friendly book materials, including recycled text paper that
is composed of at least 30 percent post-consumer waste,
whenever possible.

Contents

Contributors

William M. Block
U.S. Department of Agriculture Forest Service
Rocky Mountain Research Station

Kathi L. Borgmann
Arizona Cooperative Fish and Wildlife Research Unit
School of Natural Resources and the Environment
University of Arizona

J. Curtis Burkhalter
Ecology, Evolution, and Natural Resources
Rutgers University

Bret A. Collier
Texas A&M Institute of Renewable Natural Resources
Texas A&M University

Courtney J. Conway
U.S. Geological Survey
Idaho Cooperative Fish and Wildlife Research Unit
Department of Fish and Wildlife Sciences
University of Idaho

Clinton W. Epps
Department of Fisheries and Wildlife
Oregon State University

Clinton D. Francis
Department of Biological Sciences
California Polytechnic State University

Fred S. Guthery
Department of Natural Resource Ecology and
    Management
Oklahoma State University

Douglas H. Johnson
U.S. Geological Survey
Northern Prairie Wildlife Research Center

Julie L. Lockwood
Ecology, Evolution, and Natural Resources
Rutgers University

Heather A. Mathewson
Department of Wildlife, Sustainability, and Ecosystem
    Sciences
Tarleton State University

Kevin S. McKelvey
U.S. Department of Agriculture Forest Service
Rocky Mountain Research Station

Michael L. Morrison
Department of Wildlife and Fisheries Sciences
Texas A&M University

Amanda D. Rodewald
Cornell Lab of Ornithology and Department of
    Natural Resources
Cornell University

Jamie S. Sanderlin
U.S. Department of Agriculture Forest Service
Rocky Mountain Research Station

Michael K. Schwartz
U.S. Department of Agriculture Forest Service
Rocky Mountain Research Station

K. Shawn Smallwood
Davis, California

Bronson K. Strickland
Department of Wildlife, Fisheries and Aquaculture
Mississippi State University

Beatrice Van Horne
U.S. Department of Agriculture Forest Service
Pacific Northwest Research Station

Lisette P. Waits
Department of Fish and Wildlife Sciences
University of Idaho

John A. Wiens
School of Plant Biology
University of Western Australia
Crawley, Western Australia, Australia
Point Blue Conservation Science

Preface

There are a plethora of books that describe the many impacts that humans have on the environment, and there are many books that provide very technical (e.g., modeling) ways to analyze complex environmental impacts. Missing are books that blend an understanding of the impacts on wildlife and wildlife habitat with ways to address and hopefully start to ameliorate negative impacts. We chose the topic for each chapter in this volume to address a major issue confronting wildlife habitat. Hence the book presents in-depth coverage of individual topics while also presenting a broad coverage of topics overall.

The purpose of *Wildlife Habitat Conservation* is to deliver to a broad audience an understanding of the influences on wildlife and wildlife habitats, including an evaluation of the state-of-the-art and recommendations for a path forward that will advance management and conservation. Each chapter provides information in a form accessible to a broad audience, including but not limited to advanced undergraduate and graduate students in natural resource management and conservation (e.g., wildlife, range, conservation biology, recreation and parks); resource managers at local, state, and federal levels (e.g., state wildlife and parks departments, USFWS, BLM, Army Corps); and private land managers.

The book is organized into three parts. Part I lays the foundation for all that follows, starting with a discussion of the current and traditional use of the term *habitat* and ranging to how recent insights in demographics, spatial and temporal heterogeneity, and behavioral processes have increased the complexity of the concept of habitat. It considers how the goals of resource managers oftentimes rely on the simplified, traditional concept of habitat, resulting in tension with the habitat concept when applied to complex processes.

Part II delves into specific factors impacting wildlife habitat through more than just the removal of vegetative cover. The traditional view of habitat as a vegetation type is complicated by loss or degradation via sources other than just changes in land cover. Anthropogenic effects such as lights and sounds can degrade the environment without removing vegetation cover per se, and genetic and demographic processes are altered or accelerated with changes to land cover. Furthermore, the invasion of nonnative plants and animals alters habitat structure and processes within the system. Each chapter discusses how improving our understanding of these impacts will contribute to more effectively managing wildlife habitat.

Part III emphasizes solutions, including how to predict future changes, monitoring and planning, and restoration. These chapters investigate how wildlife managers and researchers can become more proactive in how they think about planning, monitoring, managing, and restoration in wildlife science. The final chapter synthesizes the major messages from the book.

Acknowledgments

We thank Paul Krausman, University of Montana and Editor of the Wildlife Society book series, for inviting us to assemble the contributors for this volume. Vincent Burke, Executive Editor, Johns Hopkins University Press, substantially improved the outline and direction of this project. We also thank Catherine Goldstead, Editorial Assistant, JHUP, for helping to guide the project to completion. Ross Anderson at the Texas A&M Institute for Renewable Natural Resources assisted with final manuscript preparation.

An initial draft of each chapter was reviewed by two or more independent referees and then returned to the chapter lead author for revision. We then evaluated the response of the authors to review comments and asked for additional revision. We list each reviewer here and sincerely thank them for their time and efforts in improving this book.

Chapter Reviewers

Jonathan Ballou
Smithsonian

Mark Boyce
University of Alberta

Anna Chalfoun
University of Wyoming

Melanie Colon
Texas A&M University

Bridgette Hagerty
York College of Pennsylvania

Dick Hutto
University of Montana

Bill Jensen
Emporia State University

Matt Johnson
Humboldt State University

Dylan Kesler
University of Missouri–Columbia

David King
University of Massachusetts

Ashley Long
Texas A&M University

R. William Mannan
University of Arizona

Bruce Marcot
U.S. Forest Service
Portland, OR

Scott Mills
University of Montana

Abby Powell
U.S. Geological Survey
University of Alaska–Fairbanks

C.C. St. Clair
University of Alberta

Brett Sandercock
Kansas State University

Ted Shear
North Carolina State University

Nova Silvy
Texas A&M University

Dan Simberloff
University of Tennessee

Katy Smith
Texas A&M University

John Swaddle
College of William and Mary

Peter Weisberg
University of Nevada–Reno

Patrick Zollner
Purdue University

PART I · FOUNDATION

1

The Misunderstanding of Habitat

Heather A. Mathewson
and Michael L. Morrison

Over the past century, humans have integrated *habitat* into our daily usage, creating a broadly defined term that refers to the physical location and associated conditions that an animal experiences. Oftentimes the term is qualified in various ways to express some level of suitability (e.g., good habitat, bad habitat, old habitat). It is even used in reference to people, such as the respectable Habitat for Humanity effort that builds homes for people, or to express the state of living conditions. Habitat is certainly one of the most frequently used terms in ecology. As the field of ecology has matured, so too has the application of the term, resulting in an apparent discord within the overall discipline of ecology. As research advances, through improved statistical, technological, and theoretical approaches, we have improved our understanding of the processes and mechanisms associated with an organism's habitat, yet the unfortunate result is a large collection of literature full of muddled terminology.

As thoroughly reviewed elsewhere, the term *habitat* is widely misunderstood and misused throughout ecological and management applications (see reviews by Hall et al. 1997; Morrison and Hall 2002). The failure of the scientific community to agree upon a common nomenclature contributes to misunderstandings and misapplication of other concepts such as *metapopulation theory* (Dennis et al. 2003; Baguette and Mennechez 2004; Shreeve et al. 2004) or leads to conflicts in methodology (Diaz et al. 2004), creating a plethora of unnecessary, modified habitat-based terms (Rountree and Able 2007). Thus, in this opening chapter, we thought it prudent to discuss habitat terminology, as

well as where we think we need to focus if we are to advance ecological knowledge and the pursuit of species conservation. Much of what we present here has been presented elsewhere; we feel the need to re-state some often rather basic concepts because, unfortunately, the message is not getting through to the majority of users. And by users we mean not just academic and agency scientists but also the legion of practicing biologists and ecologists that work diligently on our public and private lands.

At the most fundamental root, habitat is a binary determination. An area, and its associated conditions, either is or is not habitat. In this sense, a phrase such as "unsuitable habitat" has no meaning (Hall et al. 1997; Garshelis 2000). If a location is unsuitable for occupancy by an organism, then it, simply, is not habitat. Frequently, habitat is used to describe an area supporting a particular type of vegetation or aquatic type. This use apparently grew from the term *habitat type*, coined by Daubenmire to refer to "land units having approximately the same capacity to produce vegetation" (Daubenmire 1976, 125). Some researchers promote this definition because of the simplicity in delineation or measurements (Garshelis 2000). Furthermore, this is the definition often adopted for use in legislation or policy. This broad definition regarding "habitat" simply as space or ecological/vegetation zones leads to overuse of the term in place of more suitable and better defined terms such as *vegetation association*, *resource*, and *landscape* (see Hall et al. 1997, table 2). For example, Rountree and Able (2007) point out that part of the ambiguity of the term arises because of the perspective of the investigator. Habitat often is applied to reference an area with similar ecological conditions, such as coral reefs or mud flats, thus referencing the community of organisms reliant upon the collection of resources provided within the area. This is in contrast to an organism-focused perspective of habitat. Rountree and Able (2007) provided a list of additional terminology that differentiates between "ecological habitat type" and "organismal habitat." We argue that instead of tacking on modifiers to the term, one should adhere to a single concept for habitat and then use more appropriate terms to represent mud flats and tidal pools (e.g., biome, ecosystem; Ricklefs and Miller 2000).

For application to conservation and management of organisms, the simplicity of a location in space, or ecosite or vegetation type definition, ignores the complexity of the biological world. For management of populations, this definition is essentially useless. It eliminates consideration of spatial and temporal variation of resources and biotic interactions and how they contribute to population limitation and control (Dennis et al. 2003; Diaz et al. 2004; Mitchell 2005). Managing vegetation and other environmental features in this broad manner will likely fall short of producing the desired assemblage of wildlife by failing to adequately account for the needs of the individual species (Lindenmayer and Hunter 2010). Different taxa use resources differently across a single day, season, or life cycle, and these dependencies must be quantified in order to understand management for a single species (or population). Failing to simultaneously manage for plants and animals is a hit-or-miss strategy for animals; managing vegetation and other environmental features will address habitat for some animal species but will ignore—all or in part—habitat for other species (some of which might not be desired).

We want a concept that allows us to tackle issues such as how organisms respond to alterations in the landscape, changes in the climate, or responses to stochastic events. Thus, we follow the definition adopted by Morrison et al. (2006) in which *habitat* is an area with a combination of resources (e.g., food, cover, water) and environmental conditions (e.g., temperature, precipitation, presence or absence of predators and competitors) that promotes occupancy by individuals of a given species (or population) and allows those individuals to survive and reproduce. Habitat,

then, is a concept associated with a particular species, and sometimes even to a particular population, of plant or animal. *Habitat is thus species specific*; this is a key concept that is lost on many, many users. One cannot walk to a vista and proclaim that "this is a good" or "this is a bad" habitat without, at a bare minimum, including the specific species under consideration.

Given our definition of habitat as species-specific and encompassing necessary resources and conditions for occupancy, there are only a handful of habitat-related terms that need to be defined. The list of acceptable terms is short because only a few key terms and concepts are actually needed to understand, quantify, and manage a species' habitat (our terminology follows Morrison et al. 2006 and Morrison 2009). Unfortunately, numerous authors have applied modifiers to the basic *habitat* term, including "high quality," "marginal," "transitional," "optimal," "suitable," and many, many more (see Hall et al. 1997). Many of these modified terms arise from a failure to assign habitat to a single organism (Rountree and Able 2007), or they are inappropriately used when population demographics are not considered (Hall et al. 1997). "Suitable" or "unsuitable" habitats are most egregious in that the terms blur the fact that a lack of suitability simply means that it is not habitat in the first place. Habitat-modified terms create a jargon that is unnecessary when alternate terms are available that are defined in basic ecology textbooks (e.g., biome). This explosion of terminology has led to rampant confusion. Additionally, even casual reading of published papers shows a continual switching of terms within even the same article. Confusion reigns.

Terminology

We define *habitat use* as the way an animal uses (or consumes) physical and biological components (i.e., resources) in a habitat. Habitat use focuses on how an organism uses their habitat, not if they use an area or resource (because then it is not actually habitat). Either "habitat use" is a description of how resources are used or often it is quantified as the proportion of time that an animal spends within various components of the environment (Beyer et al. 2010).

Habitat selection is the hierarchical process of innate and learned decisions by an animal about where it should be across space and time in order to persist (Johnson 1980; Hutto 1985; Piper 2011). Habitat selection is an evolutionary process based on fitness consequences that vary with differential resource use (Morris 2011). *Habitat preference* is restricted to the consequence of the habitat selection process, resulting in the disproportional use of some resources over others. As we define these terms, habitat use and habitat preference differ from habitat selection in that the former are based on quantifiable patterns using measures of habitat availability. Unfortunately, authors oftentimes use the term *habitat selection* to represent the outcome of statistical interpretation of habitat use compared to habitat availability (i.e., habitat selection models; Mannan and Steidl 2013, 233). We argue that the term *habitat selection* should be restricted to the behavioral process (Hutto 1985; Morris 2003; Beyer et al. 2010).

Quantifying habitat use or preference requires evaluation of how much habitat occurs in the environment. Abundance, when applied to characteristics of a habitat or a resource, is restricted to the quantity of that particular component, irrespective of the number of organisms present in the habitat (Wiens 1984, 402). Abundance of a resource is not adequate for quantifying the availability of the various components of a species' habitat. Habitat availability must consider the accessibility of physical and biological components of the environment to an individual. An individual's physical limitations plus multiple interspecific, environmental, or anthropogenic factors might limit accessibility, making habitat availability a difficult measure for investigators to quantify (Beyer et al. 2010). It is, of course, difficult to assess resource availability from an animal's perspective. For example, we can measure the abundance of food for a particular species, but we cannot know that an animal can obtain all of the food in the habitat because there are many factors that restrict its availability. A simple example is plant material (leaves, fruit) that is beyond the reach of an animal; it is thus unavailable for it to feed on.

Habitat quality is the ability of the environment to provide conditions appropriate for individual and population persistence, based on the provision of resources for survival, reproduction, and population persistence, respectively (Morrison 2009, 62). Habitat quality is an assessment of a continuous gradient that necessarily is based on demographics (reproduction and survival) of individuals or populations, which fluctuate over space and time. Thus, habitat per se has no inherent "quality." Authors discuss habitat quality in paper after paper, when what they really mean, at best, is the ability or probability of an area to result in quality (i.e., the production of young or survival of individuals). One might identify a plot of land as high quality in one year, but because of myriad factors the same plot is identified as low quality in the following year (e.g., annual drought leads to low arthropod production). Furthermore, Van Horne (1983) brought to our awareness that animal density can be a misleading indicator of habitat quality. Although animal abundance can be correlated with habitat quality in some cases, the quality itself should be based on demographics of individuals or populations (Garshelis 2000; Railsback et al. 2003; Johnson 2007).

The distribution of animals is intimately tied to the concept of *niche*. The concept of niche has been defined in multiple ways over time and continues to be the subject of much discussion. As reviewed in Morrison et al. (2006, chapter 1), Grinnell (1917) introduced the term *niche* when explaining the distribution of a single species of bird. His assessments included spatial considerations (e.g., reasons for a close association with a vegetation type), dietary dimensions, and constraints placed by the need to avoid predators. Thus, in this view, the niche included both positional and functional roles in the community. Elton (1927) later described the niche as the status of an animal in the community and focused on trophic position and diet. Other, more complicated views of the niche also arose following the pioneering work by Grinnell, Elton, and others (e.g., Hutchinson 1957).

Although habitat is a core concept for developing general descriptors of the distribution of animals, it can only provide limited insight into factors responsible for animal survival and fitness, and thus population responses to changing environments. It seems apparent to us that the proliferation of terms used by authors to modify the basic habitat term, as discussed previously, is because of this failure to even think about the niche concept when studying how and why animals are distributed as they are. This limited insight from the habitat concept is because habitat alone does not usually describe the underlying mechanisms that determine survival and fecundity. Other factors, including some often related to an animal's niche, must be studied to more fully understand the mechanisms responsible for animal survival and fitness (Morrison et al. 2006, 56–57). Research has shown that a number of environmental factors can restrict survival and productivity of a species (across its range or a portion thereof), and the influence of any single factor is not necessarily additive to the influence of any other factor. That is, usually only one factor is limiting in any particular location and time, and it is unlikely that the same factor will always be limiting because natural variation causes shifting of the quality and quantity of resources.

Thus, habitat and niche are both related and core concepts in ecology and therefore wildlife and habitat management. As developed by Morrison (2009, 64–65), focusing on habitat alone is problematic because the environmental features we measure can stay the same while use of important resources by an animal within that habitat can change. For example, changes in the species or size of prey taken by a bird foraging on shrubs: the shrubs ("habitat") do not need to change in physical dimensions or appearance. If we describe habitat only as structural or floristic aspects of vegetation, we will often fail to predict organism health because we did not recognize constraints on exploitation of other resources that are critical limiting factors (Dennis et al. 2003).

What We Have Learned: It Is Time to Move On

Habitat is not a delineation of a single vegetation type in a single location across time. Although we cannot disconnect spatial location from conditions, we can and need to recognize that this is not fixed. Habitat describes the conditions surrounding the location of an animal (Morrison 2001). It is ephemeral and it changes with variations in the environment. Landscape ecologists regard habitat as a continuum across the landscape and emphasize the importance of considering the "matrix" between patches of vegetation or ecosites. Multiscale approaches are needed to characterize habitat relationships. Unfortunately, the lingering prevalence of habitat as a single-dimension, spatially delineated area hinders our advancement of understanding requirements for species or populations. For example, the

critical habitat designation that legislation and policy rely on is an oversimplification that can result in mismanagement of resources and wildlife.

We emphasize the need for fitness covariates to support habitat quality assessments because habitat selection does not necessarily correlate with habitat quality. Too often studies used for direct management or conservation purposes default to using data (i.e., density of individuals) or statistical models of habitat selection to assign value to properties. The assumption that highly selected locations are the most beneficial locations is, in reality, an iterative evolutionary process. Evolutionary habitat selection is determined by habitat quality (i.e., variation in fitness controlled by variation in the environment), but habitat selected by individuals will vary in how productive or how well it survives (i.e., habitat quality). The assumption that these processes are in equilibrium such that a linear correlation between habitat selection and habitat quality exists is ridiculous. Garshelis (2000) pointed out fundamental flaws in the reasoning behind assumptions of habitat preference modeling, which multiple research studies have supported and expanded upon (Van Horne 1983; Railsback et al. 2003). The concepts of ecological traps or source/sink habitats are in direct opposition of habitat selection as a reflection of habitat quality. Simulations of trout populations and habitat preference models revealed multiple reasons why density was low in areas with high fitness potential (Railsback et al. 2003).

We must conclude that the field of wildlife-habitat studies has progressed little over the past fifty years or more. The primary advances in habitat studies have involved technology and statistics but not anything fundamental in the way studies are conceptualized, planned, or conducted. This is due in part to a continued misuse of the habitat concept and, thus, misapplication of associated theories. But we can improve and promote progress in this field of study. We can recognize that biological systems are complex, and that it is an oversimplification to assume that relationships with host plants, vegetation types, or environmental substrates are sufficient for species persistence (e.g., Dennis et al. 2003; Diaz et al. 2004). We can encourage identification of quality habitat only when linked with population dynamics that vary across time and space. We can address habitat selection theory by including consideration of behavioral processes, such as site familiarity or the importance of public information (Boulinier and Danchin 1997; Morris 2011; Piper 2011).

Habitat studies are stagnant because of what Morrison (2012) referred to as our current habitat sampling and analysis paradigm. The scientific community perpetuates this paradigm because we have become complacent in our construction and evaluation of habitat relationship studies. Reviewers and editors tend to publish habitat relationship studies with limited scrutiny because studies often follow the generic prescription of how we have always conducted them and the inherent limitations of such work. The majority of these studies are doomed from the start because we compromise our data collection by failing to establish a cogent framework for sampling (Morrison 2012). Oftentimes our studies are largely driven by logistical constraints, related funding limitations, and limited focus at local management areas (Morrison 2012). Unfortunately, this typically results in reliance upon "convenience sampling," results of which only allow weak conclusive statements about the sample itself rather than the desired inductive inference concerning the population of interest (Anderson 2001). As noted many years ago by Romesberg (1981), we often use statistics to try to compensate for a lack of clear scientific thinking when formulating our studies. Most of our habitat studies have little relevance to the target species with regard to viability of populations and, ultimately, species persistence. Thus, attempts to translate our published results into management guidelines for a species tend to produce one-size-fits-all documents that have little to do with long-term viability and, thus, persistence.

In a world where land management and conservation are the pinnacle goals for most ecological studies, we are faced with a desire to provide general, holistic management prescriptions and conservation initiatives (Lindenmayer and Hunter 2010). However, this should not compromise the uniqueness of components within systems. Conversely, we are continuously discouraged from conducting local-scale, case studies or baseline-monitoring studies, which when planned appropriately can be integrated into future research programs (Lindenmayer and Likens 2010). Although beyond the scope of this chapter, the foundation of any habitat study is the initial identification of the relevant biological population and hence appropriate selection of sampling locations (see Morrison 2012 for a thorough discussion).

Advancing studies of habitat involves far more than terminology. Yes, we must start with a basic set of terms that recognizes the limitations that the basic concept of habitat can provide, but the ongoing proliferation of terms is an obvious indication of our broad failure to come to grips with the limitations of "habitat" alone. Regardless of the terms used, an improperly designed study does not advance understanding in a broad and scientific manner. As mentioned in this chapter and extensively reviewed elsewhere (Anderson 2001; Morrison 2012), much has been written about our failure to properly design studies of wildlife. It is our hope that eventually all of us will carry these messages to our students and other colleagues and begin the task of advancing studies of wildlife and their habitats.

LITERATURE CITED

Anderson, D. R. 2001. "The Need to Get the Basics Right in Wildlife Field Studies." *Wildlife Society Bulletin* 29:1294–97.

Baguette, M., and G. Mennechez. 2004. "Resource and Habitat Patches, Landscape Ecology and Metapopulation Biology: A Consensual Viewpoint." *Oikos* 106:399–403.

Beyer, H. L., D. T. Haydon, J. M. Morales, J. L. Frair, M. Hebblewhite, M. Mitchell, and J. Matthiopoulos. 2010. "The Interpretation of Habitat Preference Metrics Under Use—Availability Designs." *Philosophical Transactions of the Royal Society B: Biological Sciences* 365:2245–54.

Boulinier, T., and E. Danchin. 1997. "The Use of Conspecific Reproductive Success for Breeding Patch Selection in Terrestrial Migratory Species." *Evolutionary Ecology* 11:505–17.

Daubenmire, R. 1976. "The Use of Vegetation in Assessing the Productivity of Forest Lands." *Botanical Review* 42:115–43.

Dennis, R. L. H., T. G. Shreeve, and H. V. Dyck. 2003. "Towards a Functional Resource-Based Concept for Habitat: A Butterfly Biology Viewpoint." *Oikos* 102:417–26.

Diaz, R. J., M. Solan, and R. M. Valente. 2004. "A Review of Approaches for Classifying Benthic Habitats and Evaluating Habitat Quality." *Journal of Environmental Management* 73:165–81.

Elton, C. 1927. *Animal Ecology.* London: Sidgwick and Jackson.

Garshelis, D. 2000. "Delusions of Habitat Evaluation: Measuring Use, Selection, and Importance." In *Research Techniques in Animal Ecology: Controversies and Consequences,* edited by L. Boitani and T. K. Fuller, 111–64. New York: Columbia University Press.

Grinnell, J. 1917. "The Niche-Relations of the California Thrasher." *Auk* 34:427–33.

Hall, L. S., P. R. Krausman, and M. L. Morrison. 1997. "The Habitat Concept and a Plea for Standard Terminology." *Wildlife Society Bulletin* 25:173–82.

Hutchinson, G. E. 1957. "Concluding Remarks." *Cold Spring Harbor Symposium on Quantitative Biology* 22:415–27.

Hutto, R. L. 1985. "Habitat Selection by Nonbreeding, Migratory Land Birds." In *Habitat Selection in Birds*, edited by M. L. Cody, 455–76. San Diego: Academic Press.

Johnson, D. H. 1980. "The Comparison of Usage and Availability Measurements for Evaluating Resource Preference." *Ecology* 61:65–71.

Johnson, M. D. 2007. "Measuring Habitat Quality: A Review." *Condor* 109:489–504.

Kearney, M. 2006. "Habitat, Environment and Niche: What Are We Modelling?" *Oikos* 115:186–91.

Lindenmayer, D. B., and M. Hunter. 2010. "Some Guiding Concepts for Conservation Biology." *Conservation Biology* 24:1459–68.

Lindenmayer, D. B., and G. E. Likens. 2010. "The Science and Application of Ecological Monitoring." *Biological Conservation* 143:1317–28.

Mannan, R. W., and R. J. Steidl. 2013. "Habitat." In *Wildlife Management and Conservation: Contemporary Principles and Practices*, edited by P. R. Krausman and J. W. Cain, III, 229–45. Baltimore: Johns Hopkins University Press.

Mitchell, S. C. 2005. "How Useful is the Concept of Habitat? A Critique." *Oikos* 110:634–38.

Morris, D. 2003. "Toward an Ecological Synthesis: A Case for Habitat Selection." *Oecologia* 136:1–13.

Morris, D. W. 2011. "Adaptation and Habitat Selection in the Eco-evolutionary Process." *Proceedings of the National Academy of Sciences* 278:2401–11.

Morrison, M. L. 2001. "A Proposed Research Emphasis to Overcome the Limits of Wildlife-Habitat Relationship Studies." *Journal of Wildlife Management* 65:613–23.

———. 2009. *Restoring Wildlife: Ecological Concepts and Practical Applications*. Washington, D.C.: Island Press.

———. 2012. "The Habitat Sampling and Analysis Paradigm Has Limited Value in Animal Conservation: A Prequel." *Journal of Wildlife Management* 76:438–50.

Morrison, M. L., and L. S. Hall. 2002. "Standard Terminology: Toward a Common Language to Advance Ecological Understanding and Applications." In *Predicting Species Occurrences: Issues of Scale and Accuracy*, edited by J. M. Scott et al., 43–52. Washington, D.C.: Island Press.

Morrison, M. L., B. G. Marcot, and R. W. Mannan. 2006. *Wildlife-Habitat Relationships: Concepts and Applications*. 3rd edition. Washington, D.C.: Island Press.

Piper, W. 2011. "Making Habitat Selection More 'Familiar': A Review." *Behavioral Ecology and Sociobiology* 65:1329–51.

Railsback, S. F., H. B. Stauffer, and B. C. Harvey. 2003. "What Can Habitat Preference Models Tell Us? TESTS Using a Virtual Trout Population." *Ecological Applications* 13:1580–94.

Rickleffs, R. E., and G. L. Miller. 2000. *Ecology*. 4th edition. New York: W. H. Freeman and Company.

Romesberg, H. C. 1981. "Wildlife Science: Gaining Reliable Knowledge." *Journal of Wildlife Management* 45:293–313.

Rountree, R., and K. Able. 2007. "Spatial and Temporal Habitat Use Patterns for Salt Marsh Nekton: Implications for Ecological Functions." *Aquatic Ecology* 41:25–45.

Shreeve, T. G., R. L. H. Dennis, and H. Van Dyck. 2004. "Resources, Habitats and Metapopulations—Whither Reality?" *Oikos* 106:404–8.

Van Horne, B. 1983. "Density as a Misleading Indicator of Habitat Quality." *Journal of Wildlife Management* 47:893–901.

Wiens, J. A. 1984. "Resource Systems, Populations, and Communities." In *A New Ecology: Novel Approaches to Interactive Systems*, edited by P. W. Price, C. N. Slobodchikoff, and W. S. Gaud, 397–436. New York: Wiley and Sons.

2

FRED S. GUTHERY AND
BRONSON K. STRICKLAND

Exploration and Critique of Habitat and Habitat Quality

H abitat is considered a fundamental, unifying concept in wildlife biology. Yet the word seems to have suffered meaning proliferation that has sapped it of rhetorical heft. We explore and critique *habitat* and *habitat quality* based on theoretical definitions from the literature, operational definitions in the minds of practicing biologists, and implicit definitions deduced from statements in literature. Classically, habitat is a natural home or living area with resources (e.g., food) and conditions (e.g., temperature range) that permit the existence of an organism in an area (classical outlook). Operationally, habitat is an area with demographically or behaviorally meaningful subcomponents, which are also called habitats, and that also may have subcomponents called habitats (hierarchical outlook). Cover types, plant communities, physiognomic types, vegetation types, habitat types, microhabitat, and patches are, among others, operational synonyms for sub-subcomponent habitats. These meanings (area, subcomponent, sub-subcomponent) often are used without restraint and, accordingly, the word *habitat* may issue forth in technical prose like an involuntary platitude. To improve clarity and consensibility of a message in papers dealing with habitat concepts, authors can define *habitat* and related terms early in a paper and then adhere to those definitions. We recommend that authors conduct a word search for *habitat* when they complete a paper. Then, where the word occurs, usage can be checked against stated definitions and an explicit synonym can be substituted if the definition does not hold.

Introduction

Habitat is a set of concepts rather than a single, unique concept. It is difficult to write about because of its many meanings, some of which have become ambiguous. Thus, every mention of the word may, at least tacitly, entail clarification unless authors have explicitly defined it and adhered to that definition. We have noticed that authors who define the concept usually abandon the stated definition.

To experience the polysemous nature of habitat, consider its use as an adjective. What does *habitat* mean in various adjective forms? Is *habitat* a synonym for *landscape* in the phrase *habitat fragmentation*? What is *habitat* in *habitat matrix*? In *habitat quality*? In *habitat deterioration*? A number of other words modified by the adjective *habitat* include condition, type, preference, avoidance, variable, elements, units, and components.

What does *habitat* mean when modified by adjectives such as nesting, roosting, resting, foraging, and thermal? What is the *habitat* in *low-condition habitat*? Is it generically the same *habitat* as in *quality habitat* or *thermal habitat*?

What is *habitat* in Van Horne's (1983) mathematical definition of habitat quality? Is it the same *habitat* as in Wiens (2002)? "In one way or another, predicting the occurrence of species in space and time comes down to dealing with *habitats*. The issues we face in predicting occurrences therefore are issues of *habitat*—how we define [*habitat*]; how we measure, map, and model [*habitat*]; how we analyze [*habitat*]; what [*habitat*] means to organisms; and, ultimately, how we can use knowledge about *habitat* to manage natural resources in a sustainable and balanced way" (739; emphasis added).

Or consider: "One objective of *habitat relationship* research is to identify biologically important variables that have the ability to predict species occurrence . . . *Habitat quality* for many species contains a spatial component related to the arrangement and amount of *habitat elements* across the landscape. If the spatial scale of measurement alters the values of these variables, this influences our ability for predicting species occurrence and for assessing *habitat conditions*" (Trani 2002, 151; emphasis added). What does *habitat* mean in this paragraph? Does it mean the same thing each time it is mentioned? Is it the same *habitat* as in Van Horne (1983) or Wiens (2002) or both?

These preliminary comments suggest that *habitat* may have experienced meaning proliferation and overuse, such that it may be in danger of losing (or has already lost) rhetorical heft. Wildlife biologists say *habitat* the way politicians say *the American people*. Yet these problems do not override the value of the habitat concept as a fundament in the ecology and management of wildlife (Block and Brennan 1993).

Before stating objectives, we note that our subject matter is extensive. To constrain this chapter to a reasonable length, we limited surveys to include only what we believe is the most relevant information. Further, we focus on the concepts of habitat and habitat quality and give short shrift to related topics. We will not address microhabitat concepts such as nesting, roosting, bedding, foraging sites, and so forth, though these are important topics. Neither will we address regions, biomes, or other vast quantities of space. Except for passing comments, we do not discuss selection and avoidance of entities called habitats nor do we compare and contrast habitat and niche. Rather, we focus on the concept at a mesoscale that is meaningful to wildlife managers and researchers: areas the size of ranches, farms, study areas, watersheds, wildlife refuges, and other areas that serve as living spaces for organisms seasonally or annually. Organisms that use space at a continental scale (e.g., neotropical migrants) also use space at mesoscales in daily and seasonal activities.

We surveyed the literature for definitions of habitat that arise in or from dictionaries, glossaries, and review articles and surveyed research and management biologists for the operational definitions they hold. We also deduced meanings of *habitat* from wording and context when the term appeared in the literature. We discuss habitat per se relative to the concept of habitat quality. Based on these activities, we formulate recommendations on use of the word *habitat* in the ecological literature. We regard such use as optimal if it parsimoniously preserves habitat as a powerful, traditional concept (Block and Brennan 1993) and maximizes clarity and consensibility of a message in written and verbal exchanges among biologists.

Definitions of Habitat
Dictionaries and Articles

As a classical concept, habitat generally is defined in reference to "home" or "living area": "The locality in which a plant or animal naturally grows or lives" (Oxford online dictionary); a place where a plant or animal lives (Smith and Smith 2001, 778); the natural abode of an organism (Grange 1949).

In a slightly more complex version of the definition (requisites added), *habitat* is defined as a set of resources and conditions necessary for sustained occupancy (time considered) of an area by an organism: "The resources and conditions present in an area that [permit] occupancy . . . by a given organism" (Hall et al. 1997, 175); "[t]he type of place where an animal normally lives or, more specifically, the collection of resources and conditions necessary for its occupancy" (Garshelis 2000, 112); "[t]he physical space within which the animal lives, and the abiotic and biotic entities . . . in that space" (Morrison and Hall 2002, 51).

Safriel and Eliahu defined *habitat* as "the environment of a community confined to a portion of the landscape. Each habitat is made of three species-specific components: 1. 'Environmental' physical and chemical features (e.g., temperature, salinity), 2. Resources (e.g., food, space), [and] 3. Interacting organisms (other than those functioning as resources, e.g., competitors, predators, mutualists)" (1991, 349).

Garshelis expanded his definition of *habitat* to include a "set of specific environmental features that, for terrestrial animals, is often equated to a plant community, vegetative association, or cover type" (2000, 112). In this sense, alfalfa fields, corn fields, timothy

(*Phleum pratense*) meadows, marshes, bluestem (*Andropogon* sp.) prairies, and mesquite (*Prosopis glandulosa*) brushland are habitats.

These examples stray from the classical concept of "living space with requisites" but show that habitat has sundry meanings in practical use. We turn to those meanings now, including ones held explicitly and implicitly by biologists.

Practicing Biologists

For this section, we informally contacted practicing researchers and managers and asked them to answer two queries: "What is habitat for [your choice of taxon]?" and "How do you judge the quality of [taxon] habitat?" Habitat quality will be dealt with in the next major section.

Regarding "What is habitat?," we received this comment: "This question is like 'Explain World War II—use back of page if necessary.' You could write books on this . . . and many have" (K. K. Karrow, Marais des Cygnes Wildlife Area, personal communication). True. Nevertheless, we present a sampling of responses.

The northern bobwhite (*Colinus virginianus*) provides an example of variability in the definition of habitat for biologists dealing with the species. "I view [bobwhite] habitat as a residence. Just like our homes have many components (living room, bedroom, kitchen, bathroom, garage), bobwhite habitat has many types of cover (nesting, feeding, brood-rearing, loafing, escape). Just like in our homes, each cover type fulfills a different need. All cover types are in close proximity to one another" (S. J. DeMaso, US Fish and Wildlife Service, personal communication).

"Bobwhite habitat may be viewed as circumstances or areas with characteristic microstructure (bare ground, vegetation height, plant species composition to meet food and cover requirements, etc.) and macrostructure (distribution of woody and herbaceous plants)" (T. V. Dailey, Northern Bobwhite Conservation Initiative, personal communication).

D. Rollins (Texas A&M University, personal communication) defines bobwhite habitat as "a landscape frequented by quail, either past or present; such is characterized by a plant community interspersed with woody plants (especially shrubs), grasses, and early successional forbs. Ideally, one should be able to throw a softball from one shrub the size of a Volkswagen to the next." The softball metric insures that requisites are readily accessible by bobwhites, given their mobility. If the distance to a shrub exceeds a softball throw, space may become unusable, leading to a decline in average density on an area.

"I would define bobwhite habitat as a set of patches within a landscape that contains the features bobwhites require with the availability and spacing of the features set to support bobwhite life throughout the year" (C. B. Dabbert, Texas Tech University, personal communication).

"Habitat (for anything) is the place where [it] lives and includes the resources needed for survival and reproduction" (A. T. Pearse, US Geological Survey, Northern Prairie Wildlife Research Center, personal communication). Here is a detailed example using a passerine bird: Breeding habitat for a cliff swallow (*Petrochelidon pyrrhonota*) includes "relatively open areas of fields, pasture, prairie, or agricultural land in relative proximity to a nesting structure (cliff face, bridge, highway or railroad culvert, or building) that offers a vertical wall and horizontal overhang on which the mud nests can adhere. Habitat must also include a permanent or ephemeral water source that creates mud for nest building within [3–5 km] of the nesting structure" (C. R. Brown, University of Tulsa, personal communication).

Consistent with the definitions in the preceding section, time may be involved in operational concepts of habitat. Habitat "is an area or location that is able to provide for the specific life history needs of a given . . . species at a given time. I say 'at a given time' because what constitutes useful habitat changes through time and in accordance with the life history stage of the individual" (R. O'Shaughnessy, Southern Illinois Cooperative Wildlife Research Laboratory, personal communication).

"Habitat is an area that meets the needs for an animal's existence. I used the term *existence* instead of *occurrence* to exclude circumstances when an area is not habitat but an animal may be found temporarily in it (e.g., quail on [a] football field). Thus, there is an implied element of time in my use of existence" (F. Hernández, Texas A&M University–Kingsville, personal communication).

"Habitat is an *area* that supplies all of the necessary

components . . . for an animal to survive. Habitat comprises food, cover, [and] water in a suitable arrangement with sufficient space *for a particular species*. Thus, habitat is not necessarily a vegetation type, and any reference to habitat as vegetation (e.g., 'habitat type') is confusing and should be avoided" (C. A. Harper, University of Tennessee, personal communication; emphasis in original).

Here we introduce *usable space* because this term could be used as a synonym for habitat, and the concept has explicit consideration of time. "Any . . . area can be envisioned as a set of points (e.g., Cartesian coordinates) surrounded by [biotic and abiotic features with various properties]. To be fully usable, a point must by definition be associated with [biotic and abiotic features] compatible with the physical, behavioral, and physiological adaptations of [a target species] in a time-unlimited sense" (Guthery 1997, 294). The usable space concept posits maximization of usable space in time (space–time) as the goal of management. Space–time is the product of usable space and time (e.g., ha-days). The concept also introduces a binary circumstance—space is usable or not usable—which may be viewed as a severe constraint on the concepts of crude (density on an area) and specific density (density on that portion of an area that provides usable space). The classical definition of *habitat* also creates a binary circumstance (is habitat, is not habitat).

Deductions

For this section, we searched articles in journals and chapters in books for the word *habitat*. Then, we attempted to deduce the meaning of *habitat* intended (implicitly) by the author(s) based on the context in which the word was used. For brevity, we will use the term *natural abode* to denote implicit meanings that match the classical meaning. The following examples are not exhaustive but rather provide an overview of deduced meanings.

Example 1

As researchers were seeking "the truth underlying how animals were relating to their habitats, managers were seeking ways in which to use such information to address their needs" (Stauffer 2002, 56). "Animals,"

"habitats," and "managers" are plural in this sentence fragment. It is unclear whether the statement applies to different animals on different areas or different animals on the same area with multiple habitats. If the implication is different animals on different areas, then habitat would seem to be a homogeneous area in space that somehow differs, perhaps arbitrarily, from other areas in space. Here we mean "area" in the sense of study area, ranch, watershed, management area, and so on.

If the implication is different animals of a single species on the same area, the typical study circumstance, then habitat is possibly synonymous with subdivisions of the area (e.g., patch, cover type, or element). However, it could in this context be synonymous with structural properties of vegetation (small scale) that were similar in different patches. If the implication is animals of different species on the same area, then habitat could be the classical definition (i.e., natural abode). In any case, the statement is ambiguous because *habitat* could mean a place where animals live, an area, a physiognomic type within an area, or a circumstance at a small scale (e.g., a perch).

Example 2

"Demographic information on survival and fecundity will aid greatly in establishing the quality of a particular habitat" (Stauffer 2002, 58). The phrase "a particular habitat" is telling because it implies one of a set of habitats, which could imply subdivisions in an area. However, the demographic variables imply, but do not confirm, an area of sufficient size to sustain at least a segment of a population. Here again, habitat seems to be an area that differs somehow (perhaps arbitrarily) from other areas in space. It could also apply to a set of natural abodes. However, to the extent that a natural abode includes "the set of requisites of an organism," it is illogical to expect variation in habitat quality except in a binary sense: is, is not. That is, all requirements are consummated when habitat is a set of requisites. No further need exists.

Example 3

"The landscape metrics examined here are valuable descriptors of wildlife habitat" (Trani 2002, 151). We have landscape meaning area and wildlife habitat implying a suite of species. The metrics could be viewed as summary indices with or without explicit subcomponents (e.g., patches). For example, canopy coverage of trees has a single basis in the landscape. Simpson's diversity of patches deals with multiple components of the landscape. Trani (2002) seems to mix habitat as area of interest, habitat as unique subcomponents of that area, and habitat in the classical sense (natural abode).

Example 4

"One objective of habitat relationship research is to identify biologically important variables that have the ability to predict species occurrence . . . Habitat quality for many species contains a spatial component related to the arrangement and amount of habitat elements across the landscape. If the spatial scale of measurement alters the values of these variables, this influences our ability for predicting species occurrence and for assessing habitat conditions" (Trani 2002, 151).

In this paragraph (sixty-six words), we have habitat relationship, habitat quality, habitat elements, and habitat conditions.

- Habitat relationship involves relations between wildlife and one or more habitat elements. So, the "habitat" in "habitat relationship" refers to elements that could be synonymous with patches, plant communities, cover types, or components. The "habitat" in "habitat elements" probably refers to the landscape (area). For example, we could say "landscape elements" instead of "habitat elements" with no change in meaning when landscape and habitat are synonyms.
- Habitat quality involves "a spatial component," so because of a single component we assume habitat refers to the landscape (area) and some property of the habitat elements in the landscape (e.g., fractal dimension).
- Habitat conditions (plural) perhaps imply the condition of habitat elements, but this does not make sense because our interest is in a property of the landscape (habitat quality). Perhaps the phrase "habitat conditions" refers to some unstated assumption(s) about existing conditions relative to optimal conditions, however that is defined.

We see the word *habitat* used to specify a landscape (area), a subcomponent of that landscape (element), and a property of the landscape (conditions) or its elements relative to an assumed optimal condition. We are uncertain about the last meaning.

Example 5

"The distribution of a species may be limited by the behavior of individuals in selecting their habitat" (Krebs 1972, 29). Selection implies two or more habitats (selection is not an issue with one habitat). If we view selection from the standpoint of habitat as a natural abode, then we can reduce selection opportunities to two types: natural abode and not a natural abode. In this case, selection is moot because we know tautologically that all use will, by definition, be in the natural abode. Krebs might be implying that natural abodes may be classified according to their properties, e.g., oak (*Quercus* sp.) forest natural abode, hickory (*Carya* sp.) forest natural abode, and so on. Such classification is a product of human caprice, not animal adaptation. An animal might view oak and hickory abodes more generally as deciduous tree abodes. However, if we view landscapes as containers for a mixture of not-natural-abode elements and distinct-natural-abode elements, we would see selection behavior that limits the distribution of a species. At face value, Krebs's statement seems ambiguous, but perhaps that is a problem of its interpreters.

Example 6

"Semiannual population counts . . . suggest that fall habitat conditions on [the study area] may have declined relatively more than did late-winter habitat . . . This is not to imply that autumn conditions were more severe than those of late winter, only that habitat deterioration was first apparent during the season of highest density . . . Likewise, Robinson . . . considered bobwhite carrying capacity . . . to be primarily a function of habitat available in late autumn" (Roseberry and Klimstra 1984, 91–92).

In the last sentence, Robinson's statement could apply to habitat defined as usable space, i.e., area (ha) compatible with occupation, given the adaptations of bobwhites. If this conclusion is acceptable, then habitat conditions possibly mean the proportion of available space that is usable (habitat). Habitat deterioration implies loss of usable space because of concomitant loss of essential cover features such as hedgerows. We could also deduce from these statements that Roseberry and Klimstra described trends in the area (ha) of the birds' natural abode.

Example 7

"Habitat fragmentation is usually defined as a landscape-scale process involving both habitat loss and the breaking apart of habitat" (Fahrig 2003, 487). We are given a landscape dimension, but it would make little sense in context to say "landscape fragmentation"; habitat adds the wildlife consideration. Moreover, we deduce that the landscape must once have existed in a relatively unmodified state or a state with modifications of minor and innocuous extent. The statement seems to imply that *habitat* may be defined as areas within a landscape that supported sustaining wildlife populations in the absence of fragmentation.

Example 8

Storch (1993, 256) studied habitat selection by capercaillie (*Tetrao urogallus*) in summer and autumn. Habitat to Storch was a distinct patch based on vegetation and topographic variables (subjectively designated physiognomic type) on a large area (50 km²). We can further define *habitat* according to the twenty-five variables listed in Storch's table 1. Presumably, the variables listed are known or assumed components of habitat as classically defined (natural abode). "[W]e almost always provide a long list of variables for measurement that are hopefully related to what the animal perceives as habitat" (Morrison 2012, 8). Undoubtedly, there are hundreds of papers in the ecological literature that consider habitat to be a list of variables that somehow characterize the natural abode of an organism.

Example 9

Van Horne's (1983) classic paper is titled "Density as a Misleading Indicator of Habitat Quality." How might we define *habitat* to portray the meaning of the word in this paper? Again, the classical definition and permu-

tations thereof do not seem to hold because how can a conceptual entity vary in quality if it supplies all the needs of an organism? Fortunately, Van Horne's equation 1 resolves the problem because the equation gives a mathematical definition of relative habitat quality (Q_i) for area i.

Consider the numerator in Van Horne's (1983) equation 1, which may be stated as

$$D_i = ((\textit{survival plus production for all age-classes on area } i)/(a_i N_i)) \sim \lambda_i/a_i \sim \textit{fitness density in habitat } i$$

where a_i is the area in the ith habitat ($A = \sum a_i$), λ_i is the growth multiplier for the ith habitat, and N_i is the total population for the ith habitat. Van Horne's analysis implies that both the area of a subcomponent (a_i) and the total area ($\sum a_i$) are called habitat. Examples of total areas could be a ranch, management area, study area, national wildlife refuge, and so on. This outcome (a habitat composed of habitats) is consistent with the default and deduced definitions given previously, but not with the classical definition of habitat (natural abode).

Definitions of Habitat Quality

Van Horne (1983) provides a natural segue from habitat to habitat quality ($P < 0.05$). Because of the various meanings of habitat, we will point out what we deduce to be the referenced authors' meaning in the following discussion. This protocol will help to insure common understanding of concepts.

The word *quality* has at least nine meanings, including the configuration of the oral cavity when stating a vowel. We will use the word as an index of superiority or excellence based on human value. In the absence of human value, the words *excellence* and *superiority* have no meaning.

Let D_i = fitness density in subcomponent habitat i as defined previously. Then relative habitat (area) quality as defined by Van Horne (1983) is

$$Q_i = D_i/\sum D_i.$$

Relative habitat quality has no general meaning because it is specific to any particular sample. However, useful information may be in subcomponent estimates of fitness density (D_i). Use of the word "fitness" in this definition is unfortunate because, with the exception of rare circumstances, humans can only measure surrogates of fitness (e.g., survival and fecundity in instantaneous time scales relative to the time scale of fitness).

Before discussing other concepts of habitat quality, we want to further comment on Van Horne's (1983) paper. The title is slightly ambiguous because of the phrase "density *as* a misleading indicator" (emphasis added). The word *as* is quite complex (at least twenty-one meanings including idioms). The Van Horne title could be construed as "density *is* a misleading indicator" or "density *may be* a misleading indicator." In the first case, people would never knowingly use density as a correlate of habitat quality (habitat here defined in the Van Horne sense as an area of interest composed of subareas). In the second case, which Van Horne intended, one would have to justify using density as an indicator of habitat quality because it could be misleading. Some people think in terms of the first example, i.e., density *is* a misleading indicator. In fact, density may be the only reasonable estimator of habitat (some arbitrary area) quality.

Van Horne's (1983) equation implies that habitat (subcomponent) quality increases with survival and/or fecundity. This may indeed be the case in trending populations over a defined period of time. Examples may include population growth following disease outbreaks and extreme environmental events that decreased population numbers. In addition, sustainable harvest rates may be an indicator of habitat (subcomponent) quality for game species. Yet if density-dependent processes influence these demographic variables, which seems to be a universal circumstance (Brook and Bradshaw 2006), then we encounter a dilemma. In a sustaining, nontrending population, higher survival entails lower fecundity and higher fecundity entails lower survival. If these adjustments do not occur, the finite annual growth rate exceeds 1.0 ($\lambda > 1.0$) and the population increases exponentially (Guthery and Shaw 2013). If we constrain interpretation to sustaining populations, then the only useful demographic index of habitat (subcomponent) quality is density given $\lambda = 1.0$. So density may be not only a misleading indicator in some settings but the only useful indicator of habitat (area) quality in other settings.

This outcome begs the question, how can fitness density ($D_i = \lambda_i / a_i$) vary across habitat subcomponents (a_i)? One way is for carrying capacity density (K_i / a_i) to vary among subcomponents where K_i is similar to

the asymptotic limit of population growth under the logistic model. Such variation seems plausible for large herbivores in response to variation in soil fertility and plant nutrient concentration (Midgley 1937; Strickland and Demarais 2000; Jones et al. 2008).

Another way for carrying capacity density to vary is for the proportion of space that is usable on subcomponent a_i to vary among subcomponents. Suppose $K_i / a_i = pC = C$ if the proportion of space usable is $p = 1.0$. So, in general

$$K_i / a_i = p_i C$$

where p_i = the proportion of space usable on area i. Put differently, the aforementioned equation states that crude density is proportional to specific density. It also states that carrying capacity density is proportional to usable space on habitat subcomponent i. This is tantamount to saying that the quality of habitat subcomponent a_i is proportional to usable space in the subcomponent, and this would generalize to habitat as an arbitrary area of interest (e.g., $\sum a_i$).

In discussing measures of habitat quality, Johnson (2007) used *patch* and *habitat* as synonyms. He also used *arbitrary area consisting of patches* as a synonym for *habitat*. This outlook is analogous to Van Horne's view (habitat as a collection of habitats).

Johnson (2007) defined habitat quality at the level of an individual bird as "per capita contribution to population growth expected from a given habitat." What is implied by *habitat* in this definition is not clear, but habitat as a collection of patches (essentially an area composed of habitat types) seems to be a reasonable guess. The definition also raises a question about habitat quality in a sustaining population ($\lambda = 1$). We would deduce from Johnson's definition that habitat quality does not vary for sustaining populations, and we would conclude that whether a population sustains tells us all we need to know about the quality of its habitat (natural abode or arbitrary area).

There seems to be general, somewhat rote agreement that habitat (area or subcomponent) quality is poor in the case of declining populations and good in the case of sustaining and growing populations. Of course, such a perception of quality is subject to many confounding factors such as time, weather, predation, competition, standing density, and so on. Further, the perceived quality of an area (called habitat) may vary ecologically and tribally with properties such as interspersion and diversity. (By tribal variation, we mean variation imposed by human values and allegiance to like-thinkers. For example, diversity is a desired property of human social systems, but more diversity is not necessarily a meaningful property of ecosystems because population responses to diversity may become asymptotic [Guthery 2008].) We discuss some of these broader aspects of habitat (area) quality in the following text.

One respondent reported that he judges "the quality of . . . habitat [area] based upon [his] experience as a predator (hunter) and hence based on a search image . . . for where [he has] found [the quarry] in previous outings." Actually, the respondent has developed an ad hoc habitat suitability index that scores woody cover, nesting cover, food supplies, and interspersion on a 0–10 scale. Habitat suitability models are quantitative methods of assessing habitat (area) quality.

"Bobwhite habitat quality should be judged by bobwhite density given average environmental conditions for the site" (C. B. Dabbert, Texas Tech University, personal communication). This definition relates to an area that is called a habitat. The definition implicitly invokes the concept of usable space.

"Ultimately, [habitat] quality comes down to the fitness of individuals within the habitat. We use proxies for this: survival and yearly reproductive output. At certain times of the year, density also might be a good proxy for quality" (A. T. Pearse, US Geological Survey, Northern Prairie Wildlife Research Center, personal communication). This communication also seems to imply that habitat is a synonym for area of interest (e.g., density is number/area).

"I would argue that wetland (read: habitat) of 1 ha capable of providing one hundred duck use days is better quality than a 1-ha wetland capable of providing fifty duck use days" (R. O'Shaughnessy, Southern Illinois University, personal communication). Here habitat is a type of plant community or substrate, and the quality of that community or substrate is based on the food supplies it holds. Judging habitat (area) quality based on food abundance is common in wildlife biology.

Variables for evaluating habitat quality for mallards (*Anas platyrhynchos*) include "*diversity* (i.e., the array of habitat types and condition, and their juxtaposition within the area of interest), vegetation species *compo-sition* (i.e., annual seed producers, perennials, etc.), [and] vegetation *structure* (for escape cover, thermal cover, visual isolation, etc.)" (K. K. Karrow, Marais des Cygnes Wildlife Area, personal communication). In this statement, *habitat* means an area of interest and also patches within the area (types).

Discussion

As is evident from the preceding material, the word *habitat* has multiple meanings in theory and practice. To recapitulate, biologists use it primarily to mean

- where an organism lives, which includes biotic and abiotic requisites and time.
- an arbitrary area of interest (habitat) that contains demographically or behaviorally unique subareas (habitats) that may in turn contain descriptively unique subareas (habitats). For the latter component, vegetation type, ecotype, physiognomic type, habitat type, habitat (when used as synonym for habitat type), plant community, element, patch, and so on are used interchangeably with habitat.
- gibberish (i.e., meaning apparently indecipherable).

These definitions break into three natural classes: classical, hierarchical, and irrational. The classical outlook includes definitions with meanings similar to "natural abode." The hierarchical outlook is an arbitrary area (level 1) with (demographically, behaviorally) meaningful subdivisions (level 2) that are in turn subdivided (level 3), which may in turn be subdivided (level 4), and so on. We see no reason why higher level subdivisions might not be meaningful (e.g., a wetland plant community habitat could have open water and vegetated components that could be called habitats). The irrational outlook includes any use of habitat in an indecipherable manner.

Note that a unit composed of population-supporting subunits in the Van Horne (1983) sense would seem to be uncommon in the study of hierarchical habitat. Rather, the most common expression is an arbitrary area of interest composed of subunits that are unique in some respect. Such subunits are essential for habitat (subunit) preference studies, as is evident from Johnson (1980), who dealt with habitat types.

Does the classical definition hold sway over the hierarchical definition, or vice versa, as the appropriate model for research and management? Who is to say? Both seem to have a place and, besides, both will be with us forever.

Logically, the classical version of habitat seems to be rather a dead-end concept except for descriptive science. But this descriptive science is by no means trivial. Imagine how difficult it would be to identify the minimal set of factors that would describe an organism's habitat in the classical sense. Guthery (2002), for example, posited that bobwhite habitat could be fully described with five variables: canopy coverage of woody vegetation, visual exposure to ground predators, visual exposure to aerial predators, coverage of bare ground, and operative temperature. Imagine how research to reveal minimum sets of factors that characterize a species' habitat (classical sense) could lead to original and meaningful hypotheses about behaviors and processes in the field. The main purpose of science is to extract simplicity from complexity (Cohen and Stewart 1994).

The hierarchical view of habitat fits research in that it represents sequential parsing—the study of an entity composed of parts that are in turn composed of parts and so on. This approach provides a natural means of studying how animals partition time and occurrence in third-level units that we call habitats or the synonym habitat types.

Now let us turn to the concept of habitat quality. Under the classical definition of habitat, habitat quality is a meaningless concept to which we have alluded to the point of redundancy. Classical habitat is an existential matter—it either exists or it does not. Thus, in a spatial sense, there is habitat (quality = 1) and not-habitat (quality = 0).

In the hierarchical sense, a consensus prevails that habitat (area) quality varies with some index of fitness as determined by studies that are instantaneous relative to the grand sweep of time over which fitness prevails. One might conclude that habitat A (area or subcomponent) is better than habitat B (area or subcomponent) because nest success was higher in habitat A. This inference is weak regardless of the P-value or model weight because it provides no information on other demographic variables such as survival of various age-classes. Studies that document the corresponding values for recruitment and adult survival may provide support for spatial or temporal differences in habitat (subcomponent) quality, but only for a defined area and span of time. Moreover, many studies ignore the dilemma of density dependence (Guthery and Shaw 2013): in sustaining populations, increases in productivity entail decreases in survival and vice versa. Thus, higher nest success in area A could have no influence on year-to-year trends in abundance because of the compensatory mechanisms of density dependence. Over time, the dilemma of density dependence holds with annual variation in survival and production.

Habitat quality is a dearly held concept based on human value. It apparently sprang into use as an unquestioned property of nature that is based on fitness, something that is difficult to measure except for its soft correlates. The concept needs rigorous theoretical justification and clarification if it is to be useful in building knowledge about wildlife.

Recommendations

Because the word *habitat* is intrinsically ambiguous, as its classical, hierarchical, and irrational forms show, authors should state their intended meaning early in a paper and then religiously apply that meaning throughout the paper. This would serve the goal of making the concepts in papers consensible. In other words, authors and readers would have a common understanding of meaning.

To forestall overuse and misuse of the word, we recommend that authors do a word search for *habitat* after finishing a paper. Then, at each encounter of the word he or she can check to see whether any particular usage matches the stated definition. If it does not, perhaps authors can substitute a term that is more explicit than *habitat* wherever possible. For example, vegetation types are synonyms for habitat in technical parlance, so to be precise one could use prairie, oak forest, meadow, cattail community, and other explicit synonyms instead of minestrone.

ACKNOWLEDGMENTS
We thank T. E. Fulbright and C. A. Davis for reviewing a draft of this chapter. J. D. Stafford and V. A. O'Brien contributed ideas and information. We recognize that these colleagues do not necessarily agree with all the views expressed here.

LITERATURE CITED

Block, W. M., and L. A. Brennan. 1993. "The Habitat Concept in Ornithology." *Current Ornithology* 11:35–91.

Brook, B. W., and C. J. A. Bradshaw. 2006. "Strength of Evidence for Density Dependence in Abundance Time Series of 1,198 Species." *Ecology* 87:1445–61.

Cohen, J., and I. Stewart. 1994. *The Collapse of Chaos*. Penguin Books, New York, USA.

Fahrig, L. 2003. "Effects of Habitat Fragmentation on Biodiversity." *Annual Review of Ecology, Evolution, and Systematics* 34:487–515.

Garshelis, D. L. 2000. "Delusions in Habitat Evaluation: Measuring Use, Selection, and Importance." In *Research Techniques in Animal Ecology*, edited by L. Boitani and T. K. Fuller, 111–64. Columbia University Press, New York, USA.

Grange, W. B. 1949. *The Way to Game Abundance*. Charles Scribner's Sons, New York, USA.

Guthery, F. S. 1997. "A Philosophy of Habitat Management for Northern Bobwhites." *Journal of Wildlife Management* 61:291–301.

———. 2002. *The Technology of Bobwhite Management*. Iowa State University Press, Ames, USA.

———. 2008. *A Primer on Natural Resource Science*. Texas A&M University Press, College Station, USA.

Guthery, F. S., and J. H. Shaw. 2013. "Density Dependence: Applications in Wildlife Management." *Journal of Wildlife Management* 77:33–38.

Hall, L. S., P. R. Krausman, and M. L. Morrison. 1997. "The Habitat Concept and a Plea for Standard Terminology." *Wildlife Society Bulletin* 25:173–82.

Johnson, D. H. 1980. "The Comparison of Usage and Availability Measurements for Evaluating Resource Preference." *Ecology* 61:65–71.

Johnson, M. D. 2007. "Measuring Habitat Quality: A Review." *Condor* 109:489–504.

Jones, P. D., S. Demarais, B. K. Strickland, and S. L. Edwards. 2008. "Soil Region Effects on White-Tailed Deer Forage Protein." *Southeastern Naturalist* 7:595–606.

Krebs, C. J. 1972. *Ecology*. Harper and Row, New York, New York, USA.

Midgley, A. R. 1937. "Modification of Chemical Composition of Pasture Plants by Soils." *Journal of the American Society of Agronomy* 29:498–503.

Morrison, M. L. 2012. "The Habitat Sampling and Analysis Paradigm Has Limited Value in Animal Conservation: A Prequel." *Journal of Wildlife Management* 76:1–13.

Morrison, M. L., and L. S. Hall. 2002. "Standard Terminology: Towards a Common Language to Advance Ecological Understanding and Application." In *Predicting Species Occurrences: Issues of Accuracy and Scale*, edited by J. M. Scott, P. J. Heglund, M. L. Morrison, J. B Haufler, M. G. Raphael, W. A. Wall, and F. B. Sampson, 43–52. Island Press, Washington, D.C., USA.

Oxford online dictionary. http://www.oed.com/view/Entry /82988 (accessed June 25, 2014).

Roseberry, J. L., and W. D. Klimstra. 1984. *Population Ecology of the Bobwhite*. Southern Illinois University Press, Carbondale, USA.

Safriel, U. N., and M. N. Ben-Eliahu. 1991. "The Influence of Habitat Structure and Environmental Stability on the Species Diversity of Polychaetes in Vermetid Reefs." In *Habitat Structure*, edited by S. S. Bell, E. D. McCoy, and H. R. Mushinsky, 349–69. Chapman and Hall, New York, USA.

Smith, R. L., and T. M. Smith. 2001. *Ecology and Field Biology*. San Francisco: Addison Wesley Longman.

Stauffer, D. F. 2002. "Linking Populations and Habitats: Where Have We Been? Where Are We Going?" In *Predicting Species Occurrences: Issues of Accuracy and Scale*, edited by J. M. Scott, P. J. Heglund, M. L. Morrison, J. B Haufler, M. G. Raphael, W. A. Wall, and F. B. Sampson, 53–61. Island Press, Washington, D.C., USA.

Storch, I. 1993. "Habitat Selection by Capercaillie in Summer and Autumn: Is Bilberry Important?" *Oecologia* 95:257–65.

Strickland, B. K., and S. Demarais. 2000. "Age and Regional Differences in Antlers and Mass of White-Tailed Deer." *Journal of Wildlife Management* 64:903–11.

Trani, M. K. 2002. "The Influence of Spatial Scale on Landscape Pattern Description and Wildlife Habitat Assessment." In *Predicting Species Occurrences: Issues of Accuracy and Scale*, edited by J. M. Scott, P. J. Heglund, M. L. Morrison, J. B Haufler, M. G. Raphael, W. A. Wall, and F. B. Sampson, 141–55. Island Press, Washington, D.C., USA.

Van Horne, B. 1983. "Density as a Misleading Indicator of Habitat Quality." *Journal of Wildlife Management* 47:893–901.

Wiens, J. A. 2002. "Predicting Species Occurrences: Progress, Problems, and Prospects." In *Predicting Species Occurrences: Issues of Accuracy and Scale*, edited by J. M. Scott, P. J. Heglund, M. L. Morrison, J. B Haufler, M. G. Raphael, W. A. Wall, and F. B. Sampson, 739–49. Island Press, Washington, D.C., USA.

3

Amanda D. Rodewald

Demographic Consequences of Habitat

Conserving wildlife populations often warrants a habitat-based focus, given that the amount, distribution, and quality of habitat can strongly mediate population demography. The manner in which demography is affected by habitat is determined via numerous behavioral, physiological, environmental, and stochastic processes. In this chapter, I discuss how the traditional view of habitat-population relationships is being challenged by new insights from studies that illustrate the need to better (1) distinguish between individual and population-level measures of habitat quality as they relate to population ecology, (2) address habitat needs and demographic connectivity across the full life cycle, (3) recognize how behavioral processes guiding habitat selection influence population dynamics, and (4) elucidate heterogeneity in the demographic consequences of habitat use over a wide range of spatial and temporal scales. Addressing these needs will strengthen wildlife science and management by improving our ability to understand wildlife-habitat relationships and predict the demographic consequences of habitat use.

Introduction

The identification, protection, and restoration of suitable habitat are essential components of many, if not most, wildlife conservation and management efforts. Because habitat directly or indirectly mediates the rates of birth, death, immigration, and emigration for any species, most biologists and managers consider demographic information to be the gold standard for determining which elements of habitat are most important,

as well as the overall quality of habitat. Habitats are generally considered to be of "high quality" when they support and promote health, survival, and/or reproduction of individuals within a population. However, many approaches to characterizing wildlife-habitat relationships often fail to tightly link habitat parameters to population dynamics and viability.

The apparent disconnect between habitat and population dynamics may result from several shortcomings of our traditional view of the wildlife population–habitat link. First, a failure to distinguish between individual and population-level measures of habitat quality can obscure identification of the highest quality habitat and, thus, fail to identify the management approach that best achieves desired outcomes. Demographic measures framed from the two perspectives can lead to seemingly contradictory conclusions about habitat quality and its impact on populations. Second, descriptions of habitat still have a heavy bias toward a single life stage or season—usually the breeding season. This tendency overlooks the now-rich literature demonstrating that habitat requirements can vary across different life stages and that events operating across the full life cycle govern population dynamics. Third, the ways in which population dynamics can be influenced by individual behaviors related to habitat selection are generally underrecognized. The "if you build it, they will come" view of restoration and management fails to recognize that a diverse suite of environmental, biological, and social cues are used by animals to select habitats (Morrison 2009). For example, if the presence of conspecifics is the primary cue used by a species to select habitat, newly created high-quality but vacant habitats may not be occupied. In cases where cues for settlement are lacking, even the best habitats may have weak demographic signals because they are less preferred, settled later, and/or occupied by lower quality individuals. Fourth, spatiotemporal variation in habitat-specific demography is often overlooked or not sufficiently examined because many studies of wildlife-habitat relationships occur at limited spatial and temporal scales and fail to fully capture interpopulation variation in the ways animals interact with their environment. Understanding the causes and consequences of spatiotemporal variability in demography is essential to developing effective strategies that ensure the long-term viability of populations. Permeating throughout each of these points is the fact that most habitat studies give weak consideration to ecoevolutionary links and feedbacks that are especially relevant in systems where anthropogenic change has altered selective environments.

This chapter provides an overview of recent developments that are changing our understanding of the relationship between habitat quality and population dynamics. I organize the discussion around four key topics where recent advances have provided new insights into the interrelationship of habitats and populations.

Individual- versus Population-Level Measures of Habitat Quality Can Provide Different Information, and Sometimes Contrasting Perspectives, About Population Dynamics

The idea that density does not necessarily indicate habitat quality became both widely recognized and well accepted after Van Horne's (1983) classic paper suggested that habitat quality should reflect mean individual fitness per unit area, which was defined as a function of density, production of young, and survival. Taken at face value, use of demographic parameters to indicate habitat quality is intuitively attractive and straightforward. However, as discussed in previous papers (Chalfoun and Martin 2007; Johnson 2007; Mortelliti et al. 2010), descriptions of habitat quality can be vague, assessed using the wrong fitness metrics or at inappropriate spatial scales, and often conflate individual and population perspectives.

Demographic measures at individual and population levels may prove contradictory when densities do not align with fitness measures. Even in the simplest cases where life stages do not vary in fitness, demographic rates as measured by per capita fecundity may not parallel population growth rates, or the product of density and per capita fecundity (Johnson 2007; Skagen and Yackel Adams 2011). This lack of concordance between individual and population metrics is not a trivial issue from a management perspective. Should a manager give preference to a habitat that supports the greatest individual fecundity or to that which produces the greatest number of young from the local population—something that also is function of density (fig. 3.1)? For example, suppose Site A has high quality resources and produces higher per capita fitness (2.3 young per female) than Site B, which has lower quality resources and fitness (1.6 young per female). If Site A supports fewer individuals (e.g., 10 breeding pairs × 2.3 young/pair = 23 young per year), then it would contribute less to population viability than the lower quality Site B with higher densities (e.g., 20 pairs × 1.6 young/pair = 32 young produced annually). The contradiction is, then, that from an individual's perspective, Site A would be of better quality, whereas at a population level, Site B would better support a large local or regional population.

Empirical evidence shows that the relationship between individual and population measures can vary widely. For example, density was positively associated with recruitment per capita (individual level) for 72% and with "reproduction per land area" (population level) for 85% of 109 studies reviewed (Bock and Jones 2004). Negative relationships between density and fitness, on the other hand, can result from regulatory mechanisms like inverse density dependence (Brook and Bradshaw 2006). In other cases, the negative relationship can signal the presence of an ecological trap, where individuals show preferences for habitats that result in the poorest recruitment. Thus, the variable relationship between density and fitness can obscure efforts to identify the best quality habitats. For example, Pidgeon et al. (2006) empirically demonstrated that fecundity was the best indicator of individual-level habitat quality for black-throated sparrows (*Amphispiza bilineata*), but the most robust indicator of habitat quality at the population level was the combination of nest

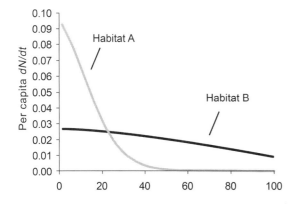

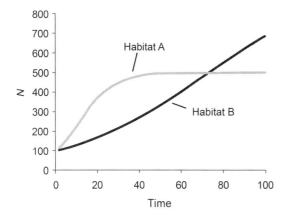

Figure 3.1 Logistic population growth curves for two hypothetical sites: site A with few high-quality resources and site B with abundant lower-quality resources. Differences between habitats result in a higher rate of population growth in A (r_A = 0.12, r_B = 0.03) and a higher carrying capacity in B (K_A = 500, K_B = 1000). Johnson (2007) simulated both populations with initial population sizes of 100 and run for 100 time intervals. If habitat quality is considered purely from an individual bird's perspective, then Habitat A is the better habitat until time 23. In contrast, if habitat quality is measured as the current population size, then Habitat A remains better until time 74. If habitat quality is considered the maximum sustained population size, as may be the perspective of many conservationists, then Habitat B is better because it has the higher carrying capacity. From Johnson 2007

amounts of low quality food and had higher carrying capacities (Hobbs and Swift 1985).

In terms of management, biologists need to explicitly consider both levels, as the most relevant likely depends upon the objective. If one is studying habitats with the aim to guide future management and restoration choices by describing resources that are most advantageous to a focal species, then individual-level measures of quality will best elucidate those "ideal" resources or conditions. After all, fitness is an individual measure and has strong evolutionary components, and managers need to identify the relative contribution to the population by individuals occupying a given habitat, also known as habitat fitness potential (*sensu* Wiens 1989a). Knowing the fitness potential of various habitats makes it possible to discern the effects of landscape composition on population dynamics and to identify optimal configurations (Pulliam 2000; Griffin and Mills 2009 Runge et al. 2006). However, if one aims to identify which existing sites or locations are most important to long-term persistence, then population-level measures, or population mean fitness, are most likely to indicate where management can have the greatest impact on population viability (fig. 3.2).

Population Dynamics Are Influenced by Habitat Used across the Full Life Cycle

Many species require multiple habitats to meet their needs across the full life cycle or across life stages, and changing needs complicate efforts to link population responses to habitat attributes. Trying to understand how habitat alteration contributes to population changes for species that rely upon multiple habitats, or even multiple locations of the same habitat type, requires determining at which location or stage habitat might limit the population. This challenge cuts across taxa, as 80% of animals have complex life cycles that involve major ecological transitions between life stages (Werner and Gilliam 1984). Even nonmigratory animals may require different habitats. Consider pelagic seabirds that spend most of their lives at sea but require terrestrial habitats, often on islands, for nesting. For other species, disparate habitat needs are met in heterogeneous habitats, as with lesser prairie chickens (*Tympanuchus pallidicinctus*) for which various reproductive behaviors, such as courtship displays, nesting,

success (or fecundity) and nest density. Another example relates to mountain shrub communities, where recently burned habitats provided small amounts of extremely high quality food to a lower carrying capacity of mule deer, whereas unburned habitats provided large

Figure 3.2 An experiment on the effects of canopy reduction on forest-breeding birds showed that sites reduced from ~28 m²/ha residual basal area (73% canopy cover) (A) to 14 m²/ha residual basal area (45% canopy cover) (B) supported 115% higher densities of Cerulean Warblers compared to unharvested control stands (Boves 2011). Although reproductive output per breeding pair was 40% lower on disturbed sites than on unharvested controls, the higher densities resulted in a much greater contribution of individuals to the local and regional population from disturbed sites than from undisturbed sites. Photos by Amanda Rodewald

and brood rearing, are each associated with different structural attributes of rangelands (Fuhlendorf and Engle 2001). Likewise, flying foxes in Australia use fruit resources in rainforests as well as nectar and pollen in coastal heath, swamps, and *Eucalyptus* forest (Law and Dickman 1998).

Variation in Habitat Use and Demography across Life Stages

Most studies define *habitat* in terms of a single stage in the annual or life cycle, with breeding being the most common. However, habitat use can vary widely across annual and life stages, particularly for migratory species whose movements span continents. Another striking habitat shift is for amphibians that undergo metamorphosis through aquatic and terrestrial life stages (Wilbur and Collins 1973; Wilbur 1980). The red-spotted newt (*Notophthalmus viridescens*) is aquatic during larval and adult stages, requiring ephemeral vernal pools and streams. However, its juvenile stage, the red eft, is terrestrial and uses upland forest. Many birds that breed in mature forest, such as wood thrush (*Hylocichla mustelina*), ovenbird (*Seiurus aurocapilla*), and worm-eating warbler (*Helmitheros vermivorum*), shift to using early-successional or shrubby habitats during postbreeding and postfledging periods (Anders et al. 1998; King et al. 2006; Vitz and Rodewald 2006). Vitz and Rodewald (2006) captured nearly 90% of birds that bred in mature forests within regenerating clearcuts during the postbreeding period in Ohio, with numbers of mature-forest species rivaling those of early-successional specialists like the prairie warbler (*Setophaga discolor*) and eastern towhee (*Piplio erythrophthalmus*). Such observed shifts in habitat use are thought to result from selection for dense cover to reduce risk of predation and/or abundant fruit resources to facilitate foraging (Vitz and Rodewald 2007). Indeed, use of habitats with dense vegetation promoted survival of fledgling ovenbirds and worm-eating warblers (Vitz and Rodewald 2011). Thus, managers are increasingly recognizing the need to address habitat requirements of both breeding and postbreeding/postfledging individuals in conservation strategies for forest birds.

Many species use multiple habitats across the annual cycle, and for migratory species the spatial scale over which these habitats occur can be staggering. Neotropical migratory birds densely occupy disturbed or second-growth forests, including agroforestry habitats, in Central and South America (Greenberg et al. 1997; Petit and Petit 2003). Shade-coffee farms are agroforestry habitats that appear to provide high quality habitat and are associated with high overwinter survival, site fidelity, and energetic condition (Johnson et al. 2006; Bakermans et al. 2009). Large-scale migrations obviously require conservation of multiple habitats contiguously distributed across large regions, as is the case for barren-ground caribou (*Rangifer tarandus*) that can migrate over five thousand km, pronghorn (*Antilocapra americana*) over five hundred km, moose (*Alces alces*) nearly four hundred km, and mule deer (*Odocoileus hemionus*) over two hundred km in their round-trip movements (reviewed in Berger 2004). Although breeding seasons are known to profoundly affect population dynamics (e.g., Hoekman et al. 2002), the demographic consequences of nonbreeding habitat use can be equally important. Sillett and Holmes (2002) showed that over 85% of the apparent annual mortality occurred during migration for black-throated blue warblers, which suggests that conservation efforts directed at migratory periods are likely to strongly influence population viability. Fortunately, population ecology tools, such as structured population models and sensitivity analyses (see Mills 2012), can be used to identify those stages and, by association, those habitats that most strongly affect population dynamics. In addition, new approaches can quantify per capita, habitat-specific contributions to population growth and, thus, allow for the examination of source-sink dynamics for animals using heterogeneous habitats or landscape mosaics (Griffin and Mills 2009).

Seasonal Interactions

Just as a habitat can strongly affect performance, reproduction, and survival, there also can be large effects that lead to seasonal interactions. Seasonal interactions, also known as carryover or lag effects, are distinguished from direct impacts on survival or reproduction within a single season by the fact that processes or events from one season affect the state or condition of individuals in subsequent seasons. Seasonal interactions at the individual level are oftentimes mediated by body condition, which tends to be closely related to habitat quality. However, there can also be population-level interactions that result from changes in population size in one season that affect per capita rates in subsequent seasons (Norris and Marra 2007). From a practical standpoint, the presence of seasonal interactions means that demographic parameters measured during one season may not indicate habitat quality if there are strong carryover effects from previous seasons and locations.

Carryover effects have been described for a wide range of taxa (reviewed in Harrison et al. 2011), including mammals (Festa-Bianchet 1998; Perryman et al. 2002; Descamps et al. 2008), reptiles (Broderick et al. 2001), and fish (Bunnell et al. 2007; Kennedy et al. 2008). In seven of eight ungulates in Kruger National Park, South Africa, juvenile survival was strongly related to rainfall, and therefore food supply, during the previous dry season (Owen-Smith et al. 2005). In roe deer (*Capreolus capreolus*), resources encountered in the spring strongly affected body mass during winter (Pettorelli et al. 2003). The size of winter food cache, in part, determined breeding condition of wolverines (*Gulo gulo*; Persson 2005). Black-throated blue warblers (*Setophaga caerulescens*) overwintering in high-quality forest were in better breeding condition than individuals occupying lower quality scrub habitats (Bearhop et al. 2005). Mass of nestling savannah sparrows (*Passerculus sandwichensis*), which was related to fledging date, affected an individual's ability to accumulate fat prior to migration (Mitchell et al. 2011). Food availability during stopover periods influenced testis growth and reproductive behavior of garden warblers (*Sylvia borin*) in ways that can affect reproduction (Bauchinger et al. 2009).

Not only can choices in one season have demographic consequences in subsequent seasons, but habitat choice in one season can constrain choices in the next by affecting the timing and speed of migration in ways that impact territory settlement and reproduction (Norris and Marra 2007). One classic example is American redstarts (*Setophaga ruticulla*), for which individuals that overwinter in high-quality mangrove habitats arrive to breeding grounds earlier, in better condition, and fledge more young than individuals that overwintered in poorer-quality scrub habitats (Marra et al. 1998; Norris et al. 2004). Indeed, the winter-habitat model created by Norris et al. (2004) predicted that females occupying high-quality winter habitat would produce two additional young fledging a month earlier compared to females from poor-quality winter habitat. Thus, the consequences of resources available to birds in their nonbreeding habitat carried over to influence their subsequent reproductive success.

The likelihood of carryover effects may vary with life history strategies. Populations of species with slow life histories, such as being long-lived with low reproductive output, should be more sensitive to events during the nonbreeding season than fast life history (e.g., low survival but high reproductive output) species, which should be most sensitive to breeding season events (Saether et al. 1996). Similarly, "capital" breeders finance reproduction from energy stores accumulated in previous months (Drent and Daan 1980; Houston et al. 2007) and should face more carryover effects than "income" breeders (Harrison et al. 2011). One example of this is gray whales (*Eschrichtius robustus*), for which winter calving success depends upon fat stores accumulated during the summer (Perryman et al. 2002). In one of the few carryover studies focusing on a year-round resident species, blue tits (*Cyanistes caeruleus*) that received supplemental food during winter initiated breeding earlier in the season and with greater reproductive success than unsupplemented individuals (Robb, McDonald, Chamberlain, and Bearhop 2008; Robb, McDonald, Chamberlain, Reynolds, Harrison, and Bearhop 2008). In addition, for species that have individuals traveling between the same specific breeding and nonbreeding areas (Esler 2000), or high migratory connectivity (Webster et al. 2002), seasonal interactions should be strong.

In their review of carryover effects across taxa ranging from marine mammals to ungulates to birds, Calvert et al. (2009) found that nonbreeding season events affected fitness in subsequent seasons at both the individual and population levels. For example, at the population level, density dependence can manifest across seasons when mortality or low reproductive rates affect the proportion of individuals able to occupy higher-quality sites in subsequent seasons (Norris and Marra 2007). The ways in which changes in density affect populations in subsequent seasons depends on how they are regulated. In the case of density-dependence due to crowding effects, increases in population size promote negative interactions among individuals (e.g., competition for food or mates) that depress reproduction (Fretwell and Lucas 1969). Alternatively, with site dependence, increases in population size result in some individuals being pushed into lower-quality territories that, in turn, depress reproduction (Rodenhouse et al. 1997). In contrast, individual-level effects happen when events in previous seasons influence individual performance in subsequent seasons via nonlethal mechanisms. Recent work shows how individual- and population-level carryover effects can be incorporated into population models (Runge and Marra 2005; Norris 2005; Norris and Taylor 2006).

Population Dynamics Are Influenced by Behavioral Processes Guiding Habitat Selection

Most habitats are heterogeneous and variable, and individuals lack complete knowledge about the state of current and future conditions. Some important determinants of fitness may not be predictable, such as predation risk (Doran and Holmes 2005), or recognized, in the case of novel alterations to habitats (e.g., invasive plants; Lloyd and Martin 2005). Further complicating matters, the consequences of settlement decisions may not be realized until later in the season and long after the decision was made (e.g., when young fledge, at time of departure for migration). Nevertheless, individuals have access to many types of information from which they can base their choice of the most appropriate strategies and/or habitats to occupy. Heterogeneous environments create the opportunity for individual behaviors to influence population-level processes because spatiotemporal variation provides options from which individuals choose. Choices have fitness consequences that ultimately can affect population dynamics. The ecology of information is the study of how organisms acquire and use information and how this affects populations, communities, landscapes, and ecosystems (Schmidt et al. 2010).

Access to information that is both public (i.e., accessible to all individuals) and private (i.e., undetectable to others and oftentimes based on the past experience of individuals, such as prior reproductive success; Wagner and Danchin 2010; Schmidt et al. 2010) can shape habitat-population links. Common cues used in habitat selection include specific environmental conditions, presence of conspecifics or heterospecifics, and reproductive success of conspecifics, such as fledgling calls or the presence of predators. When cues are related to attributes of the population itself, as with conspecific density and nest success, rather than specific habitat attributes, there is the potential for positive feedback that can dramatically affect population viability and restoration potential (Schmidt et al. 2010). Therefore, from a practical standpoint, cues matter. For instance, metapopulation persistence in patchy environments that have temporally correlated or predictable conditions was several-fold greater when fidelity was based on previous successful experience than when dispersal was uniform (Schmidt 2004; Schmidt et al. 2010). Conversely, reliance upon social cues like conspecific attraction may reduce the likelihood that birds disperse to and colonize new or restored high-quality habitats (Reed and Dobson 1993).

The optimal dispersal and settlement strategies are expected to vary based on pattern of spatial heterogeneity and temporal predictability (Doligez et al. 2003; Schmidt et al. 2010). For sites with fine-scale spatial heterogeneity (i.e., high variation among individual sites within a patch), individuals should either (1) show high fidelity to sites if temporal predictability is high or (2) disperse from unsuccessful sites and prospect within the same habitat patch. For sites with coarse-scale heterogeneity (i.e., high variation among patches), individuals should prospect for information on patch quality and base decisions on average patch quality when predictability is high but should prospect during the prebreeding season when predictability is low. Conspecific attraction, or the tendency of individuals to settle near other individuals of the same species, is only favored when variation among patches is greater than variation among sites and temporal heterogeneity is high.

An ecological trap, now a familiar phenomenon to scientists and managers, also is rooted in the ecology of information. Ecological traps form when once-reliable cues of habitat quality are no longer reliable due to altered environments (Schlaepfer et al. 2002). Loggerhead sea turtle (*Caretta caretta*) hatchlings, for example, are attracted to light from beachfront development and move inland where they face greater risk of mortality (Witherington 1997). Bobolinks (*Dolichonyx oryzivorus*) and other grassland species are attracted to hayfields for breeding, despite the fact that harvesting of hay occurs before young fledge, thereby reducing reproductive success (Bollinger et al. 1990). Decision rules can reduce or enhance the likelihood of ecological traps. Through modeling, Kokko and Sutherland (2001) showed that natal imprinting, philopatric preferences, or win-stay, lose-switch strategies tended to reduce the impact of ecological traps because individuals adjusted their habitat preferences. In general, learned cues provide more opportunity for individuals to adjust their preferences, whereas cues with strong genetic components promote the persistence of traps.

Because the cues and decision rules used in habitat selection have important consequences for population dynamics, they can provide insight into which conservation and restoration strategies are most appropriate and improve wildlife habitat models. For example, the absence of cues used in habitat selection can result in otherwise high-quality habitats going unused. Conspecific attraction can be a useful tool for conservation, particularly when there is a need to attract individuals to unoccupied, newly created, or restored habitats (Reed and Dobson 1993), as has been well established for seabirds (Podolsky and Kress 1992). Experimental playbacks of conspecific vocalizations also showed that the federally endangered black-capped vireo (*Vireo atricapilla*) was strongly attracted to conspecifics, and birds drawn to experimental sites often established territories and bred, even in subsequent years (Ward and Schlossberg 2004). Incorporating conspecific attraction into wildlife-habitat models can improve the predictive ability of habitat models (Campomizzi et al. 2008), especially when forecasting the outcome of reintroductions. Mihoub et al. (2009) showed that persistence of reintroduced populations depends heavily upon habitat selection strategies when there is high heterogeneity among habitat patch qualities. Failure of reintroductions was more likely when species used conspecific presence or avoidance strategies, both of which tended to aggregate individuals in suboptimal habitats. Direct assessment of environmental cues or attraction to areas with high conspecific reproductive success was associated with higher success rates. Given

Table 3.1 Despite a fourfold difference in density, fitness metrics (± SE) for Northern cardinals (*Cardinalis cardinalis*) were comparable in seven urban and seven rural forests in central Ohio, USA (Rodewald and Shustack 2008). In this way, cardinals appear to be resource matching and conforming to ideal free distribution. The pattern is consistent with predictions from IDF because densities are greatest in urban forests that contain greater levels of resources used and preferred by cardinals, such as fruit, bird feeders, and understory woody vegetation (Leston and Rodewald 2006).

	Urban	Rural
Density (birds per 2-ha grid)	1.30 (0.15)	0.31 (0.11)
Apparent survival, male	0.67 (0.05)	0.58 (0.06)
Apparent survival, female	0.53 (0.06)	0.63 (0.08)
Number of fledglings per pair	2.40 (0.18)	2.10 (0.18)
Nestling mass (7–9 days old; g)	24.80 (0.36)	24.70 (0.50)

these patterns, the authors suggest that reintroductions should favor release of adults, including females with dependent young, and create artificial social information for use as cues.

Decisions about habitat selection, settlement, and dispersal are among the most important behaviors that can regulate populations. Indeed, models of habitat selection nicely illustrate how behaviors governing habitat selection can have population-level consequences (Morris 2003).

Random Settlement

When breeding sites are settled at random (i.e., not related to their quality), population growth mirrors the average quality of the habitat (Pulliam and Danielson 1991; McPeek et al. 2001). Although average per-capita fecundity of the population is not related to the number of sites occupied, the variation in average per-capita fecundity increases as population size declines (McPeek et al. 2001). One consequence is that small populations might remain small for longer periods and be vulnerable to demographic or environmental stochasticity.

Ideal Free and Ideal Despotic Distributions

Under the ideal free distribution (Fretwell and Lucas 1969; Pulliam 1988), competitors are equal and select habitats that maximize fitness with the equilibrium distribution, resulting in equivalent fitness of individuals across habitats of variable quality. Population regulation is a consequence of vital rates of individuals changing with population size, with fitness-depressing

negative interactions increasing with density. In populations conforming to the ideal free distribution, density should indicate habitat quality (for empirical demonstration, see table 3.1). With ideal despotic distribution (Parker and Sutherland 1986), individuals vary in competitive ability and the highest quality habitats are occupied by the strongest competitors, such that at equilibrium, fitness is lower in poorer quality habitats. Competition and displacement are important aspects of the ideal despotic distribution model. Under this model, density is likely to be greater in poorer quality habitats and, thus, a misleading indicator of habitat quality. Consequently, before using density or distribution to indicate habitat quality, one should know the extent to which a population conforms to ideal habitat selection.

Site Dependence

Population regulation also can be accomplished solely due to settlement behaviors and without any changes in individual demographic rates, and even by way of noninteracting individuals (Pulliam and Danielson 1991; Rodenhouse et al. 1997, 1999; McPeek et al. 2001). With site dependence, the negative relationship between average per-capita fecundity of the population and population size results from the sequential use of successively poor sites, rather than from declining productivity of a site as other sites are occupied or as populations increase in size (McPeek et al. 2001). Populations are regulated despite a lack of negative density dependence at the individual level. Site dependence requires variation in the suitability of sites in terms of demography, exclusive use of sites, and adaptive choice of sites (i.e., sites are settled sequentially from best to worst or "preemption"; *sensu* Pulliam and Danielson 1991) (McPeek et al. 2001). Because the best sites are always occupied first, smaller populations should have lower variance in population growth rates than larger populations. In cases where both population size and variance in demographic rates are small, sequentially settling sites from highest to lowest quality can result in rapid population growth (McPeek et al. 2001). Species that conform to site-dependent regulation therefore should be (1) less vulnerable to stochasticity associated with small populations, (2) more likely to recover from small population size, and (3) less impacted by habitats where average suitability is relatively lower than species that randomly settle sites.

Demographic Consequences of Habitat Use Can Show High Spatiotemporal Heterogeneity

Though habitat can strongly govern population demography, a wide variety of other ecological factors can create highly variable relationships between habitat and demographic parameters—even in the absence of differences in the habitat itself. Some heterogeneity may be temporary if there are time lags following environmental change, as can be the case with species showing high levels of site fidelity (Davis and Stamps 2004). However, heterogeneity is an inherent part of many systems, and this means that focusing squarely on habitat as the key driver will result in a limited understanding of population ecology (Morrison 2001).

Population size and regulatory mechanisms can obscure the relationship between habitat quality and population demography when density-dependent mechanisms operate. If populations in high-quality habitats approach carrying capacity, then population growth rates should decline. For example, long-term work with black-throated blue warblers at Hubbard Brook LTER shows that both fecundity and the condition of young were negatively correlated with adult density within a season (Sillett and Holmes 2005), probably due to a combination of crowding (Sillett et al. 2004) and site dependence related to preemptive occupancy of territories that vary in quality (Rodenhouse et al. 1997; McPeek et al. 2001; Rodenhouse et al. 2003). What

this means is that habitats of similar quality can differ widely in terms of demography, simply as a function of population size. Small populations are more likely to have positive correlations between density and population growth rate due to a variety of mechanisms such as failure to find a mate, reduced foraging efficiency, and inability to defend against predators, which collectively are termed the Allee effect (Skagen and Yackel Adams 2011). If populations are in a density-dependent growth phrase, density and per-capita fecundity may not be correlated if settlement patterns among patches should equilibrate fitness via ideal free or ideal free despotic distribution. If populations are regulated by events occurring in different seasons and/or habitats for species with high migratory connectivity, then the association between single-season (or stage) habitat and population dynamics may be weak due to carryover effects, whereby demographic measures at one location reflect the quality of habitats used in previous seasons. For species subject to strong carryover effects, demographic consequences of habitat use may not be evident until individuals transition across seasons or years. Carryover effects can mask or create patterns in the apparent demographic consequences of habitat use and selection. For instance, if high-quality winter habitat is limiting and induces carryover effects on fitness, then there may be large demographic differences among populations or individuals occupying similar breeding habitats.

Population demography also is expected to reflect the manner in which animals interact with the abiotic and biotic components of habitat, which can change with countless ecological factors, including life stage, condition, density, food availability, and presence of other species. In other words, even if the habitat remains the same, the demographic consequences of that habitat might vary with changes in the way a species uses a habitat or interacts with other species (i.e., as its niche changes; Morrison 2001, 2012). The possibility that niche-based changes produce variation in the habitat–demography link are especially likely for ecotypes, or a distinct geographic variety, population, or race within a species that is adapted to local environmental conditions. Optimal habitat conditions are likely to differ among ecotypes, which are poorly defined and understood for most species. Recent work suggests that ecotypes can form more rapidly than previously thought in cases where there is strong habitat matching. Habitat matching is a behavioral mechanism whereby individuals select habitats that best match their phenotype and ability to use that habitat (Edelaar et al. 2008). Because habitat matching can result in spatial aggregation of individuals with similar ecological traits, rapid adaptation can follow.

When a species' niche is considered in terms of interspecific interactions, it becomes apparent how differences in community organization and species interactions also can result in spatiotemporal variation in demography in otherwise similar habitats. Not surprisingly, nest predation rate varies with abundances of many rodent nest predators, such as red squirrels (*Tamiasciurus hudsonicus*) and Eastern chipmunks (*Tamias striatus*) (Sloan et al. 1998; Clotfelter et al. 2007; Schmidt et al. 2010), which may, in turn, be a function of mast or seed production by dominant tree species. So, too, can the foraging behavior of predators change with food abundance (e.g., functional responses like prey-switching). Even the fear of predation can elicit demographic responses, as illustrated by the reduced reproductive output of song sparrows (*Me_lospiza melodia*) subjected to experimental predator calls (Zanette et al. 2011). In some cases, demographic outcomes may reflect the entire network of species interactions. Rodewald et al. (2014) showed that the structure of multispecies networks of interactions between birds and nest plants was more closely associated with avian nesting success than direct measures of habitat or landscape. Recent work suggests that species within communities (i.e., species assemblages) do not respond similarly to human-induced rapid environmental change. Some species are more likely to retain ancestral ecological attributes and preferences (i.e., niche conservatism), and this can affect the rate at which new communities form (Wiens and Graham 2005). Because many species respond individualistically, predicting future range distributions and population sizes is particularly challenging, given that there are likely to be many no-analog communities.

Demographic responses to habitat also can be dramatically affected by landscape context, or the configuration and composition of the landscape. In particular, the landscape matrix surrounding habitat patches can provide alternative habitat (Norton et al. 2000), affect movements of individuals and dispersal through the landscape (Dunning et al. 1995; Belisle et al. 2001), serve as a source of species and individuals invading habitat fragments, especially exotic species (Pysek et al. 2002), and even determine the extent of edge, area, and isolation effects on wildlife (Donovan et al. 1997; Hartley and Hunter 1998). Differing landscape contexts, or characteristics of the landscape surrounding the habitat, may explain why similar silvicultural treatments might induce strong edge-related nest predation in one landscape and no detectable effect in another (Hartley and Hunter 1998). Area requirements of the same species, too, may vary widely with the amount of fragmentation in the landscape or region (Rosenberg et al. 1999) or the amount of residential development surrounding the forest (Friesen et al. 1995; Rodewald and Bakermans 2006). In this way, landscape context is an important source of spatiotemporal heterogeneity in demographic responses to habitat attributes.

Conclusions

While biologists have long known that the quantity, quality, and distribution of habitat directly or indirectly affect wildlife populations, recent developments have improved our understanding of the relationship between habitat quality and population demography. First, habitat-population relationships need to be more explicit about the use and relevance of individual- and population-level measures of habitat quality as they relate to population dynamics and management objectives. Second, research and management strategies must address habitat needs and demographic connectivity across the full life cycle and be especially attentive to carryover effects that may create or obscure habitat-fitness relationships. Third, when interpreting distributional patterns and habitat-demography links, we need to better recognize how behavioral processes guiding habitat selection influence population dynamics. Fourth, only by describing and investigating the causes and demographic consequences of spatiotemporal heterogeneity in habitat use will we be able to implement effective management across the range of a species. By moving beyond the traditional approaches to characterizing wildlife-habitat relationships in these ways, wildlife science and management will improve its capacity to link habitat parameters to population dynamics and viability.

ACKNOWLEDGMENTS

I am grateful for the helpful reviews provided by Brett Sandercock and L. Scott Mills.

LITERATURE CITED

Anders, A., J. Faaborg, and F. Thompson. 1998. "Postfledging Dispersal, Habitat Use, and Home-Range Size of Juvenile Wood Thrushes." *Auk* 115:349–58.

Bakermans, M. H., A. C. Vitz, A. D. Rodewald, and C. G. Rengifo. 2009. "Migratory Songbird Use of Shade Coffee in the Venezuelan Andes with Implications for Conservation of Cerulean Warbler." *Biological Conservation* 142:2476–83.

Bauchinger, U., T. Van't Hof, and H. Biebach. 2009. "Food Availability during Migratory Stopover Affects Testis Growth and Reproductive Behaviour in a Migratory Passerine." *Hormones and Behavior* 55:425–33.

Bearhop, S., W. Fiedler, R. Furness, S. Votier, S. Waldron, J. Newton, G. Bowen, P. Berthold, and K. Farnsworth. 2005. "Assortative Mating as a Mechanism for Rapid Evolution of a Migratory Divide." *Science* 310:502–4.

Bearhop, S., G. Hilton, S. Votier, and S. Waldron. 2004. "Stable Isotope Ratios Indicate that Body Condition in Migrating Passerines is Influenced by Winter Habitat." *Proceedings of the Royal Society B-Biological Sciences* 271:S215–18.

Belisle, M., A. Desrochers, and M. Fortin. 2001. "Influence of Forest Cover on the Movements of Forest Birds: A Homing Experiment." *Ecology* 82:1893–1904.

Berger, J. 2004. "The Last Mile: How to Sustain Long-Distance Migration in Mammals." *Conservation Biology* 18:320–31.

Bock, C., and Z. Jones. 2004. "Avian Habitat Evaluation: Should Counting Birds Count?" *Frontiers in Ecology and the Environment* 2:403–10.

Bolger, D. T., W. D. Newmark, T. A. Morrison, and D. F. Doak. 2008. "The Need for Integrative Approaches to Understand and Conserve Migratory Ungulates." *Ecology Letters* 11:63–77.

Bollinger, E., P. Bollinger, and T. Gavin. 1990. "Effects of Hay-Cropping on Eastern Populations of the Bobolink." *Wildlife Society Bulletin* 18:142–50.

Boves, T. J. 2011. "Multiple Responses by Cerulean Warblers to Experimental Forest Disturbance in the Appalachian Mountains." PhD dissertation, University of Tennessee, Knoxville, TN.

Boulinier, T., and E. Danchin. 1997. "The Use of Conspecific Reproductive Success for Breeding Patch Selection in Terrestrial Migratory Species." *Evolutionary Ecology* 11:505–17.

Broderick, A., B. Godley, and G. Hays. 2001. "Trophic Status Drives Interannual Variability in Nesting Numbers of Marine Turtles." *Proceedings of the Royal Society B-Biological Sciences* 268:1481–87.

Brook, B. W., and C. J. A. Bradshaw. 2006. "Strength of Evidence for Density Dependence in Abundance Time Series of 1198 Species." *Ecology* 87:1445–51.

Bunnell, D. B., S. E. Thomas, and R. A. Stein. 2007. "Prey Resources before Spawning Influence Gonadal Investment of Female, But Not Male, White Crappie." *Journal of Fish Biology* 70:1838–54.

Calvert, A. M., S. J. Walde, and P. D. Taylor. 2009. "Nonbreeding-Season Drivers of Population Dynamics in Seasonal Migrants: Conservation Parallels across Taxa." *Avian Conservation and Ecology* 4:5.

Campomizzi, A. J., J. A. Butcher, S. L. Farrell, A. G. Snelgrove, B. A. Collier, K. J. Gutzwiller, M. L. Morrison, and R. N. Wilkins. 2008. "Conspecific Attraction is a Missing Component in Wildlife Habitat Modeling." *Journal of Wildlife Management* 72:331–36.

Chalfoun, A. D., and T. E. Martin. 2007. "Assessments of Habitat Preferences and Quality Depend on Spatial Scale and Metrics of Fitness." *Journal of Applied Ecology* 44:983–92.

Clotfelter, E. D., A. B. Pedersen, J. A. Cranford, N. Ram, E. A. Snajdr, V. Nolan Jr., and E. D. Ketterson. 2007. "Acorn Mast Drives Long-Term Dynamics of Rodent and Songbird Populations." *Oecologia* 154:493–503.

Danchin, E., L. Giraldeau, T. Valone, and R. Wagner. 2004. "Public Information: From Nosy Neighbors to Cultural Evolution." *Science* 305:487–91.

Davis, J. M., and J. A. Stamps. 2004. "The Effect of Natal Experience on Habitat Preferences." *Trends in Ecology and Evolution* 19:411–16.

Descamps, S., S. Boutin, D. Berteaux, A. G. McAdam, and J. Gaillard. 2008. "Cohort Effects in Red Squirrels: The Influence of Density, Food Abundance and Temperature on Future Survival and Reproductive Success." *Journal of Animal Ecology* 77:305–14.

Doligez, B., C. Cadet, E. Danchin, and T. Boulinier. 2003. "When to Use Public Information for Breeding Habitat Selection? The Role of Environmental Predictability and Density Dependence." *Animal Behaviour* 66:973–88.

Donovan, T. M., P. W. Jones, E. M. Annand, and F. R. Thompson III. 1997. "Variation in Local-Scale Edge Effects: Mechanisms and Landscape Context." *Ecology* 78:2064–75.

Doran, P., and R. Holmes. 2005. "Habitat Occupancy Patterns of a Forest Dwelling Songbird: Causes and Consequences." *Canadian Journal of Zoology* 83:1297–1305.

Drent, R., and S. Daan. 1980. "The Prudent Parent—Energetic Adjustments in Avian Breeding." *Ardea* 68:225–52.

Dunning, J., D. Stewart, B. Danielson, B. Noon, T. Root, R. Lamberson, and E. Stevens. 1995. "Spatially Explicit Population-Models—Current Forms and Future Uses." *Ecological Applications* 5:3–11.

Edelaar, P., A. M. Siepielski, and J. Clobert. 2008. "Matching Habitat Choice Causes Directed Gene Flow: A Neglected Dimension in Evolution and Ecology." *Evolution* 62:2462–72.

Esler, D. 2000. "Applying Metapopulation Theory to Conservation of Migratory Birds." *Conservation Biology* 14:366–72.

Festa-Bianchet, M. 1998. "Condition-Dependent Reproductive Success in Bighorn Ewes." *Ecology Letters* 1:91–94.

Flesch, A. D., and R. J. Steidl. 2010. "Importance of Environmental and Spatial Gradients on Patterns and Consequences of Resource Selection." *Ecological Applications* 20:1021–39.

Fretwell, S. D., and H. L. J. Lucas. 1969. "On Territorial Behavior and Other Factors Influencing Habitat Distribution in Birds. Part 1 Theoretical Development." *Acta Biotheoretica* 19:16–36.

Friesen, L., P. Eagles, and R. MacKay. 1995. "Effects of Residential Development on Forest Dwelling Neotropical Migrant Songbirds." *Conservation Biology* 9:1408–14.

Fuhlendorf, S. D., and D. M. Engle. 2001. "Restoring Heterogeneity on Rangelands: Ecosystem Management Based on Evolutionary Grazing Patterns." *BioScience* 51:625–32.

Gates, J., and L. Gysel. 1978. "Avian Nest Dispersion and Fledging Success in Field-Forest Ecotones." *Ecology* 59:871–83.

Greenberg, R., P. Bichier, A. C. Angon, and R. Reitsma. 1997. "Bird Populations in Shade and Sun Coffee Plantations in Central Guatemala." *Conservation Biology* 11:448–59.

Griffin, P. C., and L. S. Mills. 2009. "Sinks without Borders: Snowshoe Hare Dynamics in a Complex Landscape." *Oikos* 118:1487–98.

Harrison, X. A., J. D. Blount, R. Inger, D. R. Norris, and S. Bearhop. 2011. "Carry-Over Effects as Drivers of Fitness Differences in Animals." *Journal of Animal Ecology* 80:4–18.

Hartley, M., and M. Hunter. 1998. "A Meta-analysis of Forest Cover, Edge Effects, and Artificial Nest Predation Rates." *Conservation Biology* 12:465–69.

Hobbs, N., and T. Hanley. 1990. "Habitat Evaluation Do Use Availability Data Reflect Carrying Capacity." *Journal of Wildlife Management* 54:515–22.

Hobbs, N., and D. Swift. 1985. "Estimates of Habitat Carrying-Capacity Incorporating Explicit Nutritional Constraints." *Journal of Wildlife Management* 49:814–22.

Hoekman, S. T., L. S. Mills, D. W. Howerter, J. H. Devries, and I. J. Ball. 2002. "Sensitivity Analyses of the Life Cycle of Midcontinent Mallards." *Journal of Wildlife Management* 66:883–900.

Holmes, R. T. 2011. "Avian Population and Community Processes in Forest Ecosystems: Long-Term Research in the Hubbard Brook Experimental Forest." *Forest Ecology and Management* 262:20–32.

Holmes, R., P. Marra, and T. Sherry. 1996. "Habitat-Specific Demography of Breeding Black-Throated Blue Warblers (*Dendroica caerulescens*): Implications for Population Dynamics." *Journal of Animal Ecology* 65:183–95.

Homyack, J. A. 2010. "Evaluating Habitat Quality of Vertebrates Using Conservation Physiology Tools." *Wildlife Research* 37:332–42.

Houston, A. I., P. A. Stephens, I. L. Boyd, K. C. Harding, and J. M. McNamara. 2007. "Capital or Income Breeding? A Theoretical Model of Female Reproductive Strategies." *Behavioral Ecology* 18:241–50.

Johnson, M. D. 2007. "Measuring Habitat Quality: A Review." *Condor* 109:489–504.

Johnson, M. D., T. W. Sherry, R. T. Holmes, and P. P. Marra.

2006. "Assessing Habitat Quality for a Migratory Songbird Wintering in Natural and Agricultural Habitats." *Conservation Biology* 20:1433–44.

Kennedy, J., P. R. Witthames, R. D. M. Nash, and C. J. Fox. 2008. "Is Fecundity in Plaice (*Pleuronectes platessa* L.) Down-Regulated in Response to Reduced Food Intake during Autumn?" *Journal of Fish Biology* 72:78–92.

Kennedy, M., and R. Gray. 1993. "Can Ecological Theory Predict the Distribution of Foraging Animals—A Critical Analysis of Experiments on the Ideal Free Distribution." *Oikos* 68:158–66.

King, D. I., R. M. Degraaf, M.-L. Smith, and J. P. Buonaccorsi. 2006. "Habitat Selection and Habitat-Specific Survival of Fledgling Ovenbirds (*Seiurus aurocapilla*)." *Journal of Zoology* 269:414–21.

Kokko, H., and W. Sutherland. 2001. "Ecological Traps in Changing Environments: Ecological and Evolutionary Consequences of a Behaviourally Mediated Allee Effect." *Evolutionary Ecology Research* 3:537–51.

Kristan, W. B. 2003. "The Role of Habitat Selection Behavior in Population Dynamics: Source-Sink Systems and Ecological Traps." *Oikos* 103:457–68.

———. 2007. "Expected Effects of Correlated Habitat Variables on Habitat Quality and Bird Distribution." *Condor* 109:505–15.

Kristan, W. B., III, M. D. Johnson, and J. T. Rotenberry. 2007. "Choices and Consequences of Habitat Selection for Birds." *Condor* 109:485–88.

Law, B., and C. Dickman. 1998. "The Use of Habitat Mosaics by Terrestrial Vertebrate Fauna: Implications for Conservation and Management." *Biodiversity and Conservation* 7:323–33.

Leston, L. F. V., and A. D. Rodewald. 2006. "Are Urban Forests Ecological Traps for Understory Birds? An Examination Using Northern Cardinals." *Biological Conservation* 131:566–74.

Lloyd, J., and T. Martin. 2005. "Reproductive Success of Chestnut-Collared Longspurs in Native and Exotic Grassland." *Condor* 107:363–74.

Magi, M., R. Mand, H. Tamm, E. Sisask, P. Kilgas, and V. Tilgar. 2009. "Low Reproductive Success of Great Tits in the Preferred Habitat: A Role of Food Availability." *Ecoscience* 16:145–57.

Marra, P. P., K. A. Hobson, and R. T. Holmes. 1998. "Linking Winter and Summer Events in a Migratory Bird by Using Stable-Carbon Isotopes." *Science* 282(5395): 1884–86.

Mattisson, J., H. Andren, J. Persson, and P. Segerstrom. 2011. "Influence of Intraguild Interactions on Resource Use by Wolverines and Eurasian Lynx." *Journal of Mammalogy* 92:1321–30.

McPeek, M., N. Rodenhouse, R. Holmes, and T. Sherry. 2001. "A General Model of Site-Dependent Population Regulation: Population-Level Regulation without Individual-Level Interactions." *Oikos* 94:417–24.

Mihoub, J., P. Le Gouar, and F. Sarrazin. 2009. "Breeding Habitat Selection Behaviors in Heterogeneous Environments: Implications for Modeling Reintroduction." *Oikos* 118:663–74.

Mills, L. S. 2012. *Conservation of Wildlife Populations: Demography, Genetics, and Management.* 2nd edition. John Wiley & Sons, Ltd., West Sussex, UK.

Mitchell, G. W., C. G. Guglielmo, N. T. Wheelwright, C. R. Freeman-Gallant, and D. R. Norris. 2011. "Early Life Events Carry Over to Influence Pre-migratory Condition in a Free-Living Songbird." *Plos One* 6:e28838.

Morris, D. 1987. "Tests of Density-Dependent Habitat Selection in a Patchy Environment." *Ecological Monographs* 57:269–81.

———. 1988. "Habitat-Dependent Population Regulation and Community Structure." *Evolutionary Ecology* 2:253–69.

———. 2003. "How Can We Apply Theories of Habitat Selection to Wildlife Conservation and Management?" *Wildlife Research* 30:303–19.

Morrison, M. L. 2001. "A Proposed Research Emphasis to Overcome the Limits of Wildlife-Habitat Relationship Studies." *Journal of Wildlife Management* 65:613–23.

———. 2009. *Restoring Wildlife: Ecological Concepts and Practical Applications.* Island Press, Washington, D.C.

———. 2012. "The Habitat Sampling and Analysis Paradigm Has Limited Value in Animal Conservation: A Prequel." *Journal of Wildlife Management* 76:438–50.

Mortelliti, A., G. Amori, and L. Boitani. 2010. "The Role of Habitat Quality in Fragmented Landscapes: A Conceptual Overview and Prospectus for Future Research." *Oecologia* 163:535–47.

Nagy, L., and R. Holmes. 2005. "To Double-Brood or Not? Individual Variation in the Reproductive Effort in Black-Throated Blue Warblers (*Dendroica caerulescens*)." *Auk* 122:902–14.

Norris, D. 2005. "Carry-Over Effects and Habitat Quality in Migratory Populations." *Oikos* 109:178–86.

Norris, D. R., and P. P. Marra. 2007. "Seasonal Interactions, Habitat Quality, and Population Dynamics in Migratory Birds." *Condor* 109:535–47.

Norris, D., P. Marra, T. Kyser, T. Sherry, and L. Ratcliffe. 2004. "Tropical Winter Habitat Limits Reproductive Success on the Temperate Breeding Grounds in a Migratory Bird." *Proceedings of the Royal Society B-Biological Sciences* 271:59–64.

Norris, D. R., and C. M. Taylor. 2006. "Predicting the Consequences of Carry-Over Effects for Migratory Populations." *Biology Letters* 2:148–51.

Norton, M., S. Hannon, and F. Schmiegelow. 2000. "Fragments Are Not Islands: Patch vs. Landscape Perspectives on Songbird Presence and Abundance in a Harvested Boreal Forest." *Ecography* 23:209–23.

Owen-Smith, N., D. Mason, and J. Ogutu. 2005. "Correlates of Survival Rates for 10 African Ungulate Populations: Density, Rainfall and Predation." *Journal of Animal Ecology* 74:774–88.

Parejo, D., J. White, J. Clobert, A. Dreiss, and E. Danchin. 2007.

"Blue Tits Use Fledgling Quantity and Quality as Public Information in Breeding Site Choice." *Ecology* 88:2373–82.

Parker, G., and W. Sutherland. 1986. "Ideal Free Distributions When Individuals Differ in Competitive Ability: Phenotype-Limited Ideal Free Models." *Animal Behaviour* 34:1222–42.

Perryman, W., M. Donahue, P. Perkins, and S. Reilly. 2002. "Gray Whale Calf Production 1994–2000: Are Observed Fluctuations Related to Changes in Seasonal Ice Cover?" *Marine Mammal Science* 18:121–44.

Persson, J. 2005. "Female Wolverine (*Gulo gulo*) Reproduction: Reproductive Costs and Winter Food Availability." *Canadian Journal of Zoology* 83:1453–59.

Petit, L., and D. Petit. 2003. "Evaluating the Importance of Human-Modified Lands for Neotropical Bird Conservation." *Conservation Biology* 17:687–94.

Pettorelli, N., S. Dray, J. Gaillard, D. Chessel, P. Duncan, A. Illius, N. Guillon, F. Klein, and G. Van Laere. 2003. "Spatial Variation in Springtime Food Resources Influences the Winter Body Mass of Roe Deer Fawns." *Oecologia* 137:363–69.

Pidgeon, A. M., V. C. Radeloff, and N. E. Mathews. 2006. "Contrasting Measures of Fitness to Classify Habitat Quality for the Black-Throated Sparrow (*Amphispiza bilineata*)." *Biological Conservation* 132:199–210.

Piper, W. H. 2011. "Making Habitat Selection More 'Familiar': A Review." *Behavioral Ecology and Sociobiology* 65:1329–51.

Podolsky, R., and S. Kress. 1992. "Attraction of the Endangered Dark-Rumped Petrel to Recorded Vocalizations in the Galapagos Islands." *Condor* 94:448–53.

Pulliam, H. 1988. "Sources, Sinks, and Population Regulation." *American Naturalist* 132:652–61.

———. 2000. "On the Relationship between Niche and Distribution." *Ecology Letters* 3:349–61.

Pulliam, H., and B. Danielson. 1991. "Sources, Sinks, and Habitat Selection: A Landscape Perspective on Population-Dynamics." *American Naturalist* 137:S50–66.

Pysek, P., V. Jarosik, and T. Kucera. 2002. "Patterns of Invasion in Temperate Nature Reserves." *Biological Conservation* 104:13–24.

Reed, J., and A. Dobson. 1993. "Behavioral Constraints and Conservation Biology: Conspecific Attraction and Recruitment." *Trends in Ecology & Evolution* 8:253–56.

Robb, G. N., R. A. McDonald, D. E. Chamberlain, and S. Bearhop. 2008. "Food for Thought: Supplementary Feeding as a Driver of Ecological Change in Avian Populations." *Frontiers in Ecology and the Environment* 6:476–84.

Robb, G. N., R. A. McDonald, D. E. Chamberlain, S. J. Reynolds, T. J. E. Harrison, and S. Bearhop. 2008. "Winter Feeding of Birds Increases Productivity in the Subsequent Breeding Season." *Biology Letters* 4:220–23.

Rodenhouse, N., T. Sherry, and R. Holmes. 1997. "Site-Dependent Regulation of Population Size: A New Synthesis." *Ecology* 78:2025–42.

Rodenhouse, N., T. Sillett, P. Doran, and R. Holmes. 2003.

"Multiple Density-Dependence Mechanisms Regulate a Migratory Bird Population during the Breeding Season." *Proceedings of the Royal Society B-Biological Sciences* 270:2105–10.

Rodewald, A. D, and M. H. Bakermans. 2006. "What Is the Appropriate Paradigm for Riparian Forest Conservation?" *Biological Conservation* 128:193–200.

Rodewald, A. D., R. P. Rohr, M. A. Fortuna, and J. Bascompte. 2014. "Community-Level Demographic Consequences of Anthropogenic Disturbance: An Ecological Network Approach." *Journal of Animal Ecology*. doi: 10.1111/1365-2656.12224.

Rodewald, A. D., and D. P. Shustack. 2008. "Consumer Resource Matching in Urbanizing Landscapes: Are Synanthropic Species Over-Matching?" Ecology 89:515–21.

Rosenberg, K., J. Lowe, and A. Dhondt. 1999. "Effects of Forest Fragmentation on Breeding Tanagers: A Continental Perspective." *Conservation Biology* 13:568–83.

Runge, J. P., M. C. Runge, and J. D. Nichols. 2006. "The Role of Local Populations within a Landscape Context: Defining and Classifying Sources and Sinks." *American Naturalist* 167:925–38.

Runge, M. C., and P. P. Marra. 2005. "Modeling Seasonal Interactions in the Population Dynamics of Migratory Birds." In *Birds of Two Worlds: The Ecology and Evolution of Migration*, edited by R. Greenberg and P. P. Marra, 375–98. John Hopkins University Press, Baltimore, MD.

Saether, B., T. Ringsby, and E. Roskaft. 1996. "Life History Variation, Population Processes and Priorities in Species Conservation: Towards a Reunion of Research Paradigms." *Oikos* 77:217–26.

Saether, B., J. Tufto, S. Engen, K. Jerstad, O. Rostad, and J. Skatan. 2000. "Population Dynamical Consequences of Climate Change for a Small Temperate Songbird." *Science* 287:854–56.

Schlaepfer, M., M. Runge, and P. Sherman. 2002. "Ecological and Evolutionary Traps." *Trends in Ecology & Evolution* 17:474–80.

Schmidt, K. A. 2004. "Site-fidelity in Temporally Correlated Environments Enhances Population Persistence." *Ecology Letters* 7:176–84.

Schmidt, K. A., S. R. X. Dall, and J. A. van Gils. 2010. "The Ecology of Information: An Overview on the Ecological Significance of Making Informed Decisions." *Oikos* 119:304–16.

Sherry, T., and R. Holmes. 1996. "Winter Habitat Quality, Population Limitation, and Conservation of Neotropical Nearctic Migrant Birds." *Ecology* 77:36–48.

Sillett, T. S., and R. T. Holmes. 2002. "Variation in Survivorship of a Migratory Songbird throughout its Annual Cycle." *Journal of Animal Ecology* 71:296–308.

———. 2005. "Long-Term Demographic Trends, Limiting Factors, and the Strength of Density Dependence in a Breeding Population of Migratory Songbird." In *Birds of Two Worlds:*

The Ecology and Evolution of Migration, edited by R. Greenberg and P. P. Marra, 426–36. John Hopkins University Press, Baltimore, MD.

Sillett, T., R. Holmes, and T. Sherry. 2000. "Impacts of a Global Climate Cycle on Population Dynamics of a Migratory Songbird." *Science* 288:2040–42.

Sillett, T., N. Rodenhouse, and R. Holmes. 2004. "Experimentally Reducing Neighbor Density Affects Reproduction and Behavior of a Migratory Songbird." *Ecology* 85:2467–77.

Skagen, S. K., and A. A. Y. Adams. 2011. "Potential Misuse of Avian Density as a Conservation Metric." *Conservation Biology* 25:48–55.

Sloan, S., R. Holmes, and T. Sherry. 1998. "Depredation Rates and Predators at Artificial Bird Nests in an Unfragmented Northern Hardwoods Forest." *Journal of Wildlife Management* 62:529–39.

Studds, C. E., T. K. Kyser, and P. P. Marra. 2008. "Natal Dispersal Driven by Environmental Conditions Interacting across the Annual Cycle of a Migratory Songbird." *Proceedings of the National Academy of Sciences of the United States of America* 105:2929–33.

Studds, C. E., and P. P. Marra. 2005. "Nonbreeding Habitat Occupancy and Population Processes: An Upgrade Experiment with a Migratory Bird." *Ecology* 86:2380–85.

Sutherland, W., and K. Norris. 2002. "Behavioural Models of Population Growth Rates: Implications for Conservation and Prediction." *Philosophical Transactions of the Royal Society of London Series B-Biological Sciences* 357:1273–84.

Tozer, D. C., D. M. Burke, E. Nol, and K. A. Elliott. 2012. "Managing Ecological Traps: Logging and Sapsucker Nest Predation by Bears." *Journal of Wildlife Management* 76:887–98.

Van Horne, B. 1983. "Density as a Misleading Indicator of Habitat Quality." *Journal of Wildlife Management* 47:893–901.

Vitz, A. C., and A. D. Rodewald. 2006. "Can Regenerating Clearcuts Benefit Mature-Forest Songbirds? An Examination of Post-breeding Ecology." *Biological Conservation* 127:477–86.

———. 2007. "Vegetative and Fruit Resources as Determinants of Habitat Use by Mature-Forest Birds during the Post-breeding Period." *Auk* 124:494–507.

———. 2011. "Influence of Condition and Habitat Use on Survival of Post-fledging Songbirds." *Condor* 113:400–11.

Wagner, R. H., and E. Danchin. 2010. "A Taxonomy of Biological Information." *Oikos* 119:203–9.

Ward, M. P., and S. Schlossberg. 2004. "Conspecific Attraction and the Conservation of Territorial Songbirds." *Conservation Biology* 18:519–25.

Webster, M. S., P. P. Marra, S. M. Haig, S. Bensch, and R. T. Holmes. 2002. "Links Between Worlds: Unraveling Migratory Connectivity." *Trends in Ecology and Evolution* 17:76–83.

Werner, G., and J. Gilliam 1984. "The Ontogenetic Niche and Species Interactions in Size Structured Populations." *Annual Review of Ecology and Systematics* 15:393–425.

———. 1989a. *The Ecology of Bird Communities. Vol I. Foundations and Patterns*. Cambridge University Press, Cambridge UK.

———. 1989b. "Spatial Scaling in Ecology." *Functional Ecology* 3:385–97.

Wiens, J., and C. Graham. 2005. "Niche Conservatism: Integrating Evolution, Ecology, and Conservation Biology." *Annual Review of Ecology Evolution and Systematics* 36:519–39.

Wilbur, H. 1980. "Complex Life-Cycles." *Annual Review of Ecology and Systematics* 11:67–93.

Wilbur, H., and J. Collins. 1973. "Ecological Aspects of Amphibian Metamorphosis." *Science* 182:1305–14.

Witherington, B. E. 1997. "The Problem of Photopollution for Sea Turtles and Other Nocturnal Animals." In *Behavioral Approaches to Conservation in the Wild*, edited by J. R. Clemmons and R. Buchholz, 303–28. Cambridge University Press, Cambridge UK.

Zanette, L. Y., A. F. White, M. C. Allen, and M. Clinchy. 2011. "Perceived Predation Risk Reduces the Number of Offspring Songbirds Produce Per Year." *Science* 334:1398–1401.

4 Managing Habitats in a Changing World

BEATRICE VAN HORNE AND
JOHN A. WIENS

Habitats are the foundation of natural-resource management and conservation. Without habitats of the right sorts, the right sizes, the right variety, and in the right places, natural systems cannot exist.

But habitats the world over are in peril. The erosive effects of human actions on natural landscapes concerned people decades ago (e.g., Leopold 1949; Osborn 1953; Thomas 1956). These effects accelerated as burgeoning human populations converted vast areas of native terrestrial habitats into agricultural monocultures or urban sprawl and transformed natural aquatic habitats into plumbing systems. Nearly all of the North American tallgrass prairie has been converted to farmland, and over 90% of the native vegetation in the wheatbelt of Western Australia has been replaced by grain production, with the few small remnants widely scattered (Hobbs et al. 1993). The Sacramento–San Joaquin Delta in California, which historically was a vast area of tidal wetlands laced by over sixteen hundred km of tidal channels, has been engineered into a labyrinth of levees; less than 3% of the wetlands remain (Whipple et al. 2012). Habitat loss and the fragmentation of the remaining habitat have been labeled the greatest threats to biodiversity (Wilcove et al. 1998) and are the major causes of the declines of many imperiled species (Scott et al. 2010; Wiens and Gardali 2013).

The magnitude and extent of native habitat loss make it increasingly important to think about how best to manage and conserve what remains. It is not our intent here to review the panoply of papers and books that have been written about *habitat* and its many definitions and permutations (see, e.g., Cody 1985; Morrison and Hall 2002; Guthery and Strickland, this volume). Rather, we want to explore how simple conceptualizations of habitat can be compromised by the realities of nature. We conclude on a more hopeful note by considering how the effectiveness of habitat-based management and conservation can be strengthened as the pace of landscape change quickens. First, however, we consider how the aims of conservation or natural-resource management may influence how *habitat* is perceived.

Approaches to Managing Habitat

Habitat is generally taken as shorthand for the local environmental conditions in which a species of interest lives. Understanding the relationships between organisms and habitats has long been an objective of scientific research, the foundation of conservation and resource management, and the entry point for public appreciation of nature. It is by closely investigating the factors that influence survival and reproductive success that we gain a window into the processes that influence the value of a habitat in successfully supporting a species (Van Horne 1986).

Employing science to understand the processes that directly influence species of concern, however, requires focus. If we fully understand these processes, we should be able to predict how differences or changes in habitat will affect these species. Using this approach, the objectives are to learn more about a species and to understand which factors influence its selection of habitats, how differences in habitats affect mate choice or individual fitness, how the genetic structuring of populations varies with habitat conditions, and so on. This approach is often reductionist, emphasizing individuals, single species, or subdivisions within species, about which more detailed knowledge on the subject of habitats is always sought. More knowledge, in turn, generates more questions, which fuels further investigations. The goal is to minimize uncertainty. In the conservation-management arena, the emphasis is often on species that are protected by laws or regulations (e.g., the Endangered Species Act, the Marine Mammal Protection Act, the Migratory Bird Treaty) or are economically or recreationally important (e.g., salmon, deer).

In recent years, attention has been shifting to focus on multiple species, communities, ecosystems, and ecological processes. As much as we need more information on how habitats influence species viability, to extend a species-by-species approach beyond a few focal taxa is unrealistically costly in time and money. More to the point, by emphasizing single species, important trophic dynamics and ecosystem processes such as water balances or nutrient flows will be ignored. Efforts to conserve a species, even if based on detailed knowledge of its ecology and life history, may fail if the ecological processes that support its habitat are allowed to degrade. Broadening the perspective, however, inevitably means that knowledge of the multiple species is less detailed—one can't know everything about everything. Consequently, the goal frequently is to know enough to reduce uncertainty to an acceptable level for implementing management actions; more detailed knowledge may be unnecessary or difficult to justify, given the additional cost.

Of course, there is a middle ground where the species-centered and systems-centered approaches meet and blend. Because conservation and management are often directed toward particular species of concern, the information gathered in single-species studies can be critically important in framing actions. And because management and conservation frequently consider large, complex systems, they can generate and help to frame the questions (and often provide the funding) for the more detailed basic investigations.

Despite this interplay, however, conservation and management frequently emphasize a focal species or a community/ecosystem approach, rather than both. There is an inherent tension between the approaches.

There is also tension between the urge to simplify and the need to acknowledge the complexity of nature. Every situation is different in its details, but management or conservation that attempts to incorporate all the details will inevitably be so situation-specific that it has no general applicability. Some simplification is necessary. But simplification carried too far can result in generalities that are appealing but lack the reality to inform on-the-ground management. The challenge is to find a balance. To do this, one must consider several factors that, unacknowledged, can confound habitat-based research and management.

Some Realities

Whether the target of conservation or management is a particular species, a multispecies community, or a multitrophic ecosystem, *habitat* is often regarded as a fixed property of the target. Hence, the association with habitat is considered to be sufficiently tight that the species, community, and so forth can be characterized by its habitat. This is the basis for managing or conserving habitat for particular targets: protect or restore the right habitat conditions and the desired targets will thrive. Simplification is achieved by assuming constancy.

In truth, what we most often manage in the real world is a complex of areas that are being disturbed naturally or by humans at some frequency and spatial extent while simultaneously changing as a result of growth, aging, and interrelationships of the vegetation or the actions of the organisms living there (e.g., geese grazing in the tundra, predators affecting herbivore populations). Given such dynamics, it is unlikely that management will be able to maintain the full suite of habitat conditions that would be optimal for preserving biodiversity writ large. Management is further challenged because conditions in a focal landscape are influenced by factors that may originate outside its boundaries—fire, acid rain, dust, water flow, invasive species, and the like—all of which are in a continuing state of flux. And, as if this weren't enough, managing habitats is also confounded by variations within species and how they respond to habitat, variations in time, and variations in space. All of these factors thwart attempts at simplification. But how important are they?

Variation within Species and Their Responses to Habitat

In part because of laws, regulations, and the nature of legal challenges, investigations are often focused on one or several species of concern in areas that are directly under management influence or control. The emphasis on species is understandable. Species are (usually) well-defined biological entities. Most vertebrate and plant species, and many invertebrate species, are easily identified. They are useful units for management and conservation, and they resonate with the public. The value of charismatic species such as African elephants (*Loxodonta africana*), giant pandas (*Ailuropoda melanoleuca*), polar bears (*Ursus maritimus*), or California condors (*Gymnogyps californianus*) in mobilizing public support for conservation, or the importance of desired tree or fish species such as Douglas-fir (*Pseudotsuga menziesii*) or various salmonids (*Oncorhynchus* spp.) as targets for management, cannot be ignored. A good deal of scientific research carried out by graduate students in biology or wildlife departments or by single investigators rather than teams of scientists is directed at uncovering important details of the genetics, physiology, behavior, ecology, evolution, or simply the natural history of particular species. This knowledge is essential to conduct species-level management or conservation.

Studies of individual species almost always reveal fascinating and often idiosyncratic details about their relationships with habitat. Aided by technological developments (e.g., satellite-linked GPS tags, light-level geolocators; Recio et al. 2011; Lisovski et al. 2012; McKinnon et al. 2013) and advances in statistical approaches (e.g., AIC, occupancy models; Burnham and Anderson 2002; MacKenzie et al. 2006), knowledge about the habitat occupancy and use by targeted species has become increasingly detailed. Morrison (2012) has argued, however, that this greater detail may not be telling us what we really need to know about species and their habitats because the information is compromised by inappropriate study designs. By conducting studies in an area arbitrarily selected because it is convenient, measuring habitat parameters because they have been used in previous studies, bludgeoning the data with statistics, and then generalizing the results to a species over much or all of its distribution, the results and conclusions may tell us little about the factors that really affect a species' viability. Morrison suggests that we would learn more that is important by identifying biologically defined subsets of a species for study (he suggests using ecotypes that are associated with a specific type of habitat), conducting studies in areas that encompass key population processes, describing habitats using parameters that are biologically relevant to the target organisms, and bounding the scope of inference to what was actually studied. In other words, we will learn more about habitat by incorporating additional information about the biology of species into our study designs so that we measure the right variables at the right places.

Even as the term *species* encompasses variations in life-history attributes among individuals, populations, and ecotypes, necessitating a more nuanced approach, *habitat* is also subject to multiple sources of variation and interpretational challenges that make its definition elusive in any but general terms. Habitat is generally thought of as an attribute of species. Field guides usually include a section on "habitat" in their species' accounts—things like "moist coniferous forests; adjacent oaks, shade trees" or "moist woods of oaks, pines, and Douglas fir" (descriptions for chestnut-backed chickadee, *Poecile rufescens*, in Peterson 1990 and Sibley 2003, respectively). Such descriptions usually focus on major vegetation types. Habitat designations for species of conservation concern, such as those listed under federal or state endangered species acts, likewise usually begin (and often end) with documentation of the vegetation types and "critical habitats" needed to maintain the species. "Old-growth coniferous forest" for northern spotted owls (*Strix occidentalis carunia*) is an example. Thus, even though habitat is formally considered to include all important life requisites for a species, vegetation types are often used as surrogates. This simplification is understandable since general (or even specific) vegetation types are easily recognized by many people (there is little need to explain what a "moist coniferous forest" is to someone looking for a place to find chestnut-backed chickadees). Vegetation is also more easily managed to create the habitat desired for a species than are other life requisites such as food, protection from predators, appropriate thermal regimes, or suitable breeding sites. By using vegetation as a simplifying proxy for habitat, however, the linkages to habitat processes that actually affect a species can be lost.

It is frequently assumed that places where a species occurs contain good habitat and places where it is absent do not. This assumption is the basis for using correlations between environmental features and the presence or abundance of a species to describe its habitat. Species-distribution models, which are used to predict current or future distributions based on such correlations, assume that where a species occurs defines suitable habitat and that the available habitat is fully occupied ("saturated") (Wiens et al. 2009). It has long been known, however, that the abundance or density of a species in an area can influence the range of habitats occupied; Fretwell and Lucas (1969) modeled this in their "ideal free distribution" (see Rodewald, this volume). If habitat occupancy varies with density, then recording the features of an area that is occupied at a particular time to characterize a species' habitat will be sensitive to its density at that time. Moreover, whatever the density of a species in an area at a given time, it may not say much about the relationships with habitat that influence individual fitness ("habitat quality"). This is the basis for Van Horne's argument (1983) that density is a misleading indicator of habitat quality or of the concept of "ecological traps" (individuals are attracted to and occupy unsuitable habitats; Donovan and Thompson 2001). In populations having a source-sink structure (Liu et al. 2011), as many likely do, densities may sometimes be greater in the less suitable sink habitats, with population numbers maintained only by continuing immigration from the better source habitats.

Variation in Time

To simplify things, considerations of habitat in conservation and management often assume that habitat relationships documented at one time will apply to other times. The effectiveness of conservation and management actions will therefore be time-insensitive. So much of what we know about species and habitats is based on short-term studies that do not record temporal variations, which reinforces this assumption. All environments vary in time, however, and these variations can confound attempts to define the habitat conditions favoring one species or another. For example, the expansion of barred owls (*Strix varia*) into forests of the Pacific Northwest in recent years has altered the demography and habitat distribution of northern spotted owls where the species co-occur (J. D. Wiens 2012). Habitat quality may also change between years, depending on weather. Thus, a drought in the shrub-steppe of western Idaho caused a higher proportion of ground squirrels (*Urocitellus townsendii*) to survive in patches of native shrub habitat than in parched native bunchgrass habitat, where they had fared better than in shrub habitat in years of average rainfall (Van Horne et al. 1997).

Temporal dynamics over shorter periods can also influence the assessment of a species' habitat. The habitat associations of many species (especially migratory ones) change seasonally (Rodewald, this volume), or even within a season. In a Wisconsin grassland, for example, the features of habitats differed for the initial territories established by grasshopper sparrows (*Ammodramus savannarum*) and savannah sparrows (*Passerculus sandwichensis*) when birds returned in the spring. As the season progressed, however, more individuals established territories, filling the study area—habitat characteristics of the two species converged (Wiens 1973). In all of these examples, the determination of a species' habitat would differ depending on when it was assessed.

Time lags, legacy effects, or carryover effects (Rodewald, this volume) may also erode the match between habitat occupancy and habitat quality. Individuals that exhibit fidelity to previous breeding locations may continue to occupy those areas even after the original habitat has been drastically altered (Wiens and Rotenberry 1985). Unless appropriate measures of fitness (e.g., survival, reproduction) are also recorded, assessments of habitat made only after the conditions have changed will yield incorrect documentations of habitat. This is especially likely to occur for long-lived organisms that are sedentary or philopatric, such as trees or many birds.

Variation in Space

Just as all times are not the same, all places are not the same. While few would suggest that studies of habitat relations in a woodland could be extrapolated to a grassland, it is easier to assume that patterns documented in one oak woodland in one area might apply to a similar oak woodland in another area. If habitat is defined only by the dominant vegetation (e.g., "moist coniferous forest" for chestnut-backed chickadees), this simplification is easier still.

As Morrison (2012) observed, documentations of species' habitats are usually made by measuring parameters within an arbitrarily defined (and usually small) study area. Extrapolating results to a larger area assumes that the habitat preferences and relationships of a species remain constant over a range of spatial scales. This is unlikely to be the case. Studies of sage sparrows (*Artemisiospiza belli*), for example, illustrate how the associations between abundance and vegetation features change with changes in the scale of analysis (Wiens et al. 1987; Wiens 1989). At a broad, biogeographic scale, for example, densities were correlated with high shrub coverage and low grass coverage, while at a regional scale densities were greatest in areas with more grass and fewer shrubs. These changes occurred over a broad range of scales encompassing different ecoregions; within a narrower range (e.g., a few hectares to tens of square kilometers within a region), habitat relationships might be less sensitive to scaling differences.

The limits to how far one can extrapolate among areas or scales depend on one's objectives. The more general the objectives, the less important the differences among areas or across scales may be. But the inferences, and the guidance for management and conservation, also become more tenuous. At some point, differences in key factors are missed, compromising or invalidating extrapolations. Identifying the bounds of extrapolation requires that we know enough about how key habitat factors affect fitness to be confident that they remain relatively unchanged in their effects from place to place or among scales.

Documentations of the habitat associations of species are sensitive not only to place-to-place variations

and scale effects but to spatial context as well. Another simplifying assumption is that all important relationships between individuals and their habitats are contained within the defined study area. With the exception of oceanic islands, study areas are not isolated, stand-alone places but exist within a broader landscape mosaic. The boundaries of a study area are permeable to influences from beyond the boundaries (Hansen and di Castri 1992), especially if the study area is small. The features of the surrounding landscape can affect what occurs within the study area, and thus the inferences that are made about habitat. Three decades ago, Janzen (1983) called attention to how the surroundings of a protected area could alter the survival and reproduction of individuals within the area due to the actions of predators or competitors living in the adjacent landscape. Linkages with the landscape are part of the habitat within an area.

Much of what we have said about the problems with study areas applies also to management that is designed to conserve habitat in a reserve area or system of reserves. Uncertainties about the appropriate units to represent a species, flexibility in habitat selection within one or several species, variations in density and occupancy of habitat that may be unrelated to habitat suitability, time lags and temporal dynamics, and the effects of spatial variation, scale, and landscape context confound habitat management. It is important to consider the larger landscape, including multiple ownerships, in management for conservation, even if research is based on simplified portions of the landscape.

Habitat management for conservation, however, is often restricted to areas designated for biodiversity protection, especially of imperiled species that have small populations and severely restricted distributions. Inside such reserves or protected areas, habitat management for species or communities may be the primary goal, while outside, other resource-management goals or human activities (e.g., forestry, grazing, recreation) usually have primacy. Because of natural succession, climate change, or processes or threats moving across the reserve boundary, conditions within the reserve will likely change in ways that cannot be managed cost-effectively. A small reserve established for a focal species may change in ways that no longer benefit the species; the invasion of spotted owl habitat by barred owls or changes in the fire regimes that maintain habitat for

Kirtland's warbler (*Setophaga kirtlandii*) (Bocetti et al. 2012) are examples. Conversely, the larger landscape may also contribute to habitat conditions and connectivity that benefit species within the reserve.

The Challenge of a Changing World

Habitat management generally assumes that, absent major disturbances, the habitat within an area will persist more or less unchanged once the area is protected or managed or the habitat restored. Although successional processes may move the habitat away from (or toward) a desired state, these changes are generally thought to be gradual and at least to some degree predictable. Spatial variation may create heterogeneity in successional states at multiple scales, but this is often viewed as variation about a long-term equilibrium, as envisioned in shifting-mosaic steady-state concepts of succession (Borman and Likens 1979; Turner et al. 1993). Of course, ecologists and managers recognize that environments vary over time in ways unrelated to ecological succession; neither still believes in a strict equilibrium view of nature. The expectation that such variation occurs around a stable, long-term mean, however, is encapsulated in the concepts of stationarity (in aquatic environments) or historical range of variation (in terrestrial systems). Both concepts have been discredited (Milly et al. 2008; Wiens et al. 2012); even the average conditions are not stable and unvarying over any time scale relevant to conservation or management. Consequently, many of the factors determining "habitat" vary over multiple scales in time and space, burdening habitat management with cascading uncertainties. Unpredictable extreme events such as hurricanes and floods (Dale et al. 1998; CCSP 2008) only add to the uncertainty.

This is the current situation. While the future is by definition uncertain, the changes now underway are sure to create additional uncertainty in habitat-based management and conservation by altering not only the environmental context of habitats but the very nature of the habitats that people wish to manage, whatever the targets. Changes in global climate are projected to have profound effects on regional and local precipitation, temperature regimes, and the frequency and magnitude of extreme events. Changes in land use, driven by the combination of climate change, local and global economic forces, and changing societal demands for resources and commodities, will alter landscapes over multiple scales.

In some cases, ecological responses to these changes may be gradual, as species expand, contract, or shift distributions. Species-distribution modeling for a variety of taxa (e.g., Iverson et al. 2008; Lawler et al. 2009; Stralberg et al. 2009) has shown how extensive these distributional changes may be. The effects will differ among species, disrupting the composition of local communities and the trophic webs of ecosystems. In other cases, the changes may be sudden, as systems are pushed beyond thresholds of resilience or tolerance to drought, temperature, disturbance, or other factors. Such changes have already occurred, for example, in the shift from shrub dominance to grass dominance in shrubsteppe ecosystems (West and Hassan 1985) or from grassland to mesquite (*Prosopis glandulosa*) shrubland in the Chihuahuan Desert (Buffington and Herbel 1965). In both situations, grazing by domestic livestock, the invasion of exotic vegetation, and/or increased fire frequency and extent have shifted the ecosystems into an alternative state, as envisioned by state-and-transition models (Bestelmeyer 2006; Bestelmeyer et al. 2011). The habitat changes are potentially irreversible.

Whether the changes are gradual or sudden, however, they will produce a continuing disassembly and reassembly of biological assemblages, resulting in novel combinations of species and "no-analog" ecosystems that neither we nor the species have seen before (Williams and Jackson 2007; Stralberg et al. 2009; Hobbs et al. 2013). Over the ecological time scales that are relevant to conservation and management, bedrock geology may not change (unless there are major tectonic events), but everything else that we associate with habitat—vegetation composition and structure; the composition of prey, predator, and competitor assemblages; food-web dynamics; and even the soil or hydrology—will change, creating new mixtures and new habitats. The new habitats, in turn, will provide opportunities for the establishment of species new to the area, some of which may further disrupt the ecological systems while diminishing prospects for other, desirable species. The novel habitats and ecosystems of the future will require novel approaches to conservation and management.

Does the uncertainty about what habitats will exist in the future mean that the age of habitat-based management and conservation is past? Not necessarily. The habitats that exist in the future will still be habitats—places in which organisms can find the essential resources to survive and reproduce, populations can persist, and ecological processes can continue to operate. They will be habitats for *something*; we just may not know what.

A Way Forward for Habitat Research and Management

In this chapter, we have discussed some of the difficulties in simplifying habitat research and management to assure that investments are justifiable while still providing what is needed to maintain habitat and prevent further loss. The array of confounding factors we have discussed is daunting. Although not all need be included in all research designs or management plans, their potential impacts on habitat assessments should be acknowledged and given careful thought, instead of falling prey to the desire for simplification. For example, once one accepts that a landscape includes places that are changing at various rates in ways that alter habitat quality for species or communities, simple management rules that place bounds on economic and recreational activities, prescribe habitat restoration practices, or detail treatments for recognized threats such as fire must be adapted to a variety of current and expected conditions. Effective management may require using complex models to project future habitat conditions in landscapes at multiple scales, based on the gradual and/or sudden changes that are expected. Such models are always imperfect, but they may provide useful estimates of what is likely to take place in the landscape.

Resource managers and conservationists are faced with an increasingly complex and demanding task. As managers pursue their larger mission, which may involve managing lands for national defense, mining, energy development, recreation, hunting and fishing, or other activities, they must also manage habitat to avoid harming species of concern. In addition, managers must work with diverse landowners with multiple interests, which requires using an array of management practices that address habitat conditions across

the broader landscape. This work is messy, involving many societal and legal agendas and goals. Our current institutions (state and federal land management agencies, conservation organizations) are not generally organized to support and facilitate such approaches.

The ideal solution for protecting habitat would encompass large landscapes with multiple owners and with multiple management goals. It would affirm the importance of habitat and biodiversity conservation goals and support the use of science and modeling to establish and project habitat conditions over time and space, while applying decision support approaches (Marcot 2006; Heaton et al. 2008) to manage competing goals for the landscape. Tools in remote sensing, geographic information systems, and statistics now enable us to incorporate more of the complexity of nature into management and conservation practices. Cooperative approaches to habitat management (e.g., Collaborative Forest Landscape Restoration Program, Title IV of the Omnibus Public Land Management Act of 2009; Bagstad et al. 2012) are also increasing. The best approach would be to combine the focused, set-aside reserve approach with a cooperative approach to habitats over entire landscapes.

To conclude, we suggest several elements of an approach that may help to resolve the tension between simplification and complexity in habitat research and management. Although this is not a comprehensive list, we hope it will provide a useful starting point for identifying what we should be thinking about as we work to understand and conserve the elusive entity we call "good habitat."

1. Acknowledge that creating a mix of dominant vegetation types across a large landscape does not by itself ensure a diversity of animals. Additional information about ecological processes and species requirements is needed to evaluate the potential of habitat to support current and future biodiversity.

2. Maintain multiple trophic levels, including desired vegetation, herbivores, and predators. As Aldo Leopold observed long ago (1949), a functioning ecosystem requires all the parts.

3. Implement management over broad areas and multiple ownerships to consider processes occurring at broader scales while also addressing the

objectives of the multiple parties involved. Fuel buildup in western and Rocky Mountain forests, declining water tables in the desert southwest or the upper Columbia River basin, or the lack of early successional habitats for grassland birds in the northeastern United States are examples. Search for the shared objectives!

4. Consider how plant and animal species move across a landscape that is changing and create movement pathways. Habitat linkages need to be designed to facilitate seasonal and dispersive movements of a wide variety of species, while filtering out invasive species.

5. Consider the roles of protected areas in the context of larger complex landscapes. Use your best models to project conditions on these landscapes through time and consider how this context will influence biodiversity, species of concern, and conditions in the protected areas.

6. Use spatial modeling and statistics but don't get bogged down in complexity and esoterica. Always keep the objectives and the degree of certainty required in mind. Greater detail and precision are not always better. Ask what is good enough to meet your objectives with acceptable scientific certainty.

7. Recognize contingencies and thresholds that define the limits to resiliency. Resiliency may depend on replacing species that have declined or disappeared with other species with the same or similar functions, which will be dependent on processes such as water flow, soil building, and trophic interactions. Active intervention may be needed if process thresholds have been crossed, so understanding the processes is, once again, the nub of the problem.

8. Learn to live with, or even embrace, uncertainty. Weigh actions in terms of the consequences of being wrong. Couch actions in the context of risk, both ecological and economic.

9. Recognize the economic limits you face and prioritize where the costs are in line with the projected long-term benefits. Benefits should be defined by the interests of multiple stakeholders.

10. And finally, use the best current climate-change and land-use-change predictions to tailor management for the next century, not just the com-ing few years, while you ensure that the management design includes appropriately targeted scientific monitoring to allow you to learn from management activities and adapt to this new knowledge.

Including these considerations in cost-effective research and management won't be easy, but it will go a long way toward ensuring that "habitats" will persist into an uncertain future.

LITERATURE CITED

Bagstad, K. J., D. Semmens, R. Winthrop, D. Jaworski, and J. Larson. 2012. Ecosystem Services Valuation to Support Decisionmaking on Public Lands—A Case Study of the San Pedro River Watershed, Arizona. US Geological Survey Scientific Investigations Report 2012-5251, 93 pp.

Bestelmeyer, B. T. 2006. "Threshold Concepts and Their Use in Rangeland Management and Restoration: The Good, the Bad, and the Insidious." *Restoration Ecology* 14:325–29.

Bestelmeyer, B. T., A. Ellison, W. Fraser, K. Gorman, S. Holbrook, C. Laney, M. Ohman, D. C. Peters, F. C. Pillsbury, A. Rassweiler, R. Schmidt, and S. Sharma. 2011. "Analysis of Abrupt Transitions in Ecological Systems." *Ecosphere* 2(12): Article 129.

Bocetti, C. I., D. D. Goble, and J. M. Scott. 2012. "Using Conservation Management Agreements to Secure Postrecovery Perpetuation of Conservation-Reliant Species: The Kirtland's Warbler as a Case Study." *BioScience* 62: 874–79.

Bormann, F. H., and G. E. Likens. 1979. *Pattern and Process in a Forested Ecosystem.* New York, NY, Springer-Verlag.

Buffington, L. C., and C. H. Herbel. 1965. "Vegetation Changes in a Semidesert Grassland Range from 1858 to 1963." *Ecological Monographs* 35:139–64.

Burnham, K. P., and D. R. Anderson. 2002. *Model Selection and Multimodel Inference: A Practical Information-Theoretic Approach.* 2nd edition. New York, NY, Springer-Verlag.

CCSP. 2008. Weather and Climate Extremes in a Changing Climate: Regions of Focus: North America, Hawaii, Caribbean, and US Pacific Islands. A Report by the US Climate Change Science Program and the Subcommittee on Global Change Research (T. R. Karl, G. A. Meehl, C. D. Miller, S. J. Hassol, A. M. Waple, and W. L. Murray, eds.). Department of Commerce, NOAA's National Climate Data Center, Washington, D.C.

Cody, M. L., ed. 1985. *Habitat Selection in Birds.* Orlando, FL, Academic Press.

Dale, V. H., A. Lugo, J. MacMahon, and S. Pickett. 1998. "Ecosystem Management in the Context of Large, Infrequent Disturbances." *Ecosystems* 1:546–57.

Donovan, T. M., and F. R. Thompson III. 2001. "Modeling the Ecological Trap Hypothesis: A Habitat and Demographic

Analysis for Migrant Songbirds." *Ecological Applications* 11:871–82.

Fretwell, S. D., and H. L. Lucas, Jr. 1969. "On Territorial Behavior and Other Factors Influencing Habitat Distribution in Birds. I. Theoretical Development." *Acta Biotheoretica* 19:16–36.

Hansen, A. J., and F. di Castri, eds. 1992. *Landscape Boundaries: Consequences for Biotic Diversity and Ecological Flows*. New York, NY, Springer-Verlag.

Heaton, J. S., K. E. Nussear, T. C. Esque, R. D. Inman, F. M. Davenport, T. E. Leuteritz, P. A. Medica, N. W. Strout, P. A. Burgess, and L. Benvenuti. 2008. "Spatially Explicit Decision Support for Selecting Translocation Areas for Mojave Desert Tortoises." *Biodiversity Conservation* 17:575–90.

Hobbs, R. J., E. S. Higgs, and C. Hall, eds. 2013. *Novel Ecosystems: Intervening in the New Ecological World Order*. Oxford, Wiley-Blackwell.

Hobbs, R. J., D. A. Saunders, L. A. Lobry de Bruyn, and A. R. Main. 1993. "Changes in Biota." In *Reintegrating Fragmented Landscapes*, edited by R. J. Hobbs and D. A. Saunders, 65–106. New York, NY, Springer-Verlag.

Iverson, L. R., A. M. Prasad, S. N. Matthews, and M. Peters. 2008. "Estimating Potential Habitat for 134 Eastern US Tree Species under Six Climate Scenarios." *Forest Ecology and Management* 254:390–406.

Janzen, D. H. 1983. "No Park is an Island: Increase in Interference from Outside as Park Size Decreases." *Oikos* 41: 402–10.

Lawler, J. J., S. L. Shafer, D. White, P. Kareiva, E. P. Maurer, A. R. Blaustein, and P. J. Bartlein. 2009. "Projected Climate-Induced Faunal Change in the Western Hemisphere." *Ecology* 90:588–97.

Leopold, A. 1949. *A Sand County Almanac and Sketches Here and There*. New York, NY, Oxford University Press.

Lisovski, S., C. M. Hewson, R. H. G. Klaassen, F. Korner-Nievergelt, M. W. Kristensen, and S. Hahn. 2012. Geolocation by Light: Accuracy and Precision Affected by Environmental Factors. *Methods in Ecology and Evolution*: doi: 10.1111/j.2041-210X.2012.00185.x.

Liu, J., V. Hull, A. T. Morzillo, and J. A. Wiens, eds. 2011. *Sources, Sinks and Sustainability*. Cambridge: Cambridge University Press.

MacKenzie, D. I., J. D. Nichols, J. A. Royle, K. H. Pollock, L. L. Bailey, and J. E. Hines. 2006. *Occupancy Estimation and Modeling: Inferring Patterns and Dynamics of Species Occurrence*. Amsterdam, Elsevier.

Marcot, B. G., P. A. Hohenlohe, S. Morey, R. Holmes, R. Molina, M. C. Turley, M. H. Huff, and J. A. Laurence. 2006. "Characterizing Species at Risk. II. Using Bayesian Belief Networks as Decision Support Tools to Determine Species Conservation Categories under the Northwest Forest Plan." *Ecology and Society* 11(2):12.

McKinnon, E. A., K. C. Fraser, and B. J. M. Stuchbury. 2013. "New Discoveries in Landbird Migration Using Geolocators, and a Flight Plan for the Future." *Auk* 130:211–22.

Milly, P. C. D., J. Betancourt, M. Falkenmark, R. M. Hirsch, Z. W. Kundzewicz, D. P. Lettenmaier, and R. J. Stouffer. 2008. "Stationarity is Dead: Whither Water Management?" *Science* 319:573–74.

Morrison, M. L. 2012. "The Habitat Sampling and Analysis Paradigm Has Limited Value in Animal Conservation: A Prequel." *Journal of Wildlife Management* 76:438–50.

Morrison, M. L., and L. S. Hall. 2002. "Standard Terminology: Toward a Common Language to Advance Ecological Understanding and Application." In *Predicting Species Occurrences: Issues of Accuracy and Scale*, edited by J. M. Scott, P. J. Heglund, M. L. Morrison, J. B. Haufler, M. G. Raphael, W. A. Wall, and F. B. Samson, 43–52. Washington, D.C., Island Press.

Osborn, F. 1953. *The Limits of the Earth*. Boston, MA, Little, Brown and Company.

Peterson, R. T. 1990. *A Field Guide to Western Birds*. 3rd edition. Boston, MA, Houghton Mifflin Company.

Recio, M. R., R. Mathieu, P. Denys, P. Sirguey, and P. J. Seddon. 2011. "Lightweight GPS-Tags, One Giant Leap for Wildlife Tracking? An Assessment Approach." *PLoS ONE* 6(12): e28225. doi:10.1371/journal.pone.0028225.

Reynolds, K. M. 2013. http://en.wikipedia.org/wiki/Ecosystem _Management_Decision_Support.

Scott, J. M., D. D. Goble, A. M. Haines, J. A. Wiens, and M. C. Neel. 2010. "Conservation-Reliant Species and the Future of Conservation." *Conservation Letters* 3:91–97.

Sibley, D. A. 2003. *The Sibley Field Guide to Birds of Western North America*. New York, NY, Alfred A. Knopf.

Stralberg, D., D. Jongsomjit, C. A. Howell, M. A. Snyder, J. D. Alexander, J. A. Wiens, and T. L. Root. 2009. "Re-Shuffling of Species with Climate Disruption: A No-Analog Future for California Birds?" *PLoS One* 4(9): e6825.doc 10.1371/journal.pone.0006825.

Thomas Jr., W. L., ed. 1956. *Man's Role in Changing the Face of the Earth*. Chicago, IL, University of Chicago Press.

Turner, M. G., W. H. Romme, R. H. Gardner, R. V. O'Neill, and T. K. Kratz. 1993. "A Revised Concept of Landscape Equilibrium: Disturbance and Stability on Scaled Landscapes." *Landscape Ecology* 8:213–27.

Van Horne, B. 1983. "Density as a Misleading Indicator of Habitat Quality." *Journal of Wildlife Management* 47:893–901.

———. 1986. "Summary: When Habitats Fail as Predictors—The Researcher's Viewpoint." In *Wildlife 2000: Modeling Habitat Relationships of Terrestrial Vertebrates*, edited by J. Verner, M. L. Morrison, and C. J. Ralph, 257–58. Madison, WI, University of Wisconsin Press.

Van Horne, B., G. S. Olson, R. L. Schooley, J. G. Corn, and K. P. Burnham. 1997. "Effects of Drought and Prolonged Winter on Townsend's Ground Squirrel Demography in Shrub-steppe Habitats." *Ecological Monographs* 67:295–315.

West, N. E., and M. A. Hassan. 1985. "Recovery of Sagebrush-Grassland Vegetation Following Wildfire." *Journal of Range Management* 38:131–34.

Whipple, A. A., R. M. Grossinger, D. Rankin, B. Stanford,

and R. A. Askevold. 2012. Sacramento-San Joaquin Delta Historical Ecology Investigation: Exploring Pattern and Process. Prepared for the California Department of Fish and Game and Ecosystem Restoration Program. A Report of SFEI-ASC's Historical Ecology Program, SFEI-ASC Publication #672, San Francisco Estuary Institute-Aquatic Science Center, Richmond, CA.

Wiens, J. 1973. "Interterritorial Habitat Variation in Grasshopper and Savannah Sparrows." *Ecology* 54:877–84.

———. 1989. *The Ecology of Bird Communities. Vol. 2: Processes and Variations*. Cambridge, Cambridge University Press.

Wiens, J. A., and T. Gardali. 2013. "Conservation-Reliance among California's At-Risk Birds." *The Condor* 115:456–64.

Wiens, J. A., G. D. Hayward, H. D. Safford, and C. M. Giffen, eds. 2012. *Historical Environmental Variation in Conservation and Natural Resource Management*. Oxford, Wiley-Blackwell.

Wiens, J. A., and J. T. Rotenberry. 1985. "The Response of Breeding Passerine Birds to Rangeland Alteration in a North American Shrubsteppe Locality." *Journal of Applied Ecology* 22:655–68.

Wiens, J. A., J. T. Rotenberry, and B. Van Horne. 1987. "Habitat Occupancy Patterns of North American Shrubsteppe Birds: The Effects of Spatial Scale." *Oikos* 48:132–47.

Wiens, J. A., D. Stralberg, D. Jongsomjit, C. A. Howell, and M. A. Snyder. 2009. "Niches, Models, and Climate Change: Assessing the Assumptions and Uncertainties." *Proceedings of the National Academy of Sciences* 106 (Supplement 2): 19729–36.

Wiens, J. D. 2012. "Competitive Interactions and Resource Partitioning Between Northern Spotted Owls and Barred Owls in Western Oregon." PhD dissertation, Oregon State University, Corvallis, OR.

Wilcove, D. S., D. Rothstein, J. Dubow, A. Phillips, and E. Loss. 1998. "Quantifying Threats to Imperiled Species in the United States." *BioScience* 48:607–15.

Williams, J. W., and S. T. Jackson. 2007. "Novel Climates, No-Analog Communities, and Ecological Surprises." *Frontiers in Ecology and the Environment* 5:475–82.

PART II • HABITATS IN PERIL

5

CLINTON D. FRANCIS

Habitat Loss and Degradation

Understanding Anthropogenic Stressors and Their Impacts on Individuals, Populations, and Communities

Human activities are drastically changing Earth's landscapes, and the magnitude of human impacts on geophysical and biological processes fuels debate on whether we are entering a new epoch coined the *Anthropocene* (Crutzen 2002; Ellis 2011). Acknowledging this fundamental change, Ellis and coworkers retooled the standard biome classification system and mapped the anthropogenic biomes of the world (Ellis and Ramankutty 2008; Ellis et al. 2010). The results are stark; less than a quarter of ice-free land remains wild. That few landscapes across the globe remain untouched by humanity should give wildlife professionals and conservation biologists pause. Should conservation efforts focus primarily on preserving "natural" habitats and should research efforts concentrate on how wildlife persist in these habitat islands? With so little natural habitat left, these areas will undoubtedly prove insufficient for the preservation of most species. Instead, biologists are increasingly tasked with understanding how wildlife and their supporting ecological systems function in the context of sustained direct and indirect interactions with humans.

Throughout this book, readers are exposed to a variety of threats to wildlife. In this chapter, I will briefly describe more traditional views of "habitat loss and degradation" and place these concepts into the context that most wildlife already live in areas affected by human activities. In many ways, this is not a description of *particular* habitats in peril but an overview of some of the ways humans make *numerous* landscapes perilous for wildlife. As such, I aim to explain the way we should think about sources of habitat loss and degradation

through mini-reviews of three aspects of human disturbance that deserve additional attention: anthropogenic noise, artificial lighting, and recreational activities. I focus on these aspects because they are ubiquitous, growing faster than the human population, and because we still know relatively little about their impacts. Throughout the chapter, I also point out how we could do a better job of collecting meaningful data that can inform effective conservation policy and highlight the benefits of approaches that work across several levels of biological organization by examining how individual responses scale to population and community-level patterns. This includes measuring wildlife responses to habitat degradation differently, using the organism's sensory systems to guide our choice of relevant and functional environmental variables to measure, and studying communities rather than single species. I will start by describing how wildlife habitat loss has typically been viewed, then provide details on how this view must be expanded in several ways.

When Is Habitat Lost?

Habitat loss occurs when an area is converted from one state to another. Traditionally, we have almost always viewed habitat loss in terms of changes in land cover (i.e., vegetation structure or type). Take, for example, the clear-cut of a once-continuous forest to make way for agriculture, grazing, or a housing development. Importantly, habitat loss is species specific: an area that becomes uninhabitable for one species might become a new inhabitable area for another.

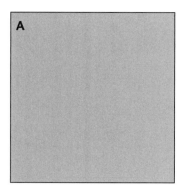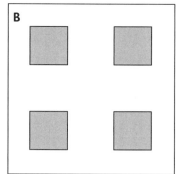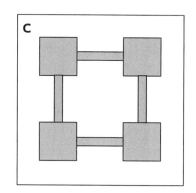

Figure 5.1 (A) A once-contiguous area (grey) undergoes both habitat loss and fragmentation, resulting in a landscape with four remaining patches (B). The use of corridors to maintain connectivity among habitat fragments (C) could be one strategy to maintain wildlife populations but may not work for all species (see main text).

Habitat Loss versus Habitat Fragmentation

Habitat loss and *habitat fragmentation* are terms often used to describe land cover change. Although these processes are different, they are not always clearly distinguished from each other in practice and have often simply been called "habitat fragmentation" in the literature (reviewed in Fahrig 2003; Ewers and Didham 2006; Collinge 2009). Habitat loss refers to the reduction in the overall area of a particular land cover type, and fragmentation entails the spatial arrangement and shape of remaining habitat patches. Why is it that habitat loss and fragmentation have proven difficult to disentangle? The short answer is that the process of habitat fragmentation is always accompanied by habitat loss, but habitat loss can occur without fragmentation (Fahrig 2003; Collinge 2009). That is, a habitat patch could be reduced in size without dividing it into separate pieces (i.e., fragmented), but it cannot be divided without losing some of the original area. These distinctions may seem trivial at first, but because wildlife biologists are interested in the viability of populations and integrity of ecological communities that support them, understanding the mechanisms responsible for observed shifts within a biological community in response to habitat change is imperative. Only by understanding *why* shifts occur might we be able to implement strategies to halt or slow population declines.

Here is a hypothetical scenario that demonstrates why the distinction between the effects of habitat loss and habitat fragmentation is important. Take two spe-cies: A & B. Individuals of Species A require a larger territory than individuals of Species B. Species A can also disperse or move farther across unfavorable en-vironments than Species B. When a once-contiguous area (fig. 5.1A) undergoes both habitat loss and habitat fragmentation (fig. 5.1B), the abundance of both spe-cies declines. Habitat loss might be the mechanism explaining a reduction in the abundance of Species A, simply because the proportion of original habitat is reduced. In contrast, this smaller amount of habitat may be sufficient to maintain the abundance of Spe-cies B, but the isolation of remnant patches precludes individuals from moving among them, also leading to a decline in population size. This is but one example of how habitat loss and fragmentation differ, and these details matter for management. For example, manage-ment efforts that create travel corridors linking sepa-rate habitat fragments might help Species B but would be ineffective for Species A (fig. 5.1C).

Thus, both habitat loss and fragmentation can func-tion in different ways to harm wildlife populations, but which mechanism is more common? Fahrig (2003) summarized the difference nicely. In general, habitat loss has strong, negative effects on biodiversity, whereas the influence of habitat fragmentation alone appears to be much weaker and may range from negative to positive. Importantly, one must carefully consider that the evidence for the influence of habitat fragmentation in the absence of habitat loss has come from relatively few studies. More details on habitat fragmentation can be found in chapter 7 of this volume, where it is discussed in detail (Smallwood, this volume), plus several other good sources (Forman 1995; Collinge and Forman 1998; Fahrig 2003; Ewers and Didham 2006; Collinge 2009).

Rethinking Measurements of Environmental Change and Wildlife Responses: Wildlife's Perspective

The traditional view of habitat loss and fragmentation that I outlined earlier neither reflects the range of environments experienced by wildlife nor the complexities that wildlife researchers need to consider when studying wildlife responses to human-caused environmental change. Aspects of the human enterprise other than land cover change can cause areas to become uninhabitable, and many landscapes are impacted by human activities in ways far more subtle than obvious cases like the transformation of a forest stand to a parking lot. Take, for example, an increase in trail density for recreation or shrub encroachment due to fire suppression or grazing. In these and other cases, it may not be an issue of habitat simply being "habitable" versus "uninhabitable." Instead, habitat quality ranges along a continuum from high quality to low quality. At some point along this continuum, an area becomes "uninhabitable," but this will depend not only on the species but also on the individual. Considerable research has shown that individuals of the same species can differ greatly in their responses to the same stimulus or environmental cue (fig. 5.2), and this variation is thought to contribute to the degree to which different individuals cope or fail to cope with environmental change (reviewed in Sih et al. 2004; Blumstein and Fernández-Juricic 2010). These individual responses have clear consequences for population persistence and community patterns (reviewed in Francis and Barber 2013; Rodewald, this volume). Less well known, however, is how nonrandom selection among individuals with differing responses to habitat degradation can influence the evolutionary trajectories of populations and either increase or decrease a population's viability.

It is also difficult to determine whether an area is "bad habitat" based only on where individuals are encountered, because the context of the environment matters greatly. An area may be unoccupied simply because better habitat exists nearby, because animals cannot readily access the patch, or for a variety of other

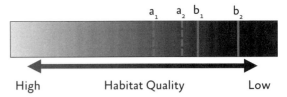

Figure 5.2 Habitat suitability should often be viewed along a continuum of quality rather than a paradigm that categorizes areas as inhabitable versus uninhabitable. Vertical lines represent the point at which a decrease in habitat quality could result in site abandonment for two different species (a and b) and individuals of the same species (denoted with subscript numbers). These thresholds will be context specific: they will vary among species, among individuals of the same species, but also according to the precise combination of environmental conditions in the area.

reasons. The reverse is also true; occupancy of an area does not necessarily mean that an area is better habitat than areas where a species is not detected. Instead, the quality of the habitat could be manifested after settlement. For example, individuals may settle in degraded and high-quality areas at similar rates, but then experience different rates of survival or reproductive success (fig. 5.3). Worse still, individuals may actually prefer degraded or altered habitat where reproductive success is very low (sometimes referred to as attractive sinks or ecological traps) relative to high-quality habitat where reproductive success is high (Robertson and Hutto 2007). These highly preferred but reproductively costly habitats can be the worst kind to preserve. For these reasons, it is unfortunate that much research involving habitat loss and degradation focuses primarily on the very coarse measures of presence/absence or abundance, especially because we know that abundance does not necessarily reflect habitat quality in terms of reproductive success (e.g., Van Horne 1983; Johnson and Temple 1986; Rodewald, this volume). These measures do tell us that there is a pattern of concern, but they usually tell us little about *why* a population declines or thrives due to a particular type of environmental change.

Part of understanding *why* organisms respond as they do to habitat degradation will require rethinking about which environmental variables should be measured in addition to quantifying distributional changes and variables reflecting reproductive success

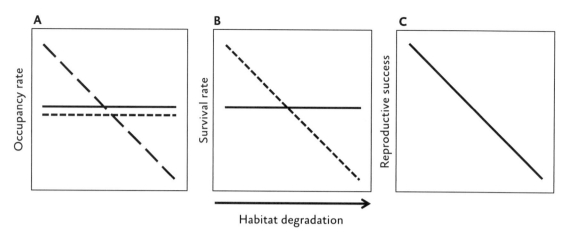

Figure 5.3 Declining habitat quality will not always cause reductions in abundance. For example, Species X (long dashed line) could decline in occupancy with degradation by avoiding degraded areas when making settlement decisions, but Species Y (short dashed line) and Z (solid line) may fail to respond to degradation while settling (A). In contrast, Species Y might experience decreased survival with degradation, whereas Species Z does not (B). Still, increased degradation may negatively affect reproductive success for Species Z (C). Importantly, over time these three different responses could all lead to the same pattern of population declines, but the mechanistic reasons responsible for the declines are different and matter for management.

and fitness. We typically measure habitat structure as it is perceived visually by humans, often failing to consider sensory systems used by other organisms that differ from our own (Van Dyck 2012). This bias is problematic because the way we perceive the structure and composition of a particular land-cover type does not necessarily reflect an area's functional significance for organisms under study. Instead, we must try to understand how organisms perceive their surroundings and frame research efforts with this knowledge to reveal how organisms respond to functional components of both natural and human-altered environments. So how might we make this shift? Because all organisms use sensory systems to obtain and respond to information in their environment, adopting a research approach grounded in sensory ecology can help reveal how and why organisms respond to natural variation in environmental conditions and anthropogenic environmental changes.

A clear example illustrating the benefit of research emphasizing the way in which wildlife perceive their environment comes from recent research on habitat selection in birds. For example, Farrell et al. (2012) used experimental playback of golden-cheeked warbler (*Setophaga chrysoparia*) vocalizations throughout an area that varied in habitat quality to determine the strength of conspecific (i.e., social) attraction for habitat selection. Surprisingly, playbacks in areas with little canopy cover, which was thought to be low-quality habitat, resulted high clustering of territories. Moreover, measures of breeding and reproductive success among males was not related to canopy cover but instead positively related to breeding density. This example illustrates that the cues used by organisms to make settlement decisions can differ quite drastically from what was assumed to be important. More importantly, however, this study reveals that a federally endangered species can successfully occupy a broader environmental range than previously thought and that playback experiments could be used in management to reestablish populations in areas where they no longer exist.

A second example comes from birds' abilities to use acoustic cues from natural enemies to make settlement choices. Playback of a common nest predator's vocalizations causes ground-nesting songbirds to avoid nesting in areas near the playback source (Emmering and Schmidt 2011). Similarly, several songbirds appear to respond in the same manner to playback of vocalizations from brown-headed cowbirds (*Molothrus ater*), which commonly reduce the reproductive success of many passerines by parasitizing their nests (Forsman and Martin 2009).

These examples illustrate how organisms use sensory systems to perceive their environment in ways different from human visual perception and standard scientific habitat quantification. Additionally, they emphasize the power of focusing research efforts on variables that are closely tied to function. This focus on function is especially important for anthropogenic stressors.

Anthropogenic Noise, Light, and Recreation: Habitat Loss without Loss of Habitat

Wildlife now face countless threats from human activities that vary both in type and severity. Three threats are discussed here because many others are covered in detail elsewhere in this book, and because these illustrate the importance of research that strives to understand how organisms obtain and respond to information in their environment.

Anthropogenic Noise

Human-generated noise is now common not only in and around cities but in rural settings exposed to industrial agriculture, dendritic transportation networks, aircraft overflight, and resource extraction and recreational activities (Barber et al. 2010). Collectively, these sources of noise amount to unprecedented ecological acoustic conditions for wildlife throughout the world and only recently has the effects of noise on wildlife gained heightened attention by ecologists and conservation biologists (Francis and Blickley 2012). Several recent reviews highlight some known effects of noise for taxonomically diverse wildlife but also emphasize that many potential consequences of noise have not yet been investigated (Patricelli and Blickley 2006; Barber et al. 2010; Kight and Swaddle 2011; Ortega 2012; Francis and Barber 2013). In the following paragraphs, several ways noise affects wildlife and some key areas that warrant more attention are discussed.

A long history of road ecology studies suggests that noise reduces habitat quality for wildlife, especially birds (e.g., Reijnen et al. 1995; Kuitunen et al. 1998; Forman et al. 2002). However, this evidence has been viewed with skepticism because many other factors might explain population declines near roads, such as mortalities from collisions with vehicles or the difference in vegetation near roads relative to comparison sites away from roads. Recently, however, several studies have dealt with the influence of confounding variables in their analyses or in their study designs and shown that noise alone can explain declines in bird species abundances and species richness (Bayne et al. 2008; Francis et al. 2009; Goodwin and Shriver 2011; Blickley et al. 2012). Interestingly, there appears to be considerable variation among species with some completely absent and others more common in noisy areas (Francis et al. 2009; Francis et al. 2012). Although we now know that noise does seem to change species distributions, it is less clear *why* some species appear to be more sensitive than others. For birds, the inability to communicate in noisy areas appears to play a large role. Several studies have shown that species with low-frequency vocalizations tend to avoid noisy areas more than species with higher frequency calls and songs (Rheindt 2003; Francis et al. 2011b; Goodwin and Shriver 2011). This pattern is directly related to the functionality of vocalizations in areas degraded by noise. Anthropogenic noise is dominated by low-frequency energy, thus species with low-frequency vocalizations experience more interference from the noise than do species with high-frequency vocalizations, making it that much more difficult for them to communicate (fig. 5.4). Because vocal frequency is negatively related to body size, one generalization might be that larger species could be more sensitive to noise than smaller species.

Why else might species avoid noisy areas? In addition to interfering with communication, continuous noise can also impair animals' abilities to use acoustic cues in a number of ways, such as interfering with an organism's ability to hunt (Schaub et al. 2008; Siemers and Schaub 2011), detect predators (Quinn et al. 2006), and navigate throughout a landscape by interfering with sounds used for orientation at a large scale (Francis and Barber 2013). However, habitat degradation due to noise can also affect wildlife after they settle in noisy areas. For several bird species, individuals breeding in noisy areas have reduced reproductive success compared to individuals in quiet areas (Halfwerk et al. 2011; Kight et al. 2012; Schroeder et al. 2012), suggesting that noise avoidance cannot be the only metric of focus for biologists seeking to quantify

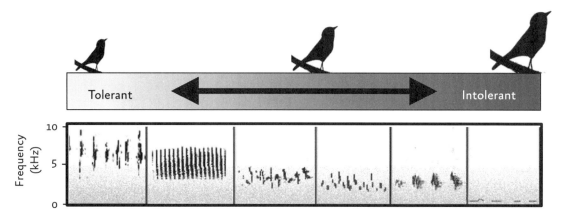

Figure 5.4 Most anthropogenic noise has increasing energy at lower frequencies (grey background in spectrograms), such that higher-frequency songs experience less acoustic interference from noise than lower-frequency songs. Additionally, because body size is negatively correlated with vocal frequency, larger species tend to be more intolerant to noise than are smaller species.

impacts from noise. Importantly, in all of these studies the reduction in reproductive success occurred in the *absence* of the influence of predation, which can complicate determining the net effect of noise on reproductive success (see the following discussion).

Halfwerk et al. (2011) proposed four nonmutually exclusive mechanisms that could explain reduced reproductive success for birds in noisy areas. First, noise may interfere with a female's ability to assess male quality and cause her to invest less energy in egg production and/or maternal care of chicks. Second, lower-quality individuals may settle in noisy areas, thus the effect of traffic noise on reproductive success would be governed by a nonrandom distribution of the quality of individuals with respect to noise. Third, noise may increase physiological stress in individuals occupying noisy areas, which can cause behavioral changes that reduce fitness and can also have direct deleterious effects. And finally, noise may compromise parent-offspring communication such that nestling provisioning is altered in some way.

Of these possibilities, there is some empirical support for a breakdown in parent-offspring communication. First, Swaddle et al. (2012) found noise significantly reduces the signal-to-noise ratio (i.e., how "hearable" a sound is compared to background noise) of chick begging calls at nearby perches used by nesting eastern bluebirds (*Sialia sialis*). In this same study system, Kight et al. (2012) reported that bluebirds fledge fewer young with increases in noise levels. A direct

link between decreased reproductive output and the influence of noise on parent-offspring communication has yet to be tested in this system. However, research on tree swallows (*Tachycineta bicolor*) found that nestlings exposed to noise beg less to the sounds made by parents arriving at the nest than did nestlings not exposed to noise (Leonard and Horn 2012). Apparently, by masking the sounds of arriving parents, noise causes nestlings to miss feeding opportunities, which could compromise growth and survival. Finally, a third study provides limited support for a communication breakdown between parents and offspring. Schroeder et al. (2012) studied the breeding biology of a house sparrow (*Passer domesticus*) population exposed to chronic noise from a large generator. Similar to the study with bluebirds, they found nests in noise-exposed areas to fledge fewer young than nests in quiet areas. Additionally, chicks reared in noisy areas were smaller than those in quiet areas, and the authors linked this trend to low provisioning rates by females. Although other explanations that could explain the inadequate provisioning rates provided by females are plausible, such as elevated stress, the authors suggest that females provision less because they are unable to hear their offspring's begging calls. Collectively, these separate studies suggest that impaired parent-offspring communication may be one mechanism that explains lower reproductive success in noisy areas. Whether the other mechanisms are at work remains to be seen.

Finally, noise can also interfere with breeding attempts well before nestlings hatch and begin to communicate with parents. Two studies have shown that territorial males inhabiting noisy areas are less successful in attaining mates than are males in less noisy areas (Habib et al. 2007; Gross et al. 2010). This may be due to several different factors, including a nonrandom distribution in the quality of individuals with respect to noise or due to females' decreased ability to detect and discriminate signals masked by noise to make mate-choice decisions. Another possibility lies in how birds that regularly inhabit noisy areas respond to noise. A growing number of birds have been shown to adaptively alter their songs and calls in response to noise (e.g., Gross et al. 2010; Francis et al. 2011a). Although this flexibility has been viewed as an important behavioral adaptation that permits species to overcome the acoustic conditions of cities and roadways that make communication difficult, the consequences of these altered songs are unknown. Because small changes in song features can influence female preference for a song or even species recognition (Nelson and Marler 1990), it is possible that noise-dependent behavioral changes to songs could render individuals unattractive to females or, worse yet, unrecognizable as species-specific signals. In other words, the flexibility that outwardly appears to be a mechanism used to cope with human noise may actually represent an evolutionary trap (Schlaepfer et al. 2002). Although the mechanisms responsible for reduced pairing success in noisy areas remain to be sorted out, the conservation implications of a smaller proportion of the population breeding has consequences of real concern to biologists.

Artificial Night Lighting

As with anthropogenic noise, ecologists and conservation biologists have only recently begun to explore the range of wildlife and ecological responses to artificial lights. This is surprising given that roughly 30% of vertebrates and 60% of all invertebrates are nocturnal (Holker et al. 2010)—clearly a large swath of diversity that may be especially sensitive to how we change nighttime conditions. An early edited volume exploring the ecological consequences of artificial light revealed that light has the potential to alter species distributions, deplete populations through direct mortality, alter species interactions, and alter animal circadian and circan-

nual rhythms (Rich and Longcore 2006). Importantly, the volume illustrates quite clearly that relatively few studies have documented the existence and severity of the many potential consequences of artificial lighting.

Perhaps the most well-known effect of light pollution involves sea turtles. Emerging hatchlings on beaches use light over the horizon of the ocean as one of several cues to find and move toward the surf (Tuxbury and Salmon 2005). However, artificial night lighting emitted from structures along the beach can cause hatchlings to travel *away* from the ocean where survival is unlikely. In this case, light pollution can be viewed as an ecological trap (Schlaepfer et al. 2002) because it mimics what had been a reliable environmental cue used for navigating toward required habitat. Fortunately, in-depth research on the sensory ecology of turtle hatchlings has permitted managers with targeted strategies to reduce inland movements. These strategies include shading lights from beach areas, or even using lights with wavelengths that do not attract hatchlings (Witherington 1997).

Unfortunately, however, light management strategies targeted for one species may not work for others. Lighting used along beaches in Florida that minimizes orientation problems for hatchling sea turtles also reduces habitat use and foraging activity in beach mice (*Peromyscus polionotus*) (Bird et al. 2004). This contrast highlights the need to understand responses to anthropogenic stressors from many species (see community-level section later on). Single-species management solutions have the potential to fail to accommodate the needs of other community members or, even worse, be counterproductive by creating additional problems.

Artificial night lighting can alter signaling behavior, too. Male green frogs (*Rana clamitans*) call less when exposed to artificial lighting (Baker and Richardson 2006). Both this decrease in calling activity and mice's avoidance of illuminated areas may be explained by the same mechanistic response—a heightened sense of predation risk with increases in ambient light. This is a likely explanation because natural light levels can influence actual predation risk across many species (reviewed in Bird et al. 2004). Whether choosing not to call for mates or minimizing foraging activities to reduce predation risk, these behavioral adjustments in response to artificial lighting could have relatively straightforward fitness consequences in the form of lowered breeding success or creating energy deficits, respectively.

In contrast, predator responses to artificial lighting appear mixed. Several early studies found some bat species to exploit the high density of insects that are attracted to and trapped by streetlamps (e.g., Rydell and Racey 1995). However, more recent studies have found other bat species to avoid artificially lit areas (Stone et al. 2009; Stone et al. 2012). How other nocturnal predators respond to artificial lighting should be a research priority.

Although not habitat per se, artificial lights located throughout avian migratory corridors contribute to countless bird mortalities each year. During their nocturnal migration, birds are drawn to and collide with structures ranging from buildings to radio towers because they are attracted to their artificial lights (Gauthreaux and Belser 2006). The precise reasons for birds' attraction to night lighting is far from clear, but explanations range from artificial lights interfering with birds' ability to orient through magnetism (Poot et al. 2008), temporarily blinding birds, or through "capture" by disorienting them and altering their flight paths (Gauthreaux and Belser 2006). Regardless of the mechanism, in efforts to reduce birds' attraction to lit structures, researchers have tested whether birds respond differently to various light colors. Poot et al. (2008) found birds to be most attracted to white and red light and less attracted to blue and green light. This initial study provides hope that mitigation efforts will become available to decrease bird collisions with man-made structures.

Although artificial lighting may pose a serious threat to birds migrating long distances, its influence on birds during other life-history stages is less clear. Several breeding European songbirds sing earlier when their territories are exposed to night lighting and at least one species' breeding biology changes drastically (Kempenaers et al. 2010). Blue tit (*Cyanistes caeruleus*) females lay eggs a few days earlier and males are two times more successful in attracting extra-pair mates in lit territories than nonlit areas. A separate study confirmed these temporal advances in reproductive physiology in the European blackbird (*Turdus merula*) but also that they molt earlier (Dominoni et al. 2013). Critically, however, in a creative experimental design

in which urban and nonurban male blackbirds were experimentally treated with light, the authors show that urban males treated with light developed their reproductive system far earlier than nonurban males, suggesting that urban light conditions have selected for different physiological phenotypes than in nonurban birds. How these changes impact fitness is unknown, but these studies suggest that artificial lighting has the potential to serve as a source of selection shaping the physiology of birds. Whether female physiology is coordinated with male physiology is unclear, and, if not, these changes could create mismatches that could affect reproductive success.

A final potential consequence of artificial lighting involves the indirect effect on wildlife mediated through the effect of lighting on invertebrate communities, which can be essential to trophic chains and for transferring materials from aquatic to terrestrial ecosystems (e.g., Perkin et al. 2011). The attraction and trapping of insects by artificial lighting is a well-known phenomenon (e.g., Eisenbeis 2006). Regardless of why insects are drawn to artificial lights, it remains unclear to what extent insect mortalities associated with artificial lights have larger consequences for the population or community. One recent study examined the attraction of moths to different wavelengths of artificial light (van Langevelde et al. 2011). Lamps with shorter wavelengths attracted more individuals and species than lamps with longer wavelengths. Additionally, larger species with large eyes were especially attracted to lamps with shorter wavelengths, presumably because they are more attracted to ultraviolet light than smaller species. Another study focused on ground-dwelling invertebrates found artificial lighting to alter the composition of these communities (Davies et al. 2012). The authors report that predatory and scavenging taxa were in higher abundances near streetlights both during the day and at night, suggesting that they were not attracted to the stimulus energy of the lights but to other changes in the community structure. It is unknown whether the increase in these taxa was due to an increased abundance of prey species that are attracted to lights or due to other factors. These two studies provide a first glimpse of how insect communities change in response to night lighting and raise important questions regarding the extent of cascading effects through mutualistic interactions, such as pollination, or predator-prey interactions that could reverberate through trophic levels.

Recreation and Presence of Humans

Management of protected areas is increasingly challenged with balancing the needs of wildlife and providing recreational opportunities for citizens. For example, in the United States, the number of people participating in outdoor recreation appears to be increasing exponentially (Reed and Merenlender 2008). Research on the influence of these activities on wildlife, and especially wildlife behavior, has a long history (e.g., Knight and Gutzwiller 1995), but much of it has fallen short of documenting population-level impacts. Consumptive uses, such as unregulated hunting and fishing, have obvious negative consequences by directly reducing wildlife population sizes, but nonconsumptive recreation can also create problems. For example, the onset of hiking in a wildlife refuge in New England triggered a sharp decline in two wood turtle (*Clemmys insculpta*) populations (Garber and Burger 1995), suggesting that even outwardly benign activities can have negative consequences for wildlife.

Some types of nonconsumptive recreation appear more likely to affect wildlife. For example, animals might respond to snowmobiles, off-road motorized vehicles, or boat traffic by fleeing or hiding, which could result in less time foraging, contributing to declining health or condition (Frid and Dill 2002). Several studies have shown wildlife to experience elevated baseline stress hormones in response to motorized recreation and vehicles. For example, both wolves (*Canis lupus*) and elk (*Cervus elaphus*) experience heightened physiological stress in response to snowmobile traffic (Creel et al. 2002). Elevated stress has also been associated with road density for elk (Millspaugh et al. 2001). From these studies, it is not clear whether elevated stress is due to the physical presence of vehicles, the association between vehicles and humans, or the noise they create. Elevated stress resulting from these responses to human activities could lead to higher risk of disease and declines in reproductive success or even survival. Motorized recreation might also influence reproduction in other ways. For example, American oystercatchers

(*Haematopus palliatus*) spend less time incubating with increased off-road motorized traffic (McGowan and Simons 2006). Typically, behavioral responses like these have been assumed to have fitness costs. However, in this particular case, McGowan and Simons (2006) found no evidence for reduced reproductive success with less time incubating. More studies need to try to quantify whether the behavioral changes we measure translate into actual costs for the individual and the population.

Perhaps more troublesome is that forms of recreation that lack the sounds and speeds of motorized forms of recreation might also have serious consequences for wildlife. For example, desert bighorn sheep (*Ovis Canadensis nelsoni*) in Canyonlands National Park, Utah, respond more strongly to hikers than to vehicle or mountain-biking traffic (Papouchis et al. 2001). Elsewhere in Utah, the flushing responses of bison (*Bison bison*), mule deer (*Odocoileus hemionus*), and pronghorn antelope (*Antilocapra americana*) did not differ when responding to hikers versus mountain bikers (Taylor and Knight 2003), suggesting that they perceived both of these intrusions as a threat worth escaping. Recreationists may not recognize flushing responses as problems for wildlife, but such responses could cause breeding animals to leave young unprotected from harsh environmental conditions or predators or have energetic consequences by forcing animals to spend less time foraging or resting.

The presence of people might also render areas unsuitable for many species and cause dramatic shifts in the composition of communities. For example, Reed and Merenlender (2008) used scat surveys to compare carnivore communities in protected areas that permitted hiking relative to those that did not in Northern California. Areas where hiking was not permitted had five times the native carnivore density than areas where hiking was permitted. Scat of domestic dogs was not detected in areas without recreation, but it was found in high densities in areas that permitted hiking. Outwardly, one might suspect that the presence of dogs might be responsible for the large differences in carnivore communities between sites, and this perception is reflected in many management practices that aim to limit the potential impacts of dogs by excluding them from protected areas or implementing leash laws (Forrest and St. Clair 2006). However, Reed and Merenlender (2011) conducted a follow-up study involving dog management practices and found no difference in the density or species richness of native carnivore species in areas where dogs were permitted off-leash, permitted on-leash, or not permitted at all. Instead, the authors argue that the presence and abundance of human visitors drive these patterns.

Although dogs may not influence the abundance and diversity of mammalian carnivores in protected areas, birds appear to be more sensitive. A study in Australia found that bird detections and estimates of species richness were more severely affected by experimental treatments of humans walking with dogs on-leash than humans walking alone (Banks and Bryant 2007). These temporary disturbances that displace birds or alter their behavior might not always have long-lasting consequences, but they can cause wildlife to abandon otherwise suitable habitats in some circumstances. For example, a recent study in Finland compared breeding bird communities along hiking trails to those in undisturbed control areas within Oulanka National Park (Kangas et al. 2010). They found that the composition of the bird community differed between these two areas but that total species richness did not. The difference appeared to be driven by high sensitivity of ground-nesting species to hikers.

Finally, recreation and other human activities have the potential to impact the evolutionary trajectories of populations. For example, estimates of bird mortalities from collisions with vehicles are expected to exceed eighty million per year in the United States (Erickson et al. 2005), and a recent study suggests that these mortalities may not only be nonrandom within a population but also declining, which is suggestive of an evolutionary response (Brown and Brown 2013). Specifically, Brown and Brown (2013) reported that cliff swallows (*Petrochelidon pyrrhonota*) killed by vehicles accessing a popular recreation site have longer wings than the population at large and that over a thirty-year period both wing length and number of road-killed swallows has declined. Although the authors could not control for additional variables that could explain some of these patterns—such as the density of scavengers that remove killed birds from the road and bias road-kill estimates or temporal changes in the sizes and shapes of vehicles—this study illustrates that human

activities could play an important role in the evolution of populations. Whether anthropogenic forces less obvious than road kill can also serve as strong sources of selection should be a priority in future research.

Collectively, these studies suggest that obviously detrimental and outwardly harmless forms of recreation can have consequences for wildlife that span behavioral adjustments, population demographic, and evolutionary change, plus community-level changes. Most of these examples demonstrate changes in behavior, such as fleeing or hiding, and some document changes in the communities present in areas used by recreationists. To what degree the findings from the first set of studies are linked to the second remains unclear, but it has often been assumed that behavioral responses to avoid humans adequately reflect vulnerability to human activity (Gill et al. 2001). Testing this link in a quantitative manner remains an important goal for wildlife-disturbance studies (Blumstein and Fernández-Juricic 2010). Still needed are studies that determine at what threshold flushing responses to humans translates into permanent site abandonment or measurable declines in individual fitness due to energetic trade-offs.

A Community-Level Lens of Habitats in Peril: Interactions and Indirect Effects

In most studies involving habitat loss or degradation, we infer that the decrease in abundance of wildlife is the direct result of the change in habitat quality. Yet we know natural communities are usually more complicated and interactions among community members can make it difficult to simply draw a straight line between a measure of degradation and a documented response. Instead, many responses documented in degraded areas are likely indirect effects that are mediated by interactions among species. Additionally, regardless of whether a species' distribution changes as a direct or indirect result of habitat degradation, it is also important to ask what the consequences of that change will be for the larger ecosystem. Community-level approaches to understanding the ecological responses to habitat degradation will provide much greater insights on not only the mechanisms driving ecological change but also the consequences for organisms with no direct link to the disturbance of interest.

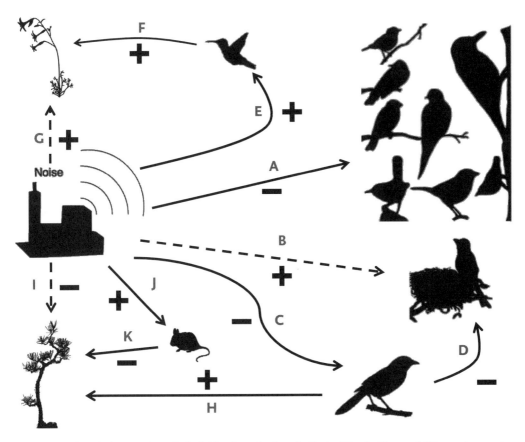

Figure 5.5 Pathway by which noise directly (solid lines) and indirectly (dashed lines) affects wildlife species, predatory-prey dynamics, and interactions among species and plants. See text for full description.

I highlight this community-level approach with work from my own long-term research involving the effects of noise on natural communities. Starting in 2005, my colleagues and I have isolated chronic anthropogenic noise as a single experimental stimulus by sampling woodlands surrounding natural gas wells with and without noise-producing compressors. Initial findings revealed that noise, in the absence of other anthropogenic stimuli, reduces the breeding bird community's species richness and alters the composition of the community such that communities in noisy areas were not simply subsets of communities in quiet areas (Francis et al. 2009; fig. 5.5A). For example, some species, such as the house finch (*Carpodacus mexicanus*) and black-chinned hummingbird (*Archilochus alexandri*), nest almost exclusively at noisy sites. This filtering appears to be largely explained by species-specific vocalizations (Francis et al. 2011b), as discussed previously. Importantly, however, this nonrandom sorting of

the avian community based on vocal frequency should explain negative (avoidance) or neutral (persistence) responses but should not explain positive (attraction) responses to noisy areas.

A possible explanation for the positive responses by some species comes from detailed study of the avian community's nesting success and nest-predation experiments using artificial nests. In contrast to the pattern of declining reproductive success with noise discussed earlier, in our study system nest success increases with noise levels (Francis et al. 2009; Francis et al. 2012b; fig. 5.5B). This pattern is evident both among and within species and is driven by decreases in predation pressure with increases in noise levels because a key nest predator, the western scrub-jay (*Aphelocoma californica*), tends to avoid noisy areas (figs. 5.5C, D). Given recent evidence that birds use acoustic cues representative of predation risk to make settlement decisions (e.g., Emmering and Schmidt 2011), it is possible that the positive responses to noise observed in finches, hummingbirds, and other species actually represent settlement decisions made based on the absence of acoustic cues representative of predation risk by jays.

Noise and other stressors can also have indirect effects on ecological systems with long-lasting consequences. The western scrub-jay and black-chinned hummingbird each provide important ecological services for plants in the form of seed dispersal and pollination, respectively. In our study area, noise appears to have both positive and negative indirect effects on plants receiving these services (Francis et al. 2012a). Experiments using artificial flowers suggest that the high abundance of hummingbirds in noisy areas leads to more visits to flowers and a greater movement of pollen among them relative to flowers in less noisy areas (figs. 5.5E–G). In contrast, experiments focused on piñon pine (*Pinus edulis*) seed dispersal suggest that the scrub-jay's avoidance of noisy areas appears to disrupt seed dispersal services and lead to reduced seedling recruitment in noisy areas (Francis et al. 2012a; figs. 5.5C, H, I). An interesting supplement to this pattern is that mice (*Peromyscus*) prey upon piñon seeds more often in noisy relative to quiet areas (figs. 5.5I–K), perhaps representing an indirect effect of noise mediated by reduced competition for food resources or less predation pressure from acoustically oriented nocturnal predators that may be sensitive to noise (e.g., owls). While these mechanistic causes for mice and other species' responses to noise still must be sorted out, the long-term consequences to ecosystem structure resulting from declining seed recruitment of a dominant tree is of considerable conservation concern.

Collectively, this example illustrates that anthropogenic stressors can have indirect effects that reverberate throughout natural communities. For our study system, the importance of the scrub-jay's response to noise for other organisms suggests that research on wildlife with important ecological roles within a community could be particularly important to understanding the full ecological costs of environmental degradation. Thus, additional research focused on predator responses could be especially revealing; their responses could cause the disruption of top-down effects, which can trigger comprehensive changes to the structure of both heterotrophic and autotrophic organisms within a community (reviewed in Estes et al. 2011). Clearly, understanding these cascades will only come from careful studies that adopt a more rounded research approach that measures responses by multiple organisms simultaneously and explicitly considers relationships among community members.

Key Research Needs for Moving Forward
Measuring Noise and Light and Quantifying the Magnitude of Their Effects

Although research on artificial lighting and noise pollution is still rather new, the early work on these anthropogenic stimuli clearly shows that they can be problematic for many wildlife species and communities. To fully understand the mechanisms responsible for various responses and unravel the cascading ecological consequences of their effects, future research on the impacts of these stimuli must be grounded in the sensory biology of the organism(s) and functional aspects of the stimulus. Why is this so? Because species have unique sensitivities to different wavelengths of light and sound, and artificial lighting and noise stimuli also vary drastically in their spectral content. These stimuli can also vary in their predictability, intensity, and duration (Francis and Barber 2013). This variation guarantees that species will differ in their responses to these stimuli. For this reason, it is critical that studies properly characterize noise or light stimuli but also pay special attention to the features that will be of functional significance to the organism. In doing so, there is substantial promise in finding appropriate and cost-effective mitigation techniques to minimize the effects of these stimuli on wildlife and the ecological systems that support them. Additional information on properly characterizing noise and wildlife responses can be found in several recent papers (Pater et al. 2009; Ellison et al. 2012; Francis and Barber 2013).

Also important is describing the spatial footprint of these two stimuli and linking their variation across space to actual ecological impacts. Most research to date, including some of my own work, has evaluated the influence of these stimuli using study designs in which the stimulus is either present or absent (i.e., lit versus dark, quiet versus noisy). Many of these studies have been instrumental in opening our eyes to the fact that these stimuli can have important consequences for wildlife, but they are of little use to management and policy because they say nothing about how wildlife responses change with incremental changes in noise or light conditions. Because policy makers and managers are interested in questions like "how much is too much?," future studies need to adopt study designs from research groups that describe how responses vary along gradients of noise and light exposure (e.g., Francis et al. 2011c; Halfwerk et al. 2011; Kight et al. 2012). Doing so has the potential to provide meaningful data that reflects how wildlife respond to subtle changes in these stimuli—information that will be essential to setting regulatory limits.

When Do Avoidance Responses to Recreational Activities Translate into Negative Consequences for the Individual or Population?

Research involving wildlife responses to recreation will greatly benefit from explicitly linking observed behavioral responses to recreational activities to fitness consequences. There are now countless examples of wildlife fleeing from approaching hikers, snowmobiles, or other common forms of recreation (reviewed in Knight and Gutzwiller 1995), and perhaps one generalization is that human presence that is both unpredictable and close to wildlife elicit stronger behavioral responses (Kight and Swaddle 2007; Blumstein and Fernández-Juricic 2010). Although many examples of these responses exist, we still need a greater understanding of whether these and other behavioral responses decrease an individual's ability to survive or reproduce. Ultimately, we want to know whether there are costs to individuals and if they lead to population-level consequences. Frameworks exist that should aid researchers in studies aiming to link behavioral responses to individual costs. For example, the risk-disturbance hypothesis explicitly views animal responses to human activities as analogous to their responses to predation risk (Frid and Dill 2002). Whether this framework is used or not, investigators will need to think about these responses in terms of energetic trade-offs and how individuals maximize fitness.

Moving beyond Occupancy and Abundance

More studies should go beyond simple measures of site occupancy or abundance. The greatest risk is when we assume that species that remain in areas exposed to human activities or other forms of anthropogenic environmental change are doing well and likely need no form of protection. This assumption is problematic because distributional changes may only occur if individuals have alternative areas to go to (Gill et al. 2001) and, as illustrated through several examples discussed previously, known impacts beyond site abandonment are plentiful and will undoubtedly grow given the number of probable effects yet to be investigated. It may be that the most vulnerable populations are those in which site abandonment is unlikely because other suitable habitats are unavailable or the costs associated with moving to them are too high (Gill et al. 2001). For populations in decline, but for reasons other than site abandonment, we certainly want to know *why*, and efforts need to clarify whether it is an issue of decreased survival, reproduction, or both. Critically, we also want to know whether responses are direct or mediated through other organisms in the community. Measuring multiple responses by many species will be logistically challenging and more expensive than many current efforts. However, these measurements are unambiguously important to informing conservation policies that have the potential to work.

Summary

Land cover change is just one way that habitats can be degraded or lost to wildlife, and research studies highlighted in this chapter emphasize why focusing solely on the amount, type, and spatial arrangement of land cover is inadequate in explaining animal responses to habitat degradation. Instead, investigators must consider forces that contribute to habitat loss and degradation without the removal or change in vegetation, such as noise and light pollution and the presence of humans and our pets. Similarly, single-species studies focused on the presence or abundance of a species provide an incomplete picture of how wildlife and their supporting ecological systems respond to human activities. Future work can benefit from an organism-centric, sensory-based approach that links the proximate mechanisms governing responses to anthropogenic stressors to their individual-, population-, and community-level consequences. As we learn more about the severity and extent of these stressors' impacts on natural systems, we may also discover unforeseen sensory-based solutions that could help maintain wildlife populations in habitats once thought to be lost.

LITERATURE CITED

Baker, B. J., and J. M. L. Richardson. 2006. "The Effect of Artificial Light on Male Breeding-Season Behaviour in Green Frogs, *Rana clamitans melanota*." *Canadian Journal of Zoology* 84:1528–32.

Banks, P. B., and J. V. Bryant. 2007. "Four-Legged Friend or Foe? Dog Walking Displaces Native Birds from Natural Areas." *Biology Letters* 3:611–13.

Barber, J. R., K. R. Crooks, and K. M. Fristrup. 2010. "The Costs of Chronic Noise Exposure for Terrestrial Organisms." *Trends in Ecology & Evolution* 25:180–89.

Bayne, E. M., L. Habib, and S. Boutin. 2008. "Impacts of Chronic Anthropogenic Noise from Energy-Sector Activity on Abundance of Songbirds in the Boreal Forest." *Conservation Biology* 22:1186–93.

Bird, B. L., L. C. Branch, and D. L. Miller. 2004. "Effects of Coastal Lighting on Foraging Behavior of Beach Mice." *Conservation Biology* 18:1435–39.

Blickley, J. L., D. Blackwood, and G. L. Patricelli. 2012. "Experimental Evidence for the Effects of Chronic Anthropogenic Noise on Abundance of Greater Sage-Grouse at Leks." *Conservation Biology* 26:461–71.

Blumstein, D. T., and E. Fernández-Juricic. 2010. *A Primer of Conservation Behavior*. Sinauer Associates, Sunderland, MA.

Brown, C. R., and M. B. Brown. 2013. "Where Has All the Road Kill Gone?" *Current Biology* 23:R233–34.

Collinge, S. K. 2009. *Ecology of Fragmented Landscapes*. John Hopkins University Press, Baltimore, MD.

Collinge, S. K., and R. T. T. Forman. 1998. "A Conceptual Model of Land Conversion Processes: Predictions and Evidence from a Microlandscape Experiment with Grassland Insects." *Oikos* 82:66–84.

Creel, S., J. E. Fox, A. Hardy, J. Sands, B. Garrott, and R. O. Peterson. 2002. "Snowmobile Activity and Glucocorticoid Stress Responses in Wolves and Elk." *Conservation Biology* 16:809–14.

Crutzen, P. J. 2002. "Geology of Mankind." *Nature* 415:23.

Davies, T. W., J. Bennie, and K. J. Gaston. 2012. "Street Lighting Changes the Composition of Invertebrate Communities." *Biology Letters* 8:764–67.

Dominoni, D., M. Quetting, and J. Partecke. 2013. "Artificial Light at Night Advances Avian Reproductive Physiology." *Proceedings of the Royal Society B: Biological Sciences* 280.

Eisenbeis, G. 2006. "Artificial Night Lighting and Insects: Attraction of Insects to Streetlamps in a Rural Setting in Germany." In *Ecological Consequences of Artificial Night Lighting*, edited by C. Rich and T. Longcore, 281–304. Island Press, Washington, D.C.

Ellis, E. C. 2011. "Anthropogenic Transformation of the Terrestrial Biosphere." *Philosophical Transactions of the Royal Society A: Mathematical Physical and Engineering Sciences* 369:1010–35.

Ellis, E. C., K. K. Goldewijk, S. Siebert, D. Lightman, and N. Ramankutty. 2010. "Anthropogenic Transformation of the Biomes, 1700 to 2000." *Global Ecology and Biogeography* 19:589–606.

Ellis, E. C., and N. Ramankutty. 2008. "Putting People in the Map: Anthropogenic Biomes of the World." *Frontiers in Ecology and the Environment* 6:439–47.

Ellison, W. T., B. L. Southall, C. W. Clark, and A. S. Frankel. 2012. "A New Context-Based Approach to Assess Marine Mammal Behavioral Responses to Anthropogenic Sounds." *Conservation Biology* 26:21–28.

Emmering, Q. C., and K. A. Schmidt. 2011. "Nesting Songbirds Assess Spatial Heterogeneity of Predatory Chipmunks by Eavesdropping on their Vocalizations." *Journal of Animal Ecology* 80:1305–12.

Erickson, W. P., G. D. Johnson, and D. P. Young. 2005. A Summary and Comparison of Bird Mortality from Antrhopogenci Causes with an Emphasis on Collisions. USDA Forest Service General Technical Report PSW-GTR 191: 1029–1042. General Technical Report PSW-GTR 191.

Estes, J. A., J. Terborgh, J. S. Brashares, M. E. Power, J. Berger, W. J. Bond, S. R. Carpenter, T. E. Essington, R. D. Holt, J. B. C. Jackson, R. J. Marquis, L. Oksanen, T. Oksanen, R. T. Paine, E. K. Pikitch, W. J. Ripple, S. A. Sandin, M. Scheffer, T. W. Schoener, J. B. Shurin, A. R. E. Sinclair, M. E. Soule, R. Virtanen, and D. A. Wardle. 2011. "Trophic Downgrading of Planet Earth." *Science* 333:301–6.

Ewers, R. M., and R. K. Didham. 2006. "Confounding Factors in the Detection of Species Responses to Habitat Fragmentation." *Biological Reviews* 81:117–42.

Fahrig, L. 2003. "Effects of Habitat Fragmentation on Biodiversity." *Annual Review of Ecology Evolution and Systematics* 34:487–515.

Farrell, S. L., M. L. Morrison, A. J. Campomizzi, and R. N. Wilkins. 2012. "Conspecific Cues and Breeding Habitat Selection in an Endangered Woodland Warbler." *Journal of Animal Ecology* 81:1056–64.

Forman, R. T. T. 1995. *Land Mosaics: The Ecology of Landscapes and Regions*. Cambridge University Press, Cambridge.

Forman, R. T. T., B. Reineking, and A. M. Hersperger. 2002. "Road Traffic and Nearby Grassland Bird Patterns in a Suburbanizing Landscape." *Environmental Management* 29:782–800.

Forrest, A., and C. St. Clair. 2006. "Effects of Dog Leash Laws and Habitat Type on Avian and Small Mammal Communities in Urban Parks." *Urban Ecosystems* 9:51–66.

Forsman, J. T., and T. E. Martin. 2009. "Habitat Selection for Parasite-Free Space by Hosts of Parasitic Cowbirds." *Oikos* 118:464–70.

Francis, C. D., and J. R. Barber. 2013. "A Framework for Understanding Noise Impacts on Wildlife: An Urgent Conservation Priority." *Frontiers in Ecology and the Environment* 11:305–13.

Francis, C. D., and J. L. Blickley. 2012. "Introduction: Research and Perspectives on the Study of Anthropogenic Noise and Birds." *Ornithological Monographs* 74:1–5.

Francis, C. D., N. J. Kleist, C. P. Ortega, and A. Cruz. 2012. "Noise Pollution Alters Ecological Services: Enhanced Pollination and Disrupted Seed Dispersal." *Proceedings of the Royal Society B: Biological Sciences* 279:2727–35.

Francis, C. D., C. P. Ortega, and A. Cruz. 2009. "Noise Pollution Changes Avian Communities and Species Interactions." *Current Biology* 19:1415–19.

———. 2011a. "Different Behavioural Responses to Anthropogenic Noise by Two Closely Related Passerine Birds." *Biology Letters* 7:850–52.

———. 2011b. "Noise Pollution Filters Bird Communities Based on Vocal Frequency." *Plos One* 6:e27052.

Francis, C. D., C. P. Ortega, R. I. Kennedy, and P. J. Nylander. 2012. "Are Nest Predators Absent from Noisy Areas or Unable to Locate Nests?" *Ornithological Monographs* 74:101–10.

Francis, C. D., J. Paritsis, C. P. Ortega, and A. Cruz. 2011. "Landscape Patterns of Avian Habitat Use and Nest Success Are Affected by Chronic Gas Well Compressor Noise." *Landscape Ecology* 26:1269–80.

Frid, A., and L. M. Dill. 2002. "Human-Caused Disturbance Stimuli as a Form of Predation Risk." *Conservation Ecology* 6:11.

Garber, S. D., and J. Burger. 1995. "A 20-Yr Study Documenting the Relationship between Turtle Decline and Human Recreation." *Ecological Applications* 5:1151–62.

Gauthreaux, S. A., and C. G. Belser. 2006. "Effects of Artificial Lighting on Migrating Birds." In *Ecological Consequences of Artificial Night Lighting*, edited by C. Rich and T. Longcore, 67–93. Island Press, Washington, D.C.

Gill, J. A., K. Norris, and W. J. Sutherland. 2001. "Why Behavioural Responses May Not Reflect the Population Consequences of Human Disturbance." *Biological Conservation* 97:265–68.

Goodwin, S. H., and W. G. Shriver. 2011. "Effects of Traffic Noise on Occupancy Patterns of Forest Birds." *Conservation Biology* 25:406–11.

Gross, K., G. Pasinelli, and H. P. Kunc. 2010. "Behavioral Plasticity Allows Short-Term Adjustment to a Novel Environment." *American Naturalist* 176:456–64.

Habib, L., E. M. Bayne, and S. Boutin. 2007. "Chronic Industrial Noise Affects Pairing Success and Age Structure of Ovenbirds *Seiurus aurocapilla*." *Journal of Applied Ecology* 44:176–84.

Halfwerk, W., L. J. M. Holleman, C. M. Lessells, and H. Slabbekoorn. 2011. "Negative Impact of Traffic Noise on Avian Reproductive Success." *Journal of Applied Ecology* 48:210–19.

Holker, F., C. Wolter, E. K. Perkin, and K. Tockner. 2010. "Light Pollution as a Biodiversity Threat." *Trends in Ecology & Evolution* 25:681–82.

Johnson, R. G., and S. A. Temple. 1986. "Assessing Habitat Quality for Birds Nesting in Fragmented Tallgrass Prairies."

In *Wildlife 2000: Modeling Habitat Relationships of Terrestrial Vertebrates*, edited by J. Verner, M. L. Morrison, and C. J. Ralph, 245–49. University of Wisconsin Press, Madison.

Kangas, K., M. Luoto, A. Ihantola, E. Tomppo, and P. Siika-maki. 2010. "Recreation-Induced Changes in Boreal Bird Communities in Protected Areas." *Ecological Applications* 20:1775–86.

Kempenaers, B., P. Borgstrom, P. Loes, E. Schlicht, and M. Valcu. 2010. "Artificial Night Lighting Affects Dawn Song, Extra-Pair Siring Success, and Lay Date in Songbirds." *Current Biology* 20:1735–39.

Kight, C. R., M. S. Saha, and J. P. Swaddle. 2012. "Anthropogenic Noise is Associated with Reductions in the Productivity of Breeding Eastern Bluebirds (*Sialia sialis*)." *Ecological Applications* 22:1989–96.

Kight, C. R., and J. P. Swaddle. 2007. "Associations of Anthropogenic Activity and distuRbance with Fitness Metrics of Eastern Bluebirds (*Sialia sialis*)." *Biological Conservation* 138:189–97.

———. 2011. "How and Why Environmental Noise Impacts Animals: An Integrative, Mechanistic Review." *Ecology Letters* 14:1052–61.

Knight, R. L., and K. J. Gutzwiller, eds. 1995. *Wildlife and Recreationists: Coexistence through Management and Research*. Island Press, Washington, D.C.

Kuitunen, M., E. Rossi, and A. Stenroos. 1998. "Do Highways Influence Density of Land Birds?" *Environmental Management* 22:297–302.

Leonard, M. L., and A. G. Horn. 2012. "Ambient Noise Increases Missed Detections in Nestling Birds." *Biology Letters* 8:530–32.

McGowan, C. P., and T. R. Simons. 2006. "Effects of Human Recreation on the Incubation Behavior of American Oyster-catchers." *Wilson Journal of Ornithology* 118:485–93.

Millspaugh, J. J., R. J. Woods, K. E. Hunt, K. J. Raedeke, G. C. Brundige, B. E. Washburn, and S. K. Wasser. 2001. "Fecal Glucocorticoid Assays and the Physiological Stress Response in Elk." *Wildlife Society Bulletin* 29:899–907.

Nelson, D. A., and P. Marler. 1990. "The Perception of Bird Song and an Ecological Concept of Signal Space." In *Comparative Perception*, edited by W. C. Stebbins and M. A. Berkley, 443–78. John Wiley, New York, New York.

Ortega, C. P. 2012. "Effects of Noise Pollution on Birds: A Brief Review of our Knowledge." *Ornithological Monographs* 74:6–22.

Papouchis, C. M., F. J. Singer, and W. B. Sloan. 2001. "Responses of Desert Bighorn Sheep to Increased Human Recreation." *Journal of Wildlife Management* 65:573–82.

Pater, L. L., T. G. Grubb, and D. K. Delaney. 2009. "Recommendations for Improved Assessment of Noise Impacts on Wildlife." *Journal of Wildlife Management* 73:788–95.

Patricelli, G. L., and J. L. Blickley. 2006. "Avian Communication in Urban Noise: Causes and Consequences of Vocal Adjustment." *Auk* 123:639–49.

Perkin, E. K., F. Hölker, J. S. Richardson, J. P. Sadler, C. Wolter, and K. Tockner. 2011. "The Influence of Artificial Light on Stream and Riparian Ecosystems: Questions, Challenges, and Perspectives." *Ecosphere* 2:art122.

Poot, H., B. J. Ens, H. de Vries, M. A. H. Donners, M. R. Wernand, and J. M. Marquenie. 2008. "Green Light for Nocturnally Migrating Birds." *Ecology and Society* 13.

Quinn, J. L., M. J. Whittingham, S. J. Butler, and W. Cresswell. 2006. "Noise, Predation Risk Compensation and Vigilance in the Chaffinch *Fringilla coelebs*." *Journal of Avian Biology* 37:601–8.

Reed, S. E., and A. M. Merenlender. 2008. "Quiet, Nonconsumptive Recreation Reduces Protected Area Effectiveness." *Conservation Letters* 1:146–54.

———. 2011. "Effects of Management of Domestic Dogs and Recreation on Carnivores in Protected Areas in Northern California." *Conservation Biology* 25:504–13.

Reijnen, R., R. Foppen, C. Terbraak, and J. Thissen. 1995. "The Effects of Car Traffic on Breeding Bird Populations in Woodland 3: Reduction of Density in Relation to the Proximity of Main Roads." *Journal of Applied Ecology* 32:187–202.

Rheindt, F. E. 2003. "The Impact of Roads on Birds: Does Song Frequency Play a Role in Determining Susceptibility to Noise Pollution?" *Journal Fur Ornithologie* 144:295–306.

Rich, C., and T. Longcore. 2006. *Ecological Consequences of Artificial Night Lighting.* Island Press, Washington, D.C.

Robertson, B. A., and R. L. Hutto. 2007. "Is Selectively Harvested Forest and Ecological Trap for Olive-Sided Flycatchers?" *The Condor* 109:109–21.

Rydell, J., and P. A. Racey. 1995. "Streetlamps and the Feeding Ecology of Insectivorous Bats." *Symposium of the Zoological Society of London* 67:291–307.

Schaub, A., J. Ostwald, and B. M. Siemers. 2008. "Foraging Bats Avoid Noise." *Journal of Experimental Biology* 211:3174–80.

Schlaepfer, M. A., M. C. Runge, and P. W. Sherman. 2002. "Ecological and Evolutionary Traps." *Trends in Ecology and Evolution* 17:474–80.

Schroeder, J., S. Nakagawa, I. R. Cleasby, and T. Burke. 2012. "Passerine Birds Breeding under Chronic Noise Experience Reduced Fitness." *Plos One* 7:e39200.

Siemers, B. M., and A. Schaub. 2011. "Hunting at the Highway: Traffic Noise Reduces Foraging Efficiency in Acoustic Predators." *Proceedings of the Royal Society B: Biological Sciences* 278:1646–52.

Sih, A., A. Bell, and J. C. Johnson. 2004. "Behavioral Syndromes: An Ecological and Evolutionary Overview." *Trends in Ecology & Evolution* 19:372–78.

Stone, E. L., G. Jones, and S. Harris. 2009. "Street Lighting Disturbs Commuting Bats." *Current Biology* 19:1123–27.

———. 2012. "Conserving Energy at a Cost to Biodiversity? Impacts of LED Lighting on Bats." *Global Change Biology* 18:2458–65.

Swaddle, J. P., C. R. Kight, S. Perera, E. Davila-Reyes, and S. Sikora. 2012. "Constraints on Acoustic Signaling among Birds Breeding in Secondary Cavities: The Effects of Weather, Cavity Material, and Noise on Sound Propagation." *Ornithological Monographs* 74:63–77.

Taylor, A. R., and R. L. Knight. 2003. "Wildlife Responses to Recreation and Associated Visitor Perceptions." *Ecological Applications* 13:951–63.

Tuxbury, S. M., and M. Salmon. 2005. "Competitive Interactions between Artificial Lighting and Natural Cues during Seafinding by Hatchling Marine Turtles." *Biological Conservation* 121:311–16.

Van Dyck, H. 2012. "Changing Organisms in Rapidly Changing Anthropogenic Landscapes: The Significance of the 'Umwelt'-Concept and Functional Habitat for Animal Conservation." *Evolutionary Applications* 5:144–53.

Van Horne, B. 1983. "Density as a Misleading Indicator of Habitat Quality." *Journal of Wildlife Management* 47:893–901.

van Langevelde, F., J. A. Ettema, M. Donners, M. F. WallisDeVries, and D. Groenendijk. 2011. "Effect of Spectral Composition of Artificial Light on the Attraction of Moths." *Biological Conservation* 144:2274–81.

Witherington, B. E. 1997. "The Problem of Photopollution for Sea Turtles and Other Nocturnal Animals." In *Behavioral Approaches to Conservation in the Wild*, edited by J. R. Clemmons and R. Buchholz, 303–28. Cambridge University Press, Cambridge.

6 — Population Genetics and Wildlife Habitat

LISETTE P. WAITS AND
CLINTON W. EPPS

Worldwide, wildlife habitat has been lost, degraded, and fragmented as humans have transformed the landscape for agriculture, transportation, and urbanization. The amount, quality, and configuration of the remaining wildlife habitat can greatly affect genetic and evolutionary processes and the viability of these populations. Decreases in the size and quality of habitat fragments lead to decreases in **effective population size**, increases in **genetic drift**, and loss of genetic diversity, while increased isolation of fragments leads to decreases in **gene flow** and loss of genetic diversity (fig. 6.1). Loss of genetic diversity is of great concern to wildlife managers because this loss increases the probability of **inbreeding depression**, decreases the adaptive potential of populations, and ultimately threatens population viability. This chapter will begin with a review of important population genetic theories, methods, and terms as they relate to habitats in peril. Specifically, we (1) describe the impacts of habitat degradation and fragmentation on population genetic processes; (2) discuss the implications for population viability, **hybridization**, and population persistence; and (3) highlight the role of molecular methods in assessing the impacts of habitat alteration on wildlife populations. Since genetic terms may be unfamiliar to some readers, we provide a glossary of key terms in box 6.1 and highlight each glossary term in bold when it is first used.

Population Genetic Methods and Theory

Genetic variation in natural populations is influenced by four main processes: mutation, gene flow, selection, and genetic drift (Hedrick 2009; Allendorf et al. 2012). Mutation is the process that creates new **alleles** due to errors in DNA replication. Gene flow occurs as a result of migration (dispersal and reproduction), and this process moves alleles between populations. Selection impacts genetic variation by increasing the frequencies of alleles that are favorable in a particular environment and decreasing the frequencies of alleles that are less favorable in that environment. Genetic drift is the change in **allele frequencies** due to random sampling effects as alleles are passed on from one generation to the next (Wright 1931). Genetic drift is thus influenced by the number of breeders and the variance in reproductive success among breeders.

Sewall Wright (1931, 1938) introduced the concept of effective population size (N_e) to mathematically and conceptually represent the effects of genetic drift. He defined N_e as "the number of breeding individuals in an idealized population that would show the same amount of dispersion of allele frequencies under random genetic drift or the same amount of **inbreeding** as the population under consideration." The N_e of a population has a large effect on genetic diversity. Populations with larger N_e will have more genetic variation because more new alleles are created by mutation due to a larger number of breeding events, and fewer alleles are lost to genetic drift since there is a larger sample of breeders each generation. N_e also influences the relative impacts of genetic drift and selection in natural populations (Fisher 1930; Wright 1931). For example, as N_e decreases, the effect of selection decreases relative to the effect of genetic drift and important adaptive genetic

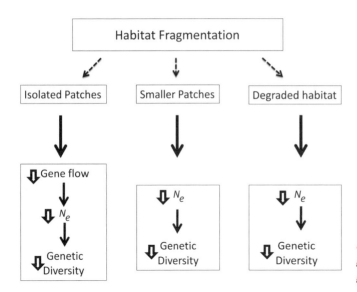

Figure 6.1 Impacts of habitat fragmentation on gene flow, effective population size (N_e), and genetic diversity.

variation can be lost (Hedrick 2009). One benchmark for conservation biology is the "50/500 rule" (Franklin 1980), which refers to minimum N_e required to prevent inbreeding (50) and offset the consequences of long-term genetic drift (500). Although this rule may have little utility as a prescription for minimum viable population sizes (Templeton 1995), it serves as a reasonable guide to determining when genetic factors should be considered (Jamieson and Allendorf 2012); thus, an N_e <500 may be regarded as small, and an N_e <50 is regarded as very small.

When studying genetic variation and gene flow in natural populations, researchers use a diversity of methods and DNA types. The collection of DNA samples from wildlife can generally be classified into three main sampling approaches: destructive sampling, nondestructive sampling, and noninvasive sampling (Taberlet et al. 1999). In destructive sampling approaches, the animal is killed to obtain the genetic sample of blood or tissue. This approach was widely used during the 1970s and 1980s when analyzing protein data (i.e., allozymes) but has been replaced by nondestructive and noninvasive genetic sampling for vertebrate animals. Nondestructive sampling involves capturing an animal and obtaining a blood, tissue, feather, mouth swab, or hair sample and then releasing the animal. Occasionally, nondestructive sampling is conducted without capturing individual animals by using a biopsy dart gun on large mammals such as whales, elephants, and cougars (Hoelzel and Amos 1988; Georgiadis et al.

1994). Noninvasive genetic sampling is achieved without capturing or handling individual animals and includes the use of a diversity of sources including hair, feces, feathers, urine, owl pellets, sloughed and shed skin, eggshells, and saliva (Waits and Paetkau 2005).

When examining genetic variation, researchers may select a location (**locus**) or locations (**loci**) in the **mitochondrial DNA** (mtDNA) or **nuclear DNA** (nDNA) genomes. MtDNA is a circular DNA molecule of the mitochondrion, an organelle that is the energy powerhouse of cells. MtDNA generally has a haploid mode of inheritance as it is passed only from mother to offspring (Birky et al. 1989). In wildlife genetics, this DNA molecule is important for species identification, phylogenetic and phylogeographic analysis and the delineation of conservation units. Two copies of nDNA are found in the chromosomes of the nucleus of a cell. Nuclear DNA has a biparental mode of inheritance as one copy is inherited from the mother and one copy is inherited from the father. In wildlife genetics, nDNA analyses are important for individual identification, kinship and parentage studies, gene flow evaluations, and the delineation of conservation and management units (Allendorf et al. 2012). When generating mtDNA or nDNA data, researchers choose to evaluate particular loci in the genome. These loci may be **neutral** or **nonneutral** (box 6.1). Neutral loci are not affected by natural selection while nonneutral or **adaptive loci** affect individual **fitness** and are under selection (Hedrick 2009).

After choosing to collect data for a particular type of

Box 6.1 Glossary of Genetic Terms

allele: A unique genetic variant observed at a particular locus.

allele frequency: A measure of the relative frequency of an allele in a population at a particular genetic locus.

bottleneck: A sharp reduction in population size, leading to a loss of genetic diversity.

census population size (N_c): The number of individuals in a population, as distinguished from N_e (see *effective population size*).

effective population size (N_e): The number of breeding individuals in an idealized population that would show the same amount of change in allele frequencies under random genetic drift or the same amount of inbreeding as the population under consideration.

fitness: The relative ability of an individual (or population) to survive, reproduce, and propagate genes in an environment.

gene flow: The movement of alleles between populations. This is also referred to as migration.

genetic drift: The change in allele frequencies due to random sampling effects as alleles are passed on from one generation to the next.

genetic rescue: Increasing genetic diversity of a population lacking genetic diversity due to drift or inbreeding by adding new individuals from a different source population.

genetic structure: Spatial variation in allele identity or frequency.

genotype: The combination of alleles observed at a particular locus for a specific individual.

haplotype: An allele or combination of alleles passed on from a single parent.

hybridization: Interbreeding of individuals from genetically distinct populations, subspecies, or species.

hybrid swarm: A population of individuals that all are hybrids.

hybrid zone: An area of contact between two genetically distinct populations where hybridization occurs.

inbreeding: Mating between closely related individuals.

inbreeding depression: A decrease in fitness due to breeding between closely related individuals (inbreeding).

introgression: Gene flow between populations whose individuals hybridize. This occurs due to backcrossing of hybrids to one or both parent groups.

locus (loci): A particular location in the genome of an organism. The term *loci* is the plural form of locus.

mitochondrial DNA (mtDNA): A circular DNA molecule of the mitochondrion, which is haploid and generally passed on only from mother to offspring.

mutation: A change in the DNA sequence of an organism. Mutations that occur in the DNA of germ cells can be passed on to the next generation.

neutral locus: A locus where the frequency of alleles is not affected by selection.

nonneutral or adaptive locus: A locus where the frequency of alleles is affected by selection.

nuclear DNA (nDNA): The DNA of the chromosomes found in the nucleus of a cell.

outbreeding depression: A reduction in fitness of hybrid individuals relative to the parental types.

population viability analysis (PVA): A quantitative analysis of population dynamics with the goal of assessing extinction risk.

positive selection: A process that increases the prevalence alleles that increases an individual's fitness.

purifying selection: A process that decreases the prevalence of alleles that diminish an individual's fitness.

Box 6.2 Overview of Methods Used in Genetic Analysis

polymerase chain reaction (PCR): The chemical process used to make millions of copies of a particular target DNA region or locus using specific primers (Mullis and Faloona 1987). PCR has revolutionized molecular genetics and is used in all following methods.

DNA sequencing: The chemical process used to read the sequence of nucleotides (DNA base pairs—A, G, C, T) at a particular DNA region. Early studies in wildlife genetics were dominated by a method known as Sanger sequencing (Sanger et al. 1977), but there are currently many new methods categorized as "next generation sequencing" approaches that parallelize the process and can produce a greater volume of sequence data in a shorter period of time (Shendure and Ji 2008).

microsatellite analysis: Also known as short tandem repeats (STRs) or simple sequence repeats (SSRs). Repeating sequences of two to six base pairs (e.g., CACACA or GTCGTCGTC). The process of genotyping microsatellite loci

and is the most commonly used approach for evaluating genetic diversity, gene flow, parentage, and hybridization in wildlife populations (Selkoe and Toonen 2006).

amplified fragment length polymorphism (AFLP) analysis: A PCR-based tool for assaying genetic variation that does not require knowledge of the DNA sequence of the target species. The amplified DNA fragments are separated by size on a polyacrylamide gel and visualized using fluorescence or autoradiography (Vos et al. 1995).

single nucleotide polymorphism (SNP) analysis: A new molecular method that surveys single base pair genetic polymorphisms in many locations throughout the genome. This method has not yet been used extensively in wildlife genetics, but the use of this method is predicted to expand rapidly in coming years because it surveys a greater proportion of the genome and can be automated for high throughput analyses (Vignal et al. 2002; Morin et al. 2004).

DNA and locus, a variety of methods are used to obtain genetic data in the form of **haplotypes** or **genotypes**. Fully describing these methods is beyond the scope of this chapter, but the most commonly used approaches are summarized in box 6.2. The most important methodological advance in wildlife and medical genetics was the development of the polymerase chain reaction or PCR (Mullis and Faloona 1987). This technique makes it possible to make many copies of particular DNA locus and to use low-quantity DNA sources like feces, saliva, or bones. We refer readers to the following sources for additional information on methodological approaches used in wildlife genetics (DeYoung and Honeycutt 2005; Beebee and Rowe 2008).

Impacts of Habitat Loss and Fragmentation on Gene Flow, N_e, and Diversity

Habitat loss, fragmentation, and degradation may strongly reduce genetic diversity and typically in-

crease **genetic structure** (i.e., differentiation between groups). This occurs because spatial distribution and quality of habitat are key determinants of population size and dispersal success, which in turn influence N_e and thus genetic drift. Therefore, rather than simply describing genetic structure, genetic diversity, or gene flow, researchers increasingly have attempted to determine how habitat distribution and quality influence those characteristics of natural populations. This interest has contributed to the increasing popularity of landscape genetic studies for both plants and animals (Manel et al. 2003; Holderegger et al. 2010; Storfer et al. 2010).

Habitat area is often assumed to be the strongest predictor of population size and thus N_e, particularly for metapopulations and other systems of fragmented populations (Hanski 1999). This concept can be traced to the theory of island biogeography, which posits that island area is the most important predictor of species extinction rate (MacArthur and Wilson 1967). However, this assumption is not always supported (Prugh et al. 2008), because habitat quality and configuration (e.g., edge effects or links to other habitat patches) may have greater importance. Genetic structure of fragmented populations is affected by both genetic drift and gene flow (Hutchison and Templeton 1999); in systems of small populations, connectivity between patches can have the greatest impact on genetic diversity within any single population or habitat patch (e.g., Epps et al. 2006). Therefore, N_e and thus genetic diversity may be eroded by (1) habitat loss; (2) habitat degradation, even if habitat area doesn't decrease; or (3) habitat fragmentation (fig. 6.1). Although habitat fragmentation is often closely associated with habitat loss (Fahrig 2003), it can also occur without any change in habitat area for naturally patchy populations if landscape changes influence the ability of organisms to disperse between patches. Transportation networks (roads, railroads, etc.) are likely sources of this type of fragmentation (Balkenhol and Waits 2009; Holderegger and Di Giulio 2010). A good example of this is the effect of fenced highways and canals on desert bighorn sheep (*Ovis canadensis nelsoni*) in California, USA. Although those structures are primarily located in the arid flats rather than the mountainous habitat used by bighorn sheep for foraging and escape terrain, Epps et al. (2005) showed that they contribute to increased genetic structure and reduced genetic diversity in those small populations. In other cases, such as Kuehn et al.'s (2007) study of roe deer populations in Switzerland separated by a fenced highway, roads cause genetic divergence but no detectable decrease in genetic diversity.

Whatever the cause, as N_e declines, genetic drift will increase and genetic diversity will be lost as noted previously. Strong genetic drift can even override balancing selection (when heterozygous individuals have a fitness advantage, such as disease resistance) or **positive selection** for an advantageous allele. Another problem that arises in small or fragmented populations is increased levels of inbreeding, which will exacerbate loss of genetic diversity from drift. Inbreeding occurs when close relatives mate and bear offspring, and it results in decreases in heterozygosity because related individuals are more likely to have the same allele at any given locus (Haldane and Waddington 1930). Decrease in fitness resulting from inbreeding is referred to as inbreeding depression (see examples in the following section,

"Implications of Changes in Gene Flow and Genetic Diversity"). Thus, the impacts of habitat loss, degradation, and fragmentation on small populations may be intensified by loss of fitness related to genetic diversity.

Other factors may interact with habitat quality and fragmentation to influence N_e and thus genetic diversity. The stability of populations matter: populations that fluctuate in size will lose more genetic diversity than stable populations (Frankham 1995a). The effect is worsened if a population declines to a small size for multiple generations (known as a **bottleneck**), although populations that expand rapidly after a short-term decline may lose little genetic diversity (Nei et al. 1975). Thus, strong temporal variation in habitat quality could reduce N_e. A variety of statistical tests can be used to deduce demographic events such as population bottlenecks from genetic data (Cornuet and Luikart 1996; Piry et al. 1999; Garza and Williamson 2001). Polygynous mating systems where variance in reproductive success is high may also result in N_e being much smaller than **census population size** (N_c). If, for instance, a single dominant male in a small population fathers a large proportion of the offspring, and other males have many fewer offspring, genetic drift and inbreeding will occur more rapidly than expected for species with monogamous mating (Frankham 1995a). Spatial structuring within a population (i.e., nonrandom association of groups of individuals) will typically reduce N_e for the population as a whole because drift will be accelerated by the comparatively smaller size of each group. As such, spatial structuring may be cryptic, especially when movement and dispersal through different habitats are poorly understood, and reductions in N_e may be more common than readily apparent, even for widely distributed species. For instance, Halbert et al. (2012) demonstrated that cryptic genetic structure exists within the Yellowstone National Park bison herd, which is one of only two bison populations that have persisted in their native habitat since European settlement of North America. They estimated that N_e of each subpopulation was ~100 and noted that this apparent isolation requires reconsideration of current management practices, including culling.

For a species as a whole, spatial structuring of populations (i.e., populations divided into subpopulations with limited dispersal) can help maintain higher genetic diversity (Templeton et al. 1990). If the number of subpopulations is large, even if drift causes fixation of a single allele (i.e., loss of all other alleles due to drift) in each subpopulation, different alleles will be fixed in different subpopulations by random chance, resulting in maintenance of diversity in the system as a whole. However, it is important to recognize that each subpopulation may still suffer inbreeding depression or reduced fitness, so fragmentation should not be viewed as an appropriate strategy for maintaining regional or species-wide genetic diversity in most cases.

Most research investigating the relationship between population genetic structure and habitat has focused on neutral genetic diversity. Neutral genetic loci offer an excellent means to infer the effects of habitat quality and configuration on demographic history, genetic diversity, and gene flow, as described earlier, and may have future adaptive potential, but inferring fitness consequences of low genetic diversity or low heterozygosity at neutral loci is difficult (Chapman et al. 2009). Although individuals or populations with lower genetic diversity have been shown to have reduced fitness (e.g., Johnson et al. 2011; Agudo et al. 2012), effect sizes are often very weak (Chapman et al. 2009), and many studies merely assume that neutral and adaptive genetic diversity are correlated; however, evidence is mixed. Studies tend to detect that correlation especially where genetic drift is strong (Radwan et al. 2010; e.g., Marsden et al. 2012). Until recently, few adaptive loci were described well enough to allow population-level assessments of adaptive variation, and many genes with adaptive significance show little variation. Exceptions can be found in research on genetic diversity related to immune-system function and disease resistance (e.g., the microhistocompatability complex or MHC genes). Studies of MHC variation show clear evidence of balancing selection in some systems (Hedrick 1994), even where neutral genetic diversity is severely depleted (Aguilar et al. 2004), but in others, the effects of genetic drift may overwhelm evidence of selection (Worley et al. 2006; Alcaide 2010).

New approaches to studying adaptive genetic variation, in some cases employing "next generation" sequencing techniques (box 6.2), are poised to reveal new insights into how adaptive genetic variation is affected by habitat quality, distribution, and population history (Kirk and Freeland 2011; Miller et al. 2012; Schoville et al. 2012). Selection may favor different genotypes in different environments, and such patterns are starting to be detectable in natural systems (Schoville et al. 2012). For example, a latitudinal gradient in allele frequencies for a gene associated with reproductive timing has been observed in Chinook salmon (*Oncorhyncus tshawytscha*) (O'Malley et al. 2010). Spatial gradients in fitness of a particular allele are one of the reasons why managers must use caution when seeking to alleviate inbreeding by bringing in individuals from other populations. If the source stock for those augmentations comes from different habitats, **outbreeding depression** is possible (Dobzhansky 1948). Outbreeding depression occurs when co-adapted gene complexes are disrupted and can lead to serious problems such as reproducing at the wrong time of year in a particular habitat. Outbreeding depression in natural populations has been observed across many taxa, including birds (Marr et al. 2002), amphibians (Jourdan-Pineau et al. 2012), fish (Roberge et al. 2008), plants (Goto et al. 2011), and mammals (Marshall and Spalton 2000). However, recent literature suggests that inbreeding depression occurs more frequently than outbreeding depression in natural populations and that concerns about outbreeding are often overstated (Edmands 2007; Frankham et al. 2011).

Implications of Changes in Gene Flow and Genetic Diversity
Effects on Population Viability

Ultimately, changes in gene flow and genetic diversity may affect population viability (Lacy 1997). Although it has long been understood that low genetic diversity has negative consequences, Lande (1988) argued that genetic drift and inbreeding are likely to pose significant threats only when census population sizes are very low and thus populations are already vulnerable. However, in naturally fragmented populations where local population sizes are small, such effects could accumulate rapidly if fragmentation prevents dispersal among habitat patches. Inbreeding has been demonstrated to be a serious problem in many captive breeding programs (Ralls et al. 1988) and, increasingly, in wild populations (e.g., Frankham 1998; Saccheri et al. 1998). However, because neutral genetic diversity is not a perfect index for adaptive genetic diversity, it has proved difficult to identify thresholds of neutral genetic diversity at which negative consequences for fitness result. Despite this difficulty, as we discuss in the following paragraphs, there is clear evidence of the detrimental consequences of inbreeding and low genetic diversity in natural populations (Frankham 1995b) and concensus that genetic factors can affect extinction risk (Jamieson and Allendorf 2012).

Many studies have demonstrated links between neutral genetic diversity, MHC diversity, and fitness. Correlations between neutral genetic diversity and MHC diversity may indicate strong drift or inbreeding (Radwan et al. 2010), suggesting that in many cases neutral genetic diversity can be used as an index for adaptive genetic diversity. In free-ranging raccoons (*Procyon lotor*), genetic diversity at neutral microsatellite loci was correlated with reduced endoparasite loads and prevalence of a single ectoparasite (Ruiz-Lopez et al. 2012). Conversely, Schwensow et al. (2007) found that neither neutral genetic diversity nor adaptive genetic diversity (MHC alleles) correlated with parasite load (a potential fitness indicator) in dwarf lemurs (*Cheirogaleus medius*). However, they were able to identify particular alleles that were associated with high or low parasite burdens.

In other cases, even when adaptive genetic diversity is not assessed, comparative estimates of very low neutral genetic diversity may be correlated with apparent inbreeding effects. For instance, Olano-Marin et al. (2011) found correlations between individual heterozygosity at neutral loci and reproductive success in blue tits (*Cyanistes caeruleus*), and argued that neutral loci are more sensitive to detecting inbreeding than adaptive loci. Johnson et al. (2011) detected a relationship between inbreeding and reproductive success for Sierra Nevada bighorn sheep (*Ovis canadensis sierrae*) but observed that the effects were too small to influence population trajectories. Chapman et al. (2009) reviewed many such studies and noted that heterozygosity–fitness correlations were often weak and that there appeared to be a publication bias against studies that failed to find heterozygosity–fitness correlations.

To counter the effects of low genetic diversity and inbreeding, "**genetic rescue**" may be employed (Tallmon et al. 2004). This technique requires bringing in less closely related individuals from other populations to inject new alleles into a genetically depauperate population. One of the best-known examples of this is the US Fish and Wildlife Service's decision to augment the endangered population of Florida panthers (*Puma concolor*) with mountain lions from Texas in an attempt to reverse the inbreeding effects (Pimm et al. 2006; Johnson et al. 2010). Genetic rescue may also occur naturally as was documented for an inbred population of Scandinavian wolves (*Canis lupus*) that showed much higher heterozygosity and rapid population growth after a single immigrant arrived and mated (Vila et al. 2003). Genetic rescue has been employed in other situations (e.g., Hogg et al. 2006; Miller et al. 2012; Olson et al. 2012; Heber et al. 2013) but must be done cautiously. Such efforts may contribute to outbreeding depression (see earlier discussion) if there are no closely related populations to use as sources, immigrants may kill resident juveniles (e.g., in large carnivores where infanticide is common), and the spread of disease is an ever-present concern (Tallmon et al. 2004). However, it can be successful even when source populations are also inbred or of low genetic diversity (Heber et al. 2013). Finally, similar to other translocations (e.g., Fischer and Lindenmayer 2000; Letty et al. 2007), any attempt to "rescue" a small or inbred population must consider whether the habitat will support a population of adequate size and whether threats to the original population have been adequately addressed (Hedrick and Fredrickson 2010).

Increasingly, we recognize that genetic diversity is affected by habitat area, quality, connectedness, and other ecological factors. Attempts to maintain population viability and genetic diversity must consider how those processes interact. For instance, the influence of connectivity on N_e becomes more important as average population size declines, particularly in metapopulations with high rates of population turnover, because of increased rates of genetic drift. A good example of this is a study by Cosentino et al. (2012) that evaluated forty-one small populations of tiger salamander (*Ambystoma tigrinum*) and demonstrated that colonization history, connectivity, wetland area, and presence of predatory fish all influenced genetic diversity and divergence. Despite this complexity, because of the primary relationship between N_e and genetic diversity, a basic understanding of typical population sizes is a first requirement when considering whether genetic diversity and gene flow will have a strong influence on population viability for a given species.

Incorporating Genetics in Population Viability Analysis (PVA)

Genetic impacts on extinction risk are rarely modeled in **population viability analyses** (Allendorf and Ryman 2002; Beissinger 2002), but attempts to include genetics in the PVA of animal populations highlight the importance of this variable (Mills and Smouse 1994; Robert et al. 2004; Haig and Allendorf 2006). Given the impacts of habitat loss and fragmentation on genetic variation and inbreeding, failure to include this variable in PVA is likely to result in overly optimistic projections of population persistence (Allendorf and Ryman 2002). For example, in a simulation study of eighteen threatened animal species with initial population sizes of fifty to one thousand individuals, Brook et al. (2002) showed that the average median time to extinction was increased 25–30% when the effects of inbreeding depression were included in PVA analyses. This study also demonstrated a wide variation in impacts as species showed little decrease (<5%) in median time to extinction, while others showed extreme impacts with decreases in time to extinction up to 95%.

Most studies that incorporate genetics into PVA generally include inbreeding depression effects only at the juvenile survival stage of the life cycle, yet it is known to affect multiple stages (Mills and Smouse 1994; Ogrady et al. 2006). When evaluating the impacts of inbreeding depression across the full life cycle of eighteen mammal and twelve bird species, O'Grady et al. (2006) found that inbreeding depression led to lower population sizes and a 31–41% median reduction in time to extinction. In conclusion, these studies indicate that the probability of survival of threatened species in fragmented and degraded habitat will be overestimated if genetic factors are not considered in PVA.

Impacts of Habitat Loss and Fragmentation on Hybridization and Introgression

Another potential genetic impact of human-caused habitat alternation is an increase in hybridization and **introgression**. Hybridization is the mating between individuals from genetically distinct populations, subspecies, or species. When hybrids backcross to one or both parental groups, this causes the introgression of genes between groups. The spatial location where hybridization and introgression occurs is known as a hybrid zone (Barton and Hewitt 1985). Hybridization is a natural evolutionary process which can lead to positive impacts on biodiversity, such as an increase in genetic diversity and fitness of hybrid individuals (hybrid vigor), adaptive radiations, and the creation of new species (Lewontin and Birch 1966; Arnold 1997; Seehausen 2004). However, it is also a conservation and management challenge with negative impacts for populations including outbreeding depression, loss of local adaptations, genetic swamping of one parental group, and, in some cases, extinction by hybridization (Rhymer and Simberloff 1996; Allendorf et al. 2001; Wolf et al. 2001).

Allendorf et al. (2001) provided a categorization of hybridization to help guide conservation and management decisions. Their classification is focused on the cause of the hybridization event (natural or anthropogenic) and the extent of introgression between hybridizing taxa. They suggest that anthropogenic hybridization events that are followed by backcrossing and introgression with parental groups are of greatest concern for conservation and management and should be avoided when possible. The three anthropogenic activities that contribute the most to hybridization are the introduction of nonnative plants and animals, habitat fragmentation, and habitat modification (Rhymer and Simberloff 1996; Allendorf et al. 2001). In a review of hybridization among threatened and endangered animals, Zamudio and Harrison (2010) found that 75% of hybridization cases could be linked to anthropogenic habitat change or introductions of nonnative species. In a review of plant and animal hybrid zones, Buggs (2007) concluded that human-induced habitat changes were driving factors in eleven of thirty-nine cases of hybrid zone movement, and climate change could be shifting hybrid zone location in multiple animal species. Thus, habitat loss and alteration are serious concerns for increasing the threat of hybridization among native taxa.

The introduction of nonnative species has led to numerous examples of hybridization in invertebrates, mammals, birds, fish, reptiles, and amphibians (reviewed in Rhymer and Simberloff 1996; Zamudio and Harrison 2010). A classic example of this is hybridization of introduced mallards (*Anas platyrhynchos*) with local, endemic duck species. Hybridization and introgression with mallards has been implicated in population declines of the New Zealand grey duck (*A. superciliosa superciliosa*; Rhymer et al. 1994), Hawaiian duck (*A. wyvilliana*; Griffin et al. 1989), Florida mottled duck (*A. fulvigula fulvigula*; Mazourek and Gray 1994), and the Australian (Pacific) black duck (*A. superciliosa rogersi*; Lever 1987). In addition, there is evidence that invasiveness can increase as a result of hybridization. Hybridization between different populations of invaders or with already-established local species may promote the establishment of the colonists while simultaneously allowing the development of new genetic combinations, which can facilitate the exploitation of underused or open niches (Seehausen et al. 2008). This has been a hypothesized mechanism for the successful and rapid spread of Cuban brown anole (*Anolis sagrei*; Kolbe et al. 2004, 2007).

Another major source of hybridization is the interbreeding of domestic or game species with local native species. Examples include domestic dogs (*Canis familiaris*) and wolves (*C. lupus*) in Europe (Randi 2008), dogs and Simien jackals (*Canis simensis*) in Ethiopia (Gotelli et al. 1994), domestic cats (*Felis silvestris catus*) and European wild cats (*Felis sylvestris sylvestris*; Beaumont et al. 2001; Randi 2008), domestic cats and the African wildcat (*F. libyca*; Wiseman et al. 2000), and domestic ferrets (*Mustela furo*) and European polecat (*M. putorius*; Davison et al. 1999). This has also been a major concern for native fish populations that hybridize with introduced species. For example, extensive introductions of rainbow trout (*Oncorhynchus mykiss*) and cutthroat trout (*O. clarki*) in western US watersheds have resulted in loss of diversity of native species because of massive introgression (Allendorf and Leary 1988; Dowling and Childs 1992; Leary et al. 1993).

Habitat fragmentation and alteration are also major factors in the current increase in hybridization and introgression among wildlife populations. Barriers to hybridization are often maintained by habitat differences, thus land conversion that homogenizes habitats and decreases landscape heterogeneity is believed to elevate the probability of hybridization by relaxing ecological divergent selection among species and/or by elevating the costs of mate choice in declining populations (Allendorf et al. 2001; Hendry et al. 2006). Habitat homogenization is believed to be a major factor in driving hybridization among grey wolves and coyotes (*C. latrans*) in the Great Lakes region of North America (Lehman et al. 1991; Koblmuller et al. 2009), Algonquin wolves (*C. lycaon*) and coyotes in Canada (Wilson et al. 2000; Kyle et al. 2006), and red wolves and coyotes in the southeast (Roy et al. 1996; Bohling and Waits 2011). The impact of habitat conversion on hybridization has also been documented for savannah (*Loxodonta africana*) and forest elephants (*L. cyclotis*) in Africa (Roca et al. 2005).

Another major impact of habitat loss and degradation is the decline of population size and density of native species that can lead to Allee (1931, 1938) effects that increase the likelihood of mating between native and invasive species as native animals have difficulty finding mates (Courchamp et al. 1999). This has been commonly cited as a factor leading to increases in hybridization among many rare or endangered species (Rhymer and Simberloff 1996; Allendorf et al. 2001; Wayne and Vila 2003; Grewal et al. 2004; Muñoz-Fuentes et al. 2010; Zamudio and Harrison 2010). One management approach that has been used to address this problem for red wolves is selective removal of coyotes and hybrid offspring (Miller et al. 2003; Stoskopf et al. 2005).

A final challenge for conservation and management of hybridization is the unclear legal status of hybrids (Allendorf et al. 2001; Haig and Allendorf 2006; Ellstrand et al. 2010). Hybrids are not classified or recognized by the International Union for Conservation of Nature, and only one country (South Africa) has endangered species legislation that specifically mentions hybrids. Historically, hybrids were not protected under the US Endangered Species Act, but the policy "that hybrids between endangered species, subspecies, or populations cannot be protected" was withdrawn in 1990 in response to growing scientific information on the complexities and benefits of hybridization (Allendorf et al. 2001). While a new "intercross" policy was drafted by the US Fish and Wildlife Service and National Marine Fisheries Service in 1996, it was never formally approved (Allendorf et al. 2001; Ellstrand et al. 2010). Thus, there is no official policy on the conservation status or protection of hybrid individuals. This makes management decisions regarding hybridization challenging, particularly in cases where purposeful introgression may be necessary to avoid inbreeding depression (Tallmon et al. 2004) and/or augment population sizes, such as occurred in the Florida panther (Pimm et al. 2006) and Washington pygmy rabbit (*Brachylagus idahoensis*) recovery efforts (USFWS 2012).

Using Genetic Methods to Assess and Predict the Impacts of Habitat Alterations

The relationship between habitat, genetic diversity, gene flow, and genetic structure may be viewed from two perspectives. First, because maintaining genetic diversity is fundamental to maintaining population viability, habitat loss, fragmentation, and degradation must be avoided if possible. In this context, researchers may seek to describe genetic structure and diversity to identify populations that are isolated or have very low genetic diversity, and may attempt to quantify how habitat configuration and quality influences those characteristics. Alternatively, because genetic diversity and structure may reflect habitat loss, fragmentation, or degradation, researchers may use genetic studies primarily as a tool to learn about the influences of habitat on distribution, movement, dispersal, and demographic history, rather than being concerned with precisely estimating genetic diversity or structure for their own sake.

Using Neutral and Adaptive Genetic Diversity to Evaluate Habitat

Methods of assessing genetic diversity (box 6.2) have proliferated rapidly because of technological advances in the last several decades, leading to many diversity metrics. Some methods, like allozymes (proteins used for early studies of genetic variation), have been replaced by newer systems assessing DNA sequence or genotypes at mutation-prone locations (box 6.2). Genetic diversity is typically measured at the population level (i.e., across individuals). At the individual level, we typically consider only heterozygosity, which is influenced by genetic diversity. Although it is beyond the scope of this chapter to fully review statistics for estimating genetic diversity at the population level, we briefly summarize some of the more common choices in box 6.3.

Genetic diversity metrics are often calculated across a relatively small number of neutral loci and then used

---

Box 6.3 Metrics of Genetic Diversity

Diversity metrics for DNA sequences:

- Haplotypic diversity (h): haploid version of expected heterozygosity (mitochondrial DNA only)
- Nucleotide diversity (π): calculated as the average number of differences in the nucleotides at each position along a sequence (mitochondrial or nuclear sequence)
- Number of different haplotypes (variants) detected (mitochondrial DNA only)

Diversity metrics for SNPs, microsatellites, and other molecular methods used to analyze variation in nuclear DNA (see also review in Hughes et al. 2008):

- Expected heterozygosity (H_e) is calculated based on the observed frequency of alleles, assuming Hardy-Weinberg equilibrium, and can be corrected for sample size (Nei 1978)
- Observed heterozygosity (H_o) is the proportion of individuals that are heterozygous at a particular locus or averaged across a set of loci
- Proportion of polymorphic loci (i.e., loci containing more than one allele)
- Allelic richness (A_r) is the average number of alleles per locus; A_r is quite sensitive to differences in sample size and therefore should be corrected using subsampling or rarefaction techniques (Leberg 2002)

---

cautiously as an index of diversity across the genome of a species. However, increasingly, new sequencing techniques make it possible to identify adaptive variants and examine diversity at those loci. In some cases, habitat-related variation in phenotype can be linked to particular alleles, raising the possibility of understanding fitness consequences of genetic diversity in different habitats. For instance, Linnen et al. (2013) identified multiple mutations that controlled coat color in deer mice (*Peromyscus maniculatus*) and varied in nearby populations depending on the color of the substrate. Mice on light-colored substrates that lacked alleles for light coat color were exposed to higher rates of predation by avian predators, and alleles for light coat color showed evidence of strong selection in those habitats.

Genetic diversity can reflect habitat quality and thus offers a potential assessment tool, although fewer studies have focused on this relationship compared to the well-known effects of habitat fragmentation on genetic diversity (Pitra et al. 2011). Pitra et al. (2011) found that genetic diversity of great bustards (*Otis tarda*) was correlated with habitat quality and population size and density. Epps et al. (2006) determined that H_e and allelic richness of microsatellite loci in desert bighorn sheep populations were correlated both with population connectivity and with population elevation. As predicted by a previous analysis of population extinctions (Epps et al. 2004), higher elevation populations appeared to have larger N_e and thus maintained more genetic diversity over the long term. However, we note that if dispersal is density dependent and gene flow is high, sink habitats (where mortality is high but offset by immigration) could potentially have higher than expected genetic diversity in some cases because new individuals are constantly entering those areas. Porlier et al. (2009) examined individual genetic diversity of tree swallows (*Tachycineta bicolor*) across a gradient of agricultural intensity and found that more genetically diverse individuals were found in poorer habitats with more intensive agriculture. This counterintuitive result occurred because more diverse individuals arrived first and settled in the earliest available habitat, which occurred in areas with intensive agriculture. Thus, any attempt to use genetic diversity as a metric of habitat quality should carefully consider the characteristics of that species and system.

Effects of Habitat on Gene Flow and Genetic Structure

The influence of habitat on gene flow and genetic structure has been of great interest for many years. Interpreting such studies requires a basic understanding of metrics of gene flow and genetic structure, which we review briefly in box 6.4. There are many metrics of genetic structure (i.e., differences in allele frequencies among different groups of individuals) and gene flow (rates of interbreeding among populations), but the best metric varies depending on the study system and is still widely debated (Jost 2008; Heller and Siegismund 2009; Ryman and Leimar 2009). Many analyses now rely on individual-based metrics of genetic difference rather than population-based estimates of gene flow or genetic structure. Such approaches are particularly useful for species that exist in relatively continuous habitat, where defining populations is problematic. Individual-based approaches can also be used to detect more recent landscape changes than many population-based approaches (Landguth et al. 2010; but see Epps et al. 2013). Common metrics include individual relatedness (reviewed in Csillery et al. 2006) and Rousset's *a* (Rousset 2000). In other cases, the spatial correlation of alleles is described across study areas and areas of discontinuity can be matched up to landscape features (Manel et al. 2007). Manel et al. (2007) used this approach to determine that brown bears in Scandinavia, despite being relatively continuously distributed, could be divided into three "populations" with genetic differences accumulating rapidly in the boundary regions.

Early analyses of genetic structure usually focused on large landscapes (even the entire range of a species) and evaluated evidence for areas of low or absent gene flow among populations. For instance, Hafner and Sullivan (1995) used allozymes to evaluate genetic structure across the range of American pikas (*Ochotona princeps*) and deduce major phylogeographic divisions, establish a likely maximum distance over which dispersal is possible among populations, and evaluate the influence of summer drought conditions on genetic diversity in at least a qualitative way. In recent years, effects of habitat on genetic structure largely have been addressed through landscape genetic approaches (see the following section), which allow spatially explicit tests of how different habitats influence gene flow. Many of these studies are occurring at much finer scales, enabled in part by newer molecular methods that target large numbers of highly variable loci and thus give good statistical power for resolving small genetic differences.

Landscape Genetics

The field of landscape genetics is a relatively new discipline that integrates population genetics and landscape

Box 6.4 Estimating Genetic Structure and Gene Flow

Metrics of Genetic Structure

The most commonly reported metrics of genetic structure are F_{ST} (Wright 1943) and its analogues such as G_{ST} (Nei 1973) and ϕ_{ST}. These statistics have values ranging from 0, meaning identical alleles are present at identical frequencies and thus there is no structure between populations, to 1, meaning no shared alleles and a complete structure. However, F_{ST} and G_{ST} have a number of problematic properties. For one, the maximum value possible declines with increasing diversity of the loci considered (Hedrick 2005). A variety of alternatives have been proposed that address this problem, including G'_{ST} (Hedrick 2005) and Jost's D (Jost 2008), but consensus about the appropriateness of these statistics remains elusive (e.g., Whitlock 2011).

Population Assignment Tests

In this analytical framework, all individuals in the study area typically are combined into a single analysis and sorted into varying numbers of groups by genetic similarity and the overall performance of each model (i.e., number of groups) evaluated to determine which model is most likely. This framework lends itself well to descriptive or exploratory analysis rather than hypothesis testing but does not require a priori definition of populations, although results can be influenced by uneven or biased sampling. Variants of this approach (e.g., geographical information) are implemented in various programs including STRUCTURE (Pritchard et al. 2000), GENELAND (Paetkau et al. 2004), and TESS (Chen et al. 2007).

Gene Flow

Gene flow can be estimated from F_{ST} as Nm (effective number of migrants × migration rate) using Wright's island model (Wright 1931), where $F_{ST} = 1/(1+4Nm)$ for nuclear loci. However, that transformation relies on unrealistic assumptions of equilibrium between genetic drift and migration and other simplifications, and so Nm estimates should not be interpreted literally (Whitlock and McCauley 1999). Alternative methods use coalescent theory (e.g., MIGRATE, Beerli and Felsenstein 2001) or assignment tests (e.g., Rannala and Mountain 1997; Wilson and Rannala 2003; Faubet and Gaggiotti 2008) and rely on maximum-likelihood or Bayesian approaches. These approaches use simulations to explore a wide variety of possible parameter values and determine the solution that best fits observed patterns in genetic structure. These methods also rely on a variety of assumptions and typically require extensive exploration of datasets; thus, their use is not always straightforward, and many researchers still use metrics of genetic structure (F_{ST}, G'_{ST}) as an index of gene flow.

ecology with the goal of providing "information about the interaction between landscape features and micro-evolutionary processes, such as gene flow, genetic drift and selection" (Manel et al. 2003). Landscape genetic approaches are particularly valuable for examining the underlying environmental factors that facilitate or hinder gene flow (Storfer et al. 2007; Holderegger and Wagner 2008; Storfer et al. 2010) and have multiple conservation and management applications (Segelbacher et al. 2010). Landscape genetic approaches can be used to evaluate or predict the impacts of landscape change on genetic diversity and gene flow, detect bar-riers and evaluate landscape resistance to gene flow, and evaluate current corridors and design future corridors (Holderegger and Wagner 2008; Balkenhol et al. 2009a; Segelbacher et al. 2010; Storfer et al. 2010). Landscape genetic analyses of adaptive loci can be used to better understand how environmental factors such as temperature or precipitation affect adaptive genetic variation and evolutionary response to landscape and environmental change (Holderegger et al. 2006; Manel et al. 2010). A wide range of statistical approaches are applied in landscape genetics (Storfer et al. 2007; Balkenhol et al. 2009b), and much effort is focused on developing and evaluating new methods (Faubet and Gaggiotti 2008; Cushman and Landguth 2010; Epperson et al. 2010; Legendre and Fortin 2010; Murphy et al. 2010; Balkenhol and Landguth 2011). A detailed overview of the specific methods applied in landscape genetics is beyond the scope of this chapter. However, we present two case studies to highlight the use of landscape genetics in studying impacts of environmental change on wildlife populations.

Although the potential for loss of gene flow and genetic diversity resulting from habitat fragmentation has been well understood for many years, those changes were initially difficult to detect in wild populations. However, advances in sampling using noninvasive genetic material and increased abilities of labs to process large numbers of samples have made it easier to sample wildlife populations systematically and over large landscapes. Epps et al. (2005) used those techniques to sample twenty-seven populations of desert bighorn sheep (*Ovis canadensis nelsoni*) across a large portion of their range in southeastern California and evaluate how natural and anthropogenic habitat fragmentation interacted to shape gene flow and genetic diversity, as previously described. In a follow-up study, Epps et al. (2007) tested least-cost models of connectivity in a GIS by determining which estimates of effective distance (distance mediated by habitat) were most strongly correlated with gene flow estimates among populations. In this manner, they determined that gene flow was facilitated by steeply sloped areas between core populations, identified the maximum effective distance at which gene flow occurred, and then used those models to evaluate how human-made barriers and population translocations affected connectivity among different populations of desert bighorn sheep.

Predicting the impact of climate change on wildlife habitat and populations is one of the most important challenges for future wildlife conservation (Heller and Zavaleta 2009). In a study of American marten populations of the northern Rocky Mountains in the United States, Wasserman et al. (2010, 2012) have provided one of the best examples of how landscape genetic methods can be applied to predict and evaluate the impact of climate change on wildlife populations. Using nDNA microsatellite loci and an individual-based causal modeling approach (Cushman et al. 2008) to explore relationships between landscape patterns

and gene flow processes, they found that gene flow in American marten was primarily driven by elevation (Wasserman et al. 2010). This finding allowed them to derive a model of landscape resistance that combined with simulation modeling (Landguth et al. 2010) could be used to predict the effects of predicted climate change on population connectivity and genetic diversity of the American martens using five potential future temperature scenarios (Wasserman et al. 2012; Wasserman et al. 2013). They found that even moderate warming scenarios resulted in very large reductions in population connectivity and genetic diversity. Wasserman et al. (2013) were also able to identify important corridors in the current landscape that would remain intact across the climate change scenarios and current corridors that would be lost as the climate warms.

Detecting Hybridization

Genetic analysis is the most definite method for detecting hybridization among taxonomic groups. In the 1980s and 1990s, mtDNA data were often used to detect hybrids (i.e., individuals with the morphology of one species but the mtDNA of an alternative taxonomic group). However, this approach is limited in resolution because of the maternal inheritance of mtDNA (hybridization will not be detected when the father is from the other taxonomic group) and the need for morphological data. More recently, nDNA microsatellite analysis (and occasionally AFLP analysis; see box 6.2) has been the molecular method of choice for detecting hybridization in wildlife populations. When identifying hybrids, it is possible to use diagnostic species-specific alleles (Fitzpatrick and Shaffer 2004; Schwartz et al. 2004), but increasingly, hybrids are detected by applying complex maximum-likelihood and Bayesian approaches (box 6.4, Pritchard et al. 2000; Anderson and Thompson 2002; Corander et al. 2003; Corander and Marttinen 2006; Corrander et al. 2006; Randi 2008) to datasets of 10–100s of microsatellite loci, AFLPs, or SNPs (box 6.2). Detecting hybridization using genetic approaches is most challenging when the hybridizing groups are closely related and when backcrossing to parental groups has occurred (Vaha and Primmer 2006; Bohling and Waits 2011). However, genetic analyses have added greatly to our ability to detect and understand hybridization in wildlife populations, and many of the challenges can be overcome by increasing the number of nDNA loci analyzed per individual. In the following discussion, we review two case studies that illustrate the use of genetic methods to detect the impact of habitat modification on hybridization in wildlife populations.

One good example of habitat-dependent hybridization occurs between the endangered California tiger salamander (*Ambystoma californiense*) and the nonnative barred tiger salamander (*Ambystoma tigrinum mavortium*) that is introduced by bait dealers and fisherman (Riley et al. 2003). Before the introduction of the barred tiger salamander fifty to sixty years ago, the two species had been geographically separated for five million years (Shaffer et al. 2004). Using mtDNA and nDNA loci, researchers evaluated hybridization of these species in the Salinas Valley of California where the landscape has been modified for cattle farming and agriculture creating perennial or ephemeral cattle ponds, both of which have longer hydroperiods than the highly ephemeral vernal pools that were historically the natural breeding habitat of *A. californiense* (Riley et al. 2003; Fitzpatrick and Shaffer 2004, 2007a, b). Overall, they found evidence of **hybrid swarms** in most of the Salinas Valley and demonstrated that hybridization and introgression rates were linked to introduction history and lowest in natural vernal pools, intermediate in ephemeral cattle ponds, and highest in perennial cattle ponds. They hypothesize that the habitat characteristics of the breeding ponds can promote some form of environment-dependent reproductive isolation, either through individual mate choice within the breeding pond or reduced population size of the introduced species (Fitzpatrick and Shaffer 2004, 2007b). They also documented increased fitness of hybrid individuals highlighting the increasing challenge of maintaining and protecting the genetically pure populations of the native endangered California tiger salamander (Fitzpatrick and Shaffer 2007a).

Studies of African elephants provide another good example of natural and human-caused habitat change that can lead to increased hybridization. African forest and savannah elephants are distinct species separated by a hybrid zone (Roca et al. 2001). To evaluate historical and current hybridization levels, Roca et al. (2005) collected genetic data from over seven hundred savannah and forest elephants in twenty-one different locations. They generated DNA sequence data for maternally inherited mtDNA, three neutral loci on the biparentally inherited X chromosome and one neutral locus on the paternally inherited Y chromosome to evaluate the frequency and directionality of hybridization between forest and savannah elephants. Their research indicated that hybridization occurs along forest/savannah ecotones and that forest elephants declined in regions that lost forest cover during the Holocene due to introgressive hybridization with savannah elephants. Additionally, there is evidence that ongoing deforestation and landscape conversion may foster genetic replacement of forest elephants, increasing the risk of extinction of this species (Roca et al. 2005).

Conclusions and Future Directions

As summarized previously, the environmental changes that destroy, degrade, and alter habitats have multiple impacts on wildlife at the genetic level. In addition, molecular genetic techniques greatly enhance our ability to study, understand, and predict the impact of such changes on wild populations. As molecular techniques and analytical methods continue to rapidly progress, we anticipate an increasing role for genetic analysis in unraveling those relationships and in providing solutions for conservation and management. For example, our increased ability to survey greater numbers of neutral loci will lead to greater precision and accuracy in estimates of N_e, gene flow, hybridization, and kin relationships (Allendorf et al. 2010).

Genomic approaches that survey adaptive genetic variation will enable us to better understand the adaptive response to the diversity of natural and human-altered habitats and provide better guidance on conservation priorities and the appropriate sources for population augmentation, reintroductions, and assisted migration (Ouborg et al. 2010; Funk et al. 2012). The use of genomic approaches will also increase our ability to understand the mechanistic basis of inbreeding depression and the relationship between genetic diversity and fitness at the individual and population levels (Allendorf et al. 2010; Ouborg et al. 2010). Ultimately, we are interested in better understanding and predicting how environmental change will affect neutral and adaptive diversity.

The rapidly growing field of landscape genetics has

emerged to help address these questions and will benefit from the larger neutral and adaptive genetic datasets generated from new genomics techniques and increasing coverage and resolution of GIS / remote sensing datasets. However, one challenge facing conservation and landscape geneticists of the future is the development and evaluation of new methods for utilizing genomic datasets and meaningfully linking them to environmental variables. The interdisciplinary integration of landscape ecology, population genetics, and spatial statistics holds much promise for understanding and predicting the impacts of habitat alternation on wildlife populations.

LITERATURE CITED

Agudo, R., M. Carrete, M. Alcaide, C. Rico, F. Hiraldo, and J. A. Donazar. 2012. "Genetic Diversity at Neutral and Adaptive Loci Determines Individual Fitness in a Long-Lived Territorial Bird." *Proceedings of the Royal Society B: Biological Sciences* 279:3241–49.

Aguilar, A., G. Roemer, S. Debenham, M. Binns, D. Garcelon, and R. K. Wayne. 2004. "High MHC Diversity Maintained by Balancing Selection in an Otherwise Genetically Monomorphic Mammal." *Proceedings of the National Academy of Sciences of the United States of America* 101:3490–94.

Alcaide, M. 2010. "On the Relative Roles of Selection and Genetic Drift in Shaping MHC Variation." *Molecular Ecology* 19:3842–44.

Allee, W. C. 1931. *Animal Aggregations: A Study in General Sociology*. The University of Chicago Press, Chicago.

———. 1938. *The Social Life of Animals*. W. W. Norton, New York.

Allendorf, F. W., P. A. Hohenlohe, and G. Luikart. 2010. "Genomics and the Future of Conservation Genetics." *Nature Reviews Genetics* 11:697–709.

Allendorf, F. W., and R. F. Leary. 1988. "Conservation and Distribution of Genetic Variation in a Polytypic Species, the Cutthroat Trout." *Conservation Biology* 2:170–84.

Allendorf, F. W., R. F. Leary, P. Spruell, and J. K. Wenburg. 2001. "The Problems with Hybrids: Setting Conservation Guidelines." *Trends in Ecology & Evolution* 16:613–22.

Allendorf, F. W., G. Luikart, and S. N. Aitken. 2012. *Conservation and the Genetics of Populations*. Wiley-Blackwell, Oxford.

Allendorf, F. W., and N. Ryman. 2002. "The Role of Genetics in Population Viability Analysis." In *Population Viability Analysis*, edited by S. R. Beissinger and D. R. McCullough, 5–17. The University of Chicago Press, Chicago, Illinois.

Anderson, E. C., and E. A. Thompson. 2002. "A Model-Based Method for Identifying Species Hybrids Using Multilocus Genetic Data." *Genetics* 160:1217–29.

Arnold, M. L. 1997. *Natural Hybridization and Evolution*. Oxford University Press, New York.

Balkenhol, N., F. Gugerli, S. A. Cushman, L. P. Waits, A. Coulon, J. W. Arntzen, R. Holderegger, and H. H. Wagner. 2009. "Identifying Future Research Needs in Landscape Genetics: Where to from Here?" *Landscape Ecology* 24:455–63.

Balkenhol, N., and E. L. Landguth. 2011. "Simulation Modelling in Landscape Genetics: On the Need to Go Further." *Molecular Ecology* 20:667–70.

Balkenhol, N., and L. P. Waits. 2009. "Molecular Road Ecology: Exploring the Potential of Genetics for Investigating Transportation Impacts on Wildlife." *Molecular Ecology* 18:4151–64.

Balkenhol, N., L. P. Waits, and R. J. Dezzani. 2009. "Statistical Approaches in Landscape Genetics: An Evaluation of Methods for Linking Landscape and Genetic Data." *Ecography* 32:818–30.

Barton, N. H., and G. M. Hewitt. 1985. "Analysis of Hybrid Zones." *Annual Review of Ecology and Systematics* 16:113–48.

Beaumont, M., E. M. Barratt, D. Gottelli, A. C. Kitchener, M. J. Daniels, J. K. Pritchard, and M. W. Bruford. 2001. "Genetic Diversity and Introgression in the Scottish Wildcat." *Molecular Ecology* 10:319–36.

Beebee, T. J. C., and G. Rowe. 2008. *An Introduction to Molecular Ecology*. 2nd edition. Oxford University Press, New York.

Beerli, P., and J. Felsenstein. 2001. "Maximum Likelihood Estimation of a Migration Matrix and Effective Population Sizes in *N* Subpopulations by Using a Coalescent Approach." *Proceedings of the National Academy of Sciences of the United States of America* 98:4563–68.

Beissinger, S. R. 2002. "Population Viability Analysis: Past, Present and Future." In *Population Viability Analysis*, edited by S. R. Beissinger and D. R. McCullough, 50–85. The University of Chicago Press, Chicago, Illinois.

Birky, C. W., P. Fuerst, and T. Maruyama. 1989. "Organelle Gene Diversity under Migration, Mutation and Drift: Equilibrium Expectations, Approach to Equilibrium, Effects of Heteroplasmic Cells, and Comparison to Nuclear Genes." *Genetics* 121:613–27.

Bohling, J. H., and L. P. Waits. 2011. "Assessing the Prevalence of Hybridization between Sympatric Canis Species Surrounding the Red Wolf (*Canis rufus*) Recovery Area in North Carolina." *Molecular Ecology* 20:2142–56.

Brook, B. W., D. W. Tonkyn, J. J. O'Grady, and R. Frankham. 2002. "Contribution of Inbreeding to Extinction Risk in Threaten Species." *Ecology and Society* 6:1–11.

Buggs, R. J. A. 2007. "Empirical Study of Hybrid Zone Movement." *Heredity* 99:301–12.

Chapman, J. R., S. Nakagawa, D. W. Coltman, J. Slate, and B. C. Sheldon. 2009. "A Quantitative Review of Heterozygosity-Fitness Correlations in Animal Populations." *Molecular Ecology* 18:2746–65.

Chen, C., E. Durand, F. Forbes, and O. Francois. 2007. "Bayesian Clustering Algorithms Ascertaining Spatial Population Structure: A New Computer Program and a Comparison Study." *Molecular Ecology Notes* 7:747–56.

Corander, J., and P. Marttinen. 2006. "Bayesian Identification of Admixture Events Using Multilocus Molecular Markers." *Molecular Ecology* 15:2833–43.

Corander, J., P. Waldmann, and M. J. Sillanpää. 2003. "Bayesian Analysis of Genetic Differentiation between Populations." *Genetics* 163:367–74.

Cornuet, J. M., and G. Luikart. 1996. "Description and Power Analysis of Two Tests for Detecting Recent Population Bottlenecks from Allele Frequency Data." *Genetics* 144:2001–14.

Corrander, J., P. Marttinen, and S. Mäntyniemi. 2006. "A Bayesian Method for Identification of Stock Mixtures from Molecular Marker Data." *Fishery Bulletin* 104:550–58.

Cosentino, B. J., C. A. Phillips, R. L. Schooley, W. H. Lowe, and M. R. Douglas. 2012. "Linking Extinction-Colonization Dynamics to Genetic Structure in a Salamander Metapopulation." *Proceedings of the Royal Society B: Biological Sciences* 279:1575–82.

Courchamp, F., T. Clutton-Brock, and B. Grenfell. 1999. "Inverse Density Dependence and the Allee Effect." *Trends in Ecology & Evolution Trends in Ecology & Evolution* 14:405–10.

Csillery, K., T. Johnson, D. Beraldi, T. Clutton-Brock, D. Coltman, B. Hansson, G. Spong, and J. M. Pemberton. 2006. "Performance of Marker-Based Relatedness Estimators in Natural Populations of Outbred Vertebrates." *Genetics* 173:2091–2101.

Cushman, S. A., and E. L. Landguth. 2010. "Spurious Correlations and Inference in Landscape Genetics." *Molecular Ecology* 19:3592–3602.

Davison, A., J. D. S. Birks, H. I. Griffiths, A. C. Kitchener, D. Biggins, and R. K. Butlin. 1999. Hybridization and the Phylogenetic Relationship between Polecats and Domestic Ferrets in Britain. *Biological Conservation Biological Conservation* 87:155–61.

DeYoung, R. W., and R. L. Honeycutt. 2005. "The Molecular Toolbox: Genetic Techniques in Wildlife Ecology and Management." *Journal of Wildlife Management* 69:1362–84.

Dobzhansky, T. 1948. "Genetics of Natural Populations. XVIII. Experiments on Chromosomes of *Drosophila pseudoobscura* from Different Geographical Regions." *Genetics* 33:588–602.

Dowling, T. E., and M. R. Childs. 1992. "Impact of Hybridization on a Threatened Trout of the Southwestern United States." *Conservation Biology* 6:355–64.

Edmands, S. 2007. "Between a Rock and a Hard Place: Evaluating the Relative Risks of Inbreeding and Outbreeding for Conservation and Management." *Molecular Ecology* 16:463–75.

Ellstrand, N. C., D. Biggs, A. Kaus, P. Lubinsky, L. A. McDade, L. M. Prince, K. Preston, H. M. Regan, V. Rorive, O. A. Ryder, and K. A. Schierenbeck. 2010. "Got Hybridization? A Multidisciplinary Approach for Informing Science Policy." *BioScience* 60:384–88.

Epperson, B. K., B. H. McRae, K. Scribner, S. A. Cushman, M. S. Rosenberg, M. J. Fortin, P. M. A. James, M. Murphy, S. Manel, P. Legendre, and M. R. T. Dale. 2010. "Utility of Computer Simulations in Landscape Genetics." *Molecular Ecology* 19:3549–64.

Epps, C. W., D. R. McCullough, J. D. Wehausen, V. C. Bleich, and J. L. Rechel. 2004. "Effects of Climate Change on Population Persistence of Desert-Dwelling Mountain Sheep in California." *Conservation Biology* 18:102–13.

Epps, C. W., P. J. Palsboll, J. D. Wehausen, G. K. Roderick, and D. R. McCullough. 2006. "Elevation and Connectivity Define Genetic Refugia for Mountain Sheep as Climate Warms." *Molecular Ecology* 15:4295–4302.

Epps, C. W., P. J. Palsboll, J. D. Wehausen, G. K. Roderick, R. R. Ramey II, and D. R. McCullough. 2005. "Highways Block Gene Flow and Cause a Rapid Decline in Genetic Diversity of Desert Bighorn Sheep." *Ecology Letters* 8:1029–38.

Epps, C. W., S. K. Wasser, J. L. Keim, B. M. Mutayoba, and J. S. Brashares. 2013. "Quantifying Past and Present Connectivity Illuminates a Rapidly Changing Landscape for the African Elephant." *Molecular Ecology* 22:1574–88.

Epps, C. W., J. D. Wehausen, V. C. Bleich, S. G. Torres, and J. S. Brashares. 2007. "Optimizing Dispersal and Corridor Models Using Landscape Genetics." *Journal of Applied Ecology* 44:714–24.

Fahrig, L. 2003. "Effects of Habitat Fragmentation on Biodiversity." *Annual Review of Ecology Evolution and Systematics* 34:487–515.

Faubet, P., and O. E. Gaggiotti. 2008. "A New Bayesian Method to Identify the Environmental Factors that Influence Recent Migration." *Genetics* 178:1491–1504.

Fischer, J., and D. B. Lindenmayer. 2000. "An Assessment of the Published Results of Animal Relocations." *Biological Conservation* 96:1–11.

Fisher, R. A. 1930. *The Genetical Theory of Natural Selection*. Oxford University Press, Clarendon.

Fitzpatrick, B. M., and H. B. Shaffer. 2004. "Environment-Dependent Admixture Dynamics in a Tiger Salamander Hybrid Zone." *Evolution* 58:1282–93.

———. 2007a. "Hybrid Vigor between Native and Introduced Salamanders Raises New Challenges for Conservation." *Proceedings of the National Academy of Sciences of the United States of America* 104:15793–98.

———. 2007b. "Introduction History and Habitat Variation Explain the Landscape Genetics of Hybrid Tiger Salamanders." *Ecological Applications* 17:598–608.

Frankham, R. 1995a. "Effective Population Size / Adult Population Size Rations in Wildlife: A Review." *Genetical Research* 66:95–107.

———. 1995b. "Inbreeding and Extinction: A Threshold Effect." *Conservation Biology* 9:792–99.

———. 1998. "Inbreeding and Extinction: Island Populations." *Conservation Biology* 12:665–75.

Frankham, R., J. D. Ballou, M. D. B. Eldridge, R. C. Lacy, K. Ralls, M. R. Dudash, and C. B. Fenster. 2011. "Predicting the Probability of Outbreeding Depression." Conservation Biology 25:465–75.

Franklin, I. R. 1980. "Evolutionary Change in Small Populations." In *Conservation Biology: An Evolutionary-Ecological Perspective*, edited by M. E. Soule and B. A. Wilcox, 135–50. Sinauer Associates, Inc., Sunderland, USA.

Funk, W. C., J. K. McKay, P. A. Hohenlohe, and F. W. Allendorf. 2012. "Harnessing Genomics for Delineating Conservation Units." *Trends in Ecology & Evolution* 27:489–96.

Garza, J. C., and E. G. Williamson. 2001. "Detection of Reduction in Population Size Using Data from Microsatellite Loci." *Molecular Ecology* 10:305–18.

Georgiadis, N., L. Bischof, A. Templeton, J. Patton, W. Karesh, and D. Western. 1994. "Structure and History of African Elephant Populations: I. Eastern and Southern Africa." *Journal of Heredity* 85:100–104.

Gotelli, D., C. Sillero-Zubiri, G. Applebaum, D, M. S. Roy, D. J. Girman, J. Garcia-Moreno, E. A. Ostrander, and R. K. Wayne. 1994. "Molecular Genetics of the Most Endangered Canid: The Ethiopian Wolf *Canis simensis*." *Molecular Ecology* 3:301–12.

Goto, S., H. Iijima, H. Ogawa, and K. Ohya. 2011. "Outbreeding Depression Caused by Intraspecific Hybridization between Local and Nonlocal Genotypes in *Abies sachalinensis*." *Restoration Ecology* 19:243–50.

Grewal, S. K., P. J. Wilson, T. K. Kung, K. Shami, M. T. Theberge, J. B. Theberge, and B. N. White. 2004. "A Genetic Assessment of the Eastern Wolf (*Canis lycaon*) in Algonquin Provincial Park." *Journal of Mammalogy* 85:625–32.

Griffin, C. R., R. J. Shallenberger, and S. I. Fefer. 1989. "Hawaii's Endangered Waterbirds: A Resource Management Challenge." In *Proceedings of Freshwater Wetlands and Wildlife Symposium*, edited by R. R. Sharitz and I. W. Gibbons, 155–69. Savannah River Ecology Laboratory, Aiken, South Carolina.

Hafner, D. J., and R. M. Sullivan. 1995. "Historical and Ecological Biogeography of Nearctic Pikas (*Lagomorpha, Ochotonidae*)." *Journal of Mammalogy* 76:302–21.

Haig, S. M., and F. W. Allendorf. 2006. "Hybrid Policies under the U.S. Endangered Species Act." In *The Endangered Species Act at Thirty, Vol. 2: Conserving Biodiversity in Human-Dominated Landscapes*, edited by J. M. Scott, D. D. Goble, and F. Davis, 150–63. Island Press, Washington, D.C.

Halbert, N. D., P. J. P. Gogan, P. W. Hedrick, J. M. Wahl, and J. N. Derr. 2012. "Genetic Population Substructure in Bison at Yellowstone National Park." *Journal of Heredity* 103:360–70.

Haldane, J. B. S., and C. H. Waddington. 1930. "Inbreeding and Linkage." *Genetics* 16:357–74.

Hanski, I. 1999. *Metapopulation Ecology*. Oxford University Press, New York.

Heber, S., A. Varsani, S. Kuhn, A. Girg, B. Kempenaers, and J. Briskie. 2013. "The Genetic Rescue of Two Bottlenecked South Island Robin Populations Using Translocations of Inbred Donors." *Proceedings of the Royal Society B: Biological Sciences* 280:20122228.

Hedrick, P. W. 1994. "Evolutionary Genetics of the Major Histocompatability Complex." *American Naturalist* 143:945–64.

———. 2005. "A Standardized Genetic Differentiation Measure." *Evolution* 59:1633–38.

———. 2009. *Genetics of Populations, Fourth Edition*. Jones & Bartlett Publishers, Sudbury, Massachusetts.

Hedrick, P. W., and R. Fredrickson. 2010. "Genetic Rescue Guidelines with Examples from Mexican Wolves and Florida Panthers." *Conservation Genetics* 11:615–26.

Heller, N. E., and E. S. Zavaleta. 2009. "Biodiversity Management in the Face of Climate Change: A Review of 22 Years of Recommendations." *Biological Conservation* 142:14–32.

Heller, R., and H. R. Siegismund. 2009. "Relationship between Three Measures of Genetic Differentiation G(ST), D-EST and G'(ST): How Wrong Have We Been?" *Molecular Ecology* 18:2080–83.

Hendry, A. P., P. R. Grant, B. R. Grant, H. A. Ford, M. J. Brewer, and J. Podos. 2006. "Possible Human Impacts on Adaptive Radiation: Beak Size Bimodality in Darwin's Finches." *Proceedings of the Royal Society B: Biological Sciences* 273:1887–94.

Hoelzel, A. R., and W. Amos. 1988. "DNA Fingerprinting and 'Scientific' Whaling." *Nature* 333:305.

Hogg, J. T., S. H. Forbes, B. M. Steele, and G. Luikart. 2006. "Genetic Rescue of an Insular Population of Large Mammals." *Proceedings of the Royal Society B: Biological Sciences* 273:1491–99.

Holderegger, R., D. Buehler, F. Gugerli, and S. Manel. 2010. "Landscape Genetics of Plants." *Trends in Plant Science* 15:675–83.

Holderegger, R., and M. Di Giulio. 2010. "The Genetic Effects of Roads: A Review of Empirical Evidence." *Basic and Applied Ecology* 11:522–31.

Holderegger, R., U. Kamm, and F. Gugerli. 2006. "Adaptive vs. Neutral Genetic Diversity: Implications for Landscape Genetics." *Landscape Ecology* 21:797–807.

Holderegger, R., and H. H. Wagner. 2008. "Landscape Genetics." *BioScience* 58:199–207.

Hughes, A. R., B. D. Inouye, M. T. J. Johnson, N. Underwood, and M. Vellend. 2008. "Ecological Consequences of Genetic Diversity." *Ecology Letters* 11:609–23.

Hutchison, D. W., and A. R. Templeton. 1999. "Correlation of Pairwise Genetic and Geographic Distance Measures: Inferring the Relative Influences of Gene Flow and Drift on the Distribution of Genetic Variability." *Evolution* 53:1898–1914.

Jamieson, I. G., and F. W. Allendorf. 2012. "How Does the 50/500 Rule Apply to MVPs?" *Trends in Ecology & Evolution* 27:578–84.

Johnson, H. E., L. S. Mills, J. D. Wehausen, T. R. Stephenson, and G. Luikart. 2011. "Translating Effects of Inbreeding Depression on Component Vital Rates to Overall Population Growth in Endangered Bighorn Sheep." *Conservation Biology* 25:1240–49.

Johnson, W. E., D. P. Onorato, M. E. Roelke, E. D. Land, M. Cunningham, R. C. Belden, R. McBride, D. Jansen, M. Lotz, D. Shindle, J. Howard, D. E. Wildt, L. M. Penfold,

J. A. Hostetler, M. K. Oli, and S. J. O'Brien. 2010. "Genetic Restoration of the Florida Panther." *Science* 329:1641–45.

Jost, L. 2008. "G(ST) and Its Relatives Do Not Measure Differentiation." *Molecular Ecology* 17:4015–26.

Jourdan-Pineau, H., J. Folly, P. A. Crochet, and P. David. 2012. "Testing the Influence of Family Structure and Outbreeding Depression on Heterozygosity-Fitness Correlations in Small Populations." *Evolution* 66:3624–31.

Kirk, H., and J. R. Freeland. 2011. "Applications and Implications of Neutral versus Non-neutral Markers in Molecular Ecology." *International Journal of Molecular Sciences* 12:3966–88.

Koblmuller, S., M. Nord, R. K. Wayne, and J. A. Leonard. 2009. "Origin and Status of the Great Lakes Wolf." *Molecular Ecology* 18:2313–26.

Kolbe, J. J., R. E. Glor, L. Rodríguez Schettino, A. C. Lara, A. Larson, and J. B. Losos. 2004. "Genetic Variation Increases during Biological Invasion by a Cuban Lizard." *Nature* 431:177–81.

Kolbe, J. J., A. Larson, and J. B. Losos. 2007. "Differential Admixture Shapes Morphological Variation among Invasive Populations of the Lizard *Anolis sagrei*." *Molecular Ecology* 16:1579–91.

Kuehn, R., K. E. Hindenlang, O. Holzgang, J. Senn, B. Stoeckl, and C. Sperisen. 2007. "Genetic Effect of Transportation Infrastructure on Roe Deer Populations (*Capreolus capreolus*)." *Journal of Heredity* 98:13–22.

Kyle, C. J., A. R. Johnson, B. R. Patterson, P. J. Wilson, K. Shami, S. K. Grewal, and B. N. White. 2006. "Genetic Nature of Eastern Wolves: Past, Present and Future." *Conservation Genetics* 7:273–87.

Lacy, R. C. 1997. "Importance of Genetic Variation to the Viability of Mammalian Populations." *Journal of Mammalogy* 78:320–35.

Lande, R. 1988. "Genetics and Demography in Biological Conservation." *Science* 241:1455–59.

Landguth, E. L., S. A. Cushman, M. K. Schwartz, K. S. McKelvey, M. Murphy, and G. Luikart. 2010. "Quantifying the Lag Time to Detect Barriers in Landscape Genetics." *Molecular Ecology* 19:4179–91.

Leary, R. F., F. W. Allendorf, and S. H. Forbes. 1993. "Conservation Genetics of Bull Trout in the Columbia and Klamath River Drainages." *Conservation Biology* 7:857–65.

Leberg, P. L. 2002. "Estimating Allelic Richness: Effects of Sample Size and Bottlenecks." *Molecular Ecology* 11:2445–49.

Legendre, P., and M. J. Fortin. 2010. "Comparison of the Mantel Test and Alternative Approaches for Detecting Complex Multivariate Relationships in the Spatial Analysis of Genetic Data." *Molecular Ecology Resources* 10:831–44.

Lehman, N., A. Eisenhawer, K. Hansen, L. D. Mech, R. O. Peterson, P. J. P. Gogan, and R. K. Wayne. 1991. "Introgression of Coyote Mitochondrial DNA into Sympatric North American Gray Wolf Populations." *Evolution* 45:104–19.

Letty, J., S. Marchandeau, and J. Aubineau. 2007. "Problems En-

countered by Individuals in Animal Translocations: Lessons from Field Studies." *Ecoscience* 14:420–31.

Lever, C. 1987. *Naturalized Birds of the World*. Longmans, London.

Lewontin, R. C., and L. C. Birch. 1966. "Hybridization as a Source of Variation for Adaptation to New Environments." *Evolution* 20:315–36.

Linnen, C. R., Y. P. Poh, B. K. Peterson, R. D. H. Barrett, J. G. Larson, J. D. Jensen, and H. E. Hoekstra. 2013. "Adaptive Evolution of Multiple Traits through Multiple Mutations at a Single Gene." *Science* 339:1312–16.

MacArthur, R. H., and E. O. Wilson. 1967. *The Theory of Island Biogeography*. Princeton University Press, Princeton, New Jersey.

Manel, S., F. Berthoud, E. Bellemain, G. Gaudeul, G. Luikart, J. E. Swenson, L. P. Waits, P. Taberlet, and I. Consortium. 2007. "A New Individual-Based Spatial Approach for Identifying Genetic Discontinuities in Natural Populations." *Molecular Ecology* 16:2031–43.

Manel, S., S. Joost, B. K. Epperson, R. Holderegger, A. Storfer, M. S. Rosenberg, K. T. Scribner, A. Bonin, and M. J. Fortin. 2010. "Perspectives on the Use of Landscape Genetics to Detect Genetic Adaptive Variation in the Field." *Molecular Ecology* 19:3760–72.

Manel, S., M. K. Schwartz, G. Luikart, and P. Taberlet. 2003. "Landscape Genetics: Combining Landscape Ecology and Population Genetics." *Trends in Ecology & Evolution* 18:189–97.

Marr, A. B., L. F. Keller, and P. Arcese. 2002. "Heterosis and Outbreeding Depression in Descendants of Natural Immigrants to an Inbred Population of Song Sparrows (*Melospiza melodia*)." *Evolution* 56:131–42.

Marsden, C. D., R. Woodroffe, M. G. L. Mills, J. W. McNutt, S. Creel, R. Groom, M. Emmanuel, S. Cleaveland, P. Kat, G. S. A. Rasmussen, J. Ginsberg, R. Lines, J. M. Andre, C. Begg, R. K. Wayne, and B. K. Mable. 2012. "Spatial and Temporal Patterns of Neutral and Adaptive Genetic Variation in the Endangered African Wild Dog (*Lycaon pictus*)." *Molecular Ecology* 21:1379–93.

Marshall, T. C., and J. A. Spalton. 2000. "Simultaneous Inbreeding and Outbreeding Depression in Reintroduced Arabian Oryx." *Animal Conservation* 3:241–48.

Mazourek, J. C., and P. N. Gray. 1994. "The Florida Duck or the Mallard." *Florida Wildlife* 48:29–31.

Miller, C. R., J. R. Adams, and L. P. Waits. 2003. "Pedigree-Based Assignment Tests for Reversing Coyote (*Canis latrans*) Introgression Into the Wild Red Wolf (*Canis rufus*) Population." *Molecular Ecology* 12:3287–3301.

Miller, J. M., J. Poissant, J. T. Hogg, and D. W. Coltman. 2012. "Genomic Consequences of Genetic Rescue in an Insular Population of Bighorn Sheep (*Ovis canadensis*)." *Molecular Ecology* 21:1583–96.

Mills, L. S., and P. E. Smouse. 1994. "Demographic Consequences of Inbreeding in Remnant Populations." *The American Naturalist* 144:412–31.

Morin, P. A., G. Luikart, R. K. Wayne, and S. N. P. W. Grp. 2004. "SNPs in Ecology, Evolution and Conservation." *Trends in Ecology & Evolution* 19:208–16.

Mullis, K. B., and F. A. Faloona. 1987. "Specific Synthesis of DNA in Vitro via a Polymerase-Catalyzed Chain Reaction." *Methods in Enzymology* 155:335–50.

Muñoz-Fuentes, V., C. T. Darimont, P. C. Paquet, and J. A. Leonard. 2010. "The Genetic Legacy of Extirpation and Re-colonization in Vancouver Island Wolves." *Conservation Genetics* 11:547–56.

Murphy, M. A., J. S. Evans, and A. Storfer. 2010. "Quantifying Bufo Boreas Connectivity in Yellowstone National Park with Landscape Genetics." *Ecology* 91:252–61.

Nei, M. 1973. "Analysis of Gene Diversity in Subdivided Populations." *Proceedings of the National Academy of Sciences of the United States of America* 70:3321–23.

———. 1978. "Estimation of Average Heterozygosity and Genetic Distance from a Small Number of Individuals." *Genetics* 89:583–90.

Nei, M., T. Maruyama, and R. Chakraborty. 1975. "Bottleneck Effect and Genetic Variability in Populations." *Evolution* 29:1–10.

Ogrady, J. J., B. W. Brook, D. H. Reed, J. D. Ballou, D. W. Tonkyn, and R. Frankham. 2006. "Realistic Levels of Inbreeding Depression Strongly Affect Extinction Risk in Wild Populations." *BIOC Biological Conservation* 133:42–51.

Olano-Marin, J., J. C. Mueller, and B. Kempenaers. 2011. "Correlations between Heterozygosity and Reproductive Success in the Blue Tit (*Cyanistes caeruleus*): An Analysis of Inbreeding and Single Locus Effects." *Evolution* 65:3175–94.

Olson, Z. H., D. G. Whittaker, and O. E. Rhodes. 2012. "Evaluation of Experimental Genetic Management in Reintroduced Bighorn Sheep." *Ecology and Evolution* 2:429–43.

O'Malley, K. G., M. J. Ford, and J. J. Hard. 2010. "Clock Polymorphism in Pacific Salmon: Evidence for Variable Selection Along a Latitudinal Gradient." *Proceedings of the Royal Society B: Biological Sciences* 277:3703–14.

Ouborg, N. J., C. Pertoldi, V. Loeschcke, R. Bijlsma, and P. W. Hedrick. 2010. "Conservation Genetics in Transition to Conservation Genomics." *Trends in Genetics* 26:177–87.

Paetkau, D., R. Slade, M. Burden, and A. Estoup. 2004. "Genetic Assignment Methods for the Direct, Real-Time Estimation of Migration Rate: A Simulation-Based Exploration of Accuracy and Power." *Molecular Ecology* 13:55–65.

Pimm, S. L., L. Dollar, and O. L. Bass. 2006. "The Genetic Rescue of the Florida Panther." *Animal Conservation* 9:115–22.

Piry, S., G. Luikart, and J. M. Cornuet. 1999. "BOTTLENECK: A Computer Program for Detecting Recent Reductions in the Effective Population Size Using Allele Frequency Data." *Journal of Heredity* 90:502–3.

Pitra, C., S. Suarez-Seoane, C. A. Martin, W.-J. Streich, and J. C. Alonso. 2011. "Linking Habitat Quality with Genetic Diversity: A Lesson from Great Bustards in Spain." *European Journal of Wildlife Research* 57:411–19.

Porlier, M., M. Belisle, and D. Garant. 2009. "Non-random Distribution of Individual Genetic Diversity Along an Environmental Gradient." *Philosophical Transactions of the Royal Society B: Biological Sciences* 364:1543–54.

Pritchard, J. K., M. Stephens, and P. Donnelly. 2000. "Inference of Population Structure Using Multilocus Genotype Data." *Genetics* 155:945–59.

Prugh, L. R., K. E. Hodges, A. R. E. Sinclair, and J. S. Brashares. 2008. "Effect of Habitat Area and Isolation on Fragmented Animal Populations." *Proceedings of the National Academy of Sciences of the United States of America* 105:20770–75.

Radwan, J., A. Biedrzycka, and W. Babik. 2010. "Does Reduced MHC Diversity Decrease Viability of Vertebrate Populations?" *Biological Conservation* 143:537–44.

Ralls, K., J. D. Ballou, and A. Templeton. 1988. "Estimates of Lethal Equivalents and the Cost of Inbreeding in Mammals." *Conservation Biology* 2:185–93.

Randi, E. 2008. "Detecting Hybridization between Wild Species and Their Domesticated Relatives." *Molecular Ecology* 17:285–93.

Rannala, B., and J. L. Mountain. 1997. "Detecting Immigration by Using Multilocus Genotypes." *Proceedings of the National Academy of Sciences of the United States of America* 94:9197–9201.

Rhymer, J. M., and D. Simberloff. 1996. "Extinction by Hybridization and Introgression." *Annual Review of Ecology and Systematics* 27:83–109.

Rhymer, J. M., M. J. Williams, and M. J. Braun. 1994. "Mitochondrial Analysis of Gene Flow between New Zealand Mallards (*Anas platyrhynchos*) and Grey Ducks (*A. superciliosa*)." *The Auk* 111:970–78.

Riley, S. P. D., H. Bradley Shaffer, S. Randal Voss, and B. M. Fitzpatrick. 2003. "Hybridization between a Rare, Native Tiger Salamander (*Ambystoma californiense*) and Its Introduced Congener." *Ecological Applications* 13:1263–75.

Roberge, C., E. Normandeau, S. Einum, H. Guderley, and L. Bernatchez. 2008. "Genetic Consequences of Interbreeding between Farmed and Wild Atlantic Salmon: Insights from the Transcriptome." *Molecular Ecology* 17:314–24.

Robert, A., F. Sarrazin, D. Couvet, and S. Legendre. 2004. "Releasing Adults versus Young in Reintroductions: Interactions between Demographics and Genetics." *Conservation Biology* 18:1078–87.

Roca, A. L., N. Georgiadis, and S. J. O'Brien. 2005. "Cytonuclear Genomic Dissociation in African Elephant Species." *Nature Genetics* 37:96–100.

Roca, A. L., N. Georgiadis, J. Pecon-Slattery, and S. J. O'Brien. 2001. "Genetic Evidence for Two Species of Elephant in Africa." *Science* 293:1473–77.

Rousset, F. 2000. "Genetic Differentiation between Individuals." *Journal of Evolutionary Biology* 13:58–62.

Roy, M. S., E. Geffen, D. Smith, and R. K. Wayne. 1996. "Molecular Genetics of Pre-1940 Red Wolves." *Conservation Biology* 10:1413–24.

Ruiz-Lopez, M. J., R. J. Monello, M. E. Gompper, and L. S. Eggert. 2012. "The Effect and Relative Importance of Neutral

Genetic Diversity for Predicting Parasitism Varies across Parasite Taxa." *Plos One* 7(9): e45404.

Ryman, N., and O. Leimar. 2009. "G(ST) is Still a Useful Measure of Genetic Differentiation: A Comment on Jost's D." *Molecular Ecology* 18:2084–87.

Saccheri, I., M. Kuussaari, M. Kankare, P. Vikman, W. Fortelius, and I. Hanski. 1998. "Inbreeding and Extinction in a Butterfly Metapopulation." *Nature* 392:491–94.

Sanger, F., G. M. Air, B. G. Barrell, N. L. Brown, A. R. Coulson, C. A. Fiddes, C. A. Hutchison, P. M. Slocombe, and M. Smith. 1977. "Nucleotide Sequence of Bacteriophage Phi X174 DNA." *Nature* 265:687–95.

Schoville, S. D., A. Bonin, O. Francois, S. Lobreaux, C. Melodelima, and S. Manel. 2012. "Adaptive Genetic Variation on the Landscape: Methods and Cases." In *Annual Review of Ecology, Evolution, and Systematics, Vol 43*, edited by D. J. Futuyma, 23–43.

Schwartz, M. K., K. L. Pilgrim, K. S. McKelvey, E. L. Lindquist, J. J. Claar, S. Loch, and L. F. Ruggiero. 2004. "Hybridization between Canada Lynx and Bobcats: Genetic Results and Management Implications." *Conservation Genetics* 5:349–55.

Schwensow, N., J. Fietz, K. H. Dausmann, and S. Sommer. 2007. "Neutral versus Adaptive Genetic Variation in Parasite Resistance: Importance of Major Histocompatibility Complex Supertypes in a Free-Ranging Primate." *Heredity* 99:265–77.

Seehausen, O. 2004. "Hybridization and Adaptive Radiation." *Trends in Ecology & Evolution* 19:198–207.

Seehausen, O., G. Takimoto, D. Roy, and J. Jokela. 2008. "Speciation Reversal and Biodiversity Dynamics with Hybridization in Changing Environments." *Molecular Ecology* 17:30–44.

Segelbacher, G., S. A. Cushman, B. K. Epperson, M. J. Fortin, O. Francois, O. J. Hardy, R. Holderegger, P. Taberlet, L. P. Waits, and S. Manel. 2010. "Applications of Landscape Genetics in Conservation Biology: Concepts and Challenges." *Conservation Genetics* 11:375–85.

Selkoe, K. A., and R. J. Toonen. 2006. "Microsatellites for Ecologists: A Practical Guide to Using and Evaluating Microsatellite Markers." *Ecology Letters* 9:615–29.

Shaffer, H. B., G. B. Pauly, J. C. Oliver, and P. C. Trenham. 2004. "The Molecular Phylogenetics of Endangerment: Cryptic Variation and Historical Phylogeography of the California Tiger Salamander, *Ambystoma californiense.*" *Molecular Ecology* 13:3033–49.

Shendure, J., and H. L. Ji. 2008. "Next-Generation DNA Sequencing." *Nature Biotechnology* 26:1135–45.

Storfer, A., M. A. Murphy, J. S. Evans, C. S. Goldberg, S. Robinson, S. F. Spear, R. Dezzani, E. Delmelle, L. Vierling, and L. P. Waits. 2007. "Putting the 'Landscape' in Landscape Genetics." *Heredity* 98:128–42.

Storfer, A., M. A. Murphy, S. F. Spear, R. Holderegger, and L. P. Waits. 2010. "Landscape Genetics: Where Are We Now?" *Molecular Ecology* 19:3496–3514.

Stoskopf, M. K., K. Beck, B. B. Fazio, T. K. Fuller, E. M. Gese, B. T. Kelly, F. F. Knowlton, D. L. Murray, W. Waddell, and L. Waits. 2005. "Implementing Recovery of the Red Wolf-Integrating Research Scientists and Managers." *Wildlife Society Bulletin* 33:1145–52.

Taberlet, P., L. P. Waits, and G. Luikart. 1999. "Noninvasive Genetic Sampling: Look Before You Leap." *Trends in Ecology & Evolution* 14:323–27.

Tallmon, D. A., G. Luikart, and R. S. Waples. 2004. "The Alluring Simplicity and Complex Reality of Genetic Rescue." *Trends in Ecology & Evolution* 19:489–96.

Templeton, A. R. 1995. "Measuring Genetic Diversity in Macroorganisms." In *Biodiversity Measurement and Estimation*, edited by D. L. Hawksworth, 59–64. Chapman & Hall, London.

Templeton, A. R., K. Shaw, E. Routman, and S. K. Davis. 1990. "The Genetic Consequences of Habitat Fragmentation." *Annals of the Missouri Botanical Garden* 77:13–27.

US Fish and Wildlife Service. 2012. "Recovery Plan for the Columbia Basin Distinct Population Segment of the Pygmy Rabbit (*Brachylagus idahoensis*)." U. S. Fish and Wildlife Service, Portland, Oregon.

Vaha, J. P., and C. R. Primmer. 2006. "Efficiency of Model-Based Bayesian Methods for Detecting Hybrid Individuals Under Different Hybridization Scenarios and with Different Numbers of Loci." *Molecular Ecology* 15:63–72.

Vignal, A., D. Milan, M. SanCristobal, and A. Eggen. 2002. "A Review on SNP and Other Types of Molecular Markers and Their Use in Animal Genetics." *Genetics Selection Evolution* 34:275–305.

Vila, C., A. K. Sundqvist, O. Flagstad, J. Seddon, S. Bjornerfeldt, I. Kojola, A. Casulli, H. Sand, P. Wabakken, and H. Ellegren. 2003. "Rescue of a Severely Bottlenecked Wolf (Canis lupus) Population by a Single Immigrant." *Proceedings of the Royal Society B: Biological Sciences* 270:91–97.

Vos, P., R. Hogers, M. Bleeker, M. Reijans, T. van de Lee, M. Hornes, A. Frijters, J. Pot, J. Peleman, M. Kuiper, et al. 1995. "AFLP: A New Technique for DNA Fingerprinting." *Nucleic Acids Research* 23:4407–14.

Waits, L. P., and D. Paetkau. 2005. "Noninvasive Genetic Sampling Tools for Wildlife Biologists: A Review of Applications and Recommendations for Accurate Data Collection." *Journal of Wildlife Management* 69:1419–33.

Wasserman, T. N., S. A. Cushman, J. S. Littell, A. J. Shirk, and E. L. Landguth. 2013. "Population Connectivity and Genetic Diversity of American Marten (*Martes americana*) in the United States Northern Rocky Mountains in a Climate Change Context." *Conservation Genetics* 14:529–41.

Wasserman, T. N., S. A. Cushman, M. K. Schwartz, and D. O. Wallin. 2010. "Spatial Scaling and Multi-model Inference in Landscape Genetics: *Martes americana* in Northern Idaho." *Landscape Ecology* 25:1601–12.

Wasserman, T. N., S. A. Cushman, A. S. Shirk, E. L. Landguth, and J. S. Littell. 2012. "Simulating the Effects of Climate Change on Population Connectivity of American Marten

(*Martes americana*) in the Northern Rocky Mountains, USA." *Landscape Ecology* 27:211–25.

Wayne, R. K., and C. Vila. 2003. "Molecular Genetics Studies of Wolves." In *Wolves: Behavior, Ecology, and Conservation*, edited by L. D. Mech and L. Boitani, 218–38. University of Chicago Press, Chicago, Illinois.

Whitlock, M. C. 2011. "G'_{ST} and D Do Not Replace F_{ST}." *Molecular Ecology* 20:1083–91.

Whitlock, M. C., and D. E. McCauley. 1999. "Indirect Measures of Gene Flow and Migration: F-ST Not Equal 1/(4Nm+1)." *Heredity* 82:117–25.

Wilson, G. A., and B. Rannala. 2003. "Bayesian Inference of Recent Migration Rates Using Multilocus Genotypes." *Genetics* 163:1177–91.

Wilson, P. J., S. Grewal, I. D. Lawford, J. N. M. Heal, A. G. Granacki, D. Pennock, J. B. Theberge, M. T. Theberge, D. R. Voigt, W. Waddell, R. E. Chambers, P. C. Paquet, G. Goulet, D. Cluff, and B. N. White. 2000. "DNA Profiles of the Eastern Canadian Wolf and the Red Wolf Provide Evidence for a Common Evolutionary History Independent of the Gray Wolf." *Canadian Journal of Zoology* 78:2156–66.

Wiseman, R., C. O'Ryan, and E. H. Harley. 2000. "Microsatellite Analysis Reveals that Domestic Cat (*Felis catus*) and Southern African Wild Cat (*F. lybica*) Are Genetically Distinct." *Animal Conservation* 3:221–28.

Wolf, D. E., N. Takebayashi, and L. H. Rieseberg. 2001. "Predicting the Risk of Extinction through Hybridization." *Conservation Biology* 15:1039–53.

Worley, K., J. Carey, A. Veitch, and D. W. Coltman. 2006. "Detecting the Signature of Selection on Immune Genes in Highly Structured Populations of Wild Sheep (*Ovis dalli*)." *Molecular Ecology* 15:623–37.

Wright, S. 1931. "Evolution in Mendelian Populations." *Genetics* 16:97–159.

———. 1938. "Size of Population and Breeding Structure in Relation to Evolution." *Science* 87:430–31.

———. 1943. "Isolation by Distance." *Genetics* 28:114–38.

Zamudio, K. R., and R. G. Harrison. 2010. "Hybridization in Threatened and Endangered Animal Taxa: Implications for Conservation and Management of Biodiversity." In *Molecular Approaches in Natural Resource Conservation ("the Work")*, edited by J. A. DeWoody et al., 169–89. Cambridge University Press, Cambridge.

7 — Habitat Fragmentation and Corridors

K. Shawn Smallwood

Since its conceptualization, habitat fragmentation has factored prominently in academic discussions on threats to biodiversity and species' conservation. Habitat fragmentation has also been incorporated into environmental law; e.g., one of the standards of the California Environmental Quality Act is to assess a project's potential interference with the movement of fish and wildlife. A principal countermeasure to habitat fragmentation—maintaining and improving habitat connectivity—has also factored into many academic and legal discussions. Natural or constructed corridors have often been promoted to mitigate the effects of habitat fragmentation by maintaining or improving animal movement and gene flow between habitat patches. Given its prominence in conservation biology and environmental law, one would expect that the term *habitat fragmentation* is well understood and based on a strong theoretical foundation.

However, given the widespread confusion over the meaning of the term *habitat* (Hall et al. 1997; Guthery and Strickland, this volume), which is a conceptual foundation of habitat fragmentation, it might be worth examining how well habitat fragmentation and corridors are understood. At its inception, some degree of vagueness around the concept was reasonable. Wilcox and Murphy (1985) argued, "That current theory is inadequate for resolving many of the details should not detract from what is obvious and accepted by most ecologists: habitat fragmentation is the most serious threat to biological diversity and is the primary cause of the present extinction crisis." Nearly three decades after Wilcox and Murphy (1985), and after humans have converted vast additional areas for human use, it is time to ask whether the theory has advanced sufficiently to resolve the details that matter. More importantly, it is time to examine whether the concepts of habitat fragmentation and corridors have mattered where these interrelated concepts needed to be implemented.

My objectives for this chapter are first to compare definitions of habitat fragmentation and corridors and synthesize these interrelated concepts. Next, I review environmental documents to reveal how practitioners perceive these concepts and to what degrees they apply them in impact analyses and conservation planning. Finally, I suggest how these concepts could be more consistently understood and effectively applied.

Habitat Fragmentation Defined

Wilcox and Murphy (1985) described habitat fragmentation as habitat loss and insularization. They described the risk of fragmentation as (1) destruction, reduction, or subdivision of demographic units; (2) loss of potential sources of immigrants; and (3) impedance of immigration caused by conversion of habitat between habitat patches. Saunders et al. (1991) described it as clearing of natural vegetation resulting in isolated, remnant patches of vegetation. Yahner (1996) recommended that habitat fragmentation be considered a process of diminishing size and increasing isolation of habitat fragments, where the result is habitat loss. Villard et al. (1999) described habitat fragmentation as a process in which a focal habitat type is partially or completely removed, resulting in an alternate configuration and reduced population persistence on the landscape. Bender et al. (1998) defined habitat fragmentation as an event that creates a greater number of habitat patches that are smaller in size than the original contiguous habitat. Francis (this volume) similarly described it as the spatial arrangement and shape of remaining habitat patches remaining after habitat loss. Bender et al. went on to define it as an event that produces an even greater population decline than would occur due to simple habitat loss.

Karr (1994) described fragmentation as a disruption of the linkages among patches that exchange ecologically important resources. Wilcox et al. (2002) described it as the loss of contiguity of accessible landscape from the perspective of the organism or some other ecologically important element, where landscape context, the spatial extent of the organisms' ecological interactions, and its demographic organization are critical factors.

Saunders et al. (1991) argued that habitat fragmentation not only causes biogeographic changes, i.e., increasing isolation of habitat patches, but also physical changes within the patches due to alterations in microclimate. Similarly, Wiens (1997) argued that the land conversions causing habitat fragmentation often sharpen habitat patch boundaries, alter connectivity, and shift the cost-benefit contours on the landscape. In other words, land conversions often inject into the remaining habitat fragments invasive species (Alberts et al. 1993), light pollution (Rich and Longcore 2006), noise pollution, atmospheric pollution, water pollution (Longcore et al. 1993), soil erosion, and added mortality factors such as automobile collisions, line collisions, electrocutions, poaching, and animal damage control. Furthermore, the net length of boundary increases between habitat fragments and the nonhabitat matrix, thereby increasing the number and variety of ways in which the land conversions can degrade the remaining habitat fragments. For example, large carnivores, due to their greater ranging behavior, will more often experience conflict with humans when the habitat fragments they occupy include greater edge-to-interior ratios (Woodroffe and Ginsberg 1998). Loss of the large predators can result in mesopredator release, thereby putting more predation pressure on other wildlife species residing in the fragments (Zembal 1993). Habitat fragmentation can expose patch-interior spe-

cies to competitors, predators, and parasites that were more prevalent along the patch edges (Laurence and Yensen 1991; McCollin 1993; Porneluzi et al. 1993), even leading to pest outbreaks that can further degrade or remove the vegetation in habitat fragments (Roland 1993).

Ultimately, what separates habitat fragmentation from simple habitat loss is the disproportionate reduction in numerical capacity of the remaining habitat of the same net area (also see Francis, this volume). In other words, habitat reduced to a contiguous area of one hundred ha might support one hundred individuals of a certain species, but one hundred ha of habitat fragmented by impassable barriers (e.g., irrigation ditches, a freeway, residential or commercial development) and contaminated by physical and biological pollutants might support sixty individuals, or even zero individuals. Hypothetically, the protection or creation of corridors can maintain or restore some of the numerical capacity that would otherwise be lost to fragmentation. Applied to the previous example, the creation of one or more movement corridors might restore the capacity of the one hundred ha from sixty individuals to ninety individuals. Fragmentation is thought to disproportionately reduce numerical capacity, whereas corridors are thought to counteract this effect.

Why the disproportionate reduction in fragmented habitat? Genetic isolation is often cited, and so are other processes from island biogeography theory. Another factor is the suite of altered physical and biological conditions caused by greater edge-to-interior ratios and by the activities in the emergent nonhabitat matrix (Saunders et al. 1991; Bolger et al. 1997; Wiens 1997). But a third factor is social organization, which is a much stronger force often given credit in discussions of habitat. In an extreme example, our one hundred ha of fragmented habitat might consist of one hundred one-ha patches of isolated habitat, each patch capable of supporting one individual of a particular species on the basis of resources but incapable of supporting any individuals beyond a single generation due to insufficient space for breeding pairs, offspring, or larger demographic units. The spatial extent of habitat must comport with the social organization and associated behaviors of the species.

Long-term ecological relationships have established trajectories of how individuals of each species interface with their habitats. For example, some species have developed territorial behaviors to stabilize numerical responses to resource variability and therefore cannot tolerate crowding that might initially be forced by habitat fragmentation. Many species routinely shift locations of high activity, i.e., high density, either generationally or multigenerationally. Taylor and Taylor (1979) hypothesized that this spatial shifting of abundance enables species to rest local food resources or to escape predator or parasite loads. They also suggested that dispersed young might naturally congregate in new locations while the natal colonies senesce. Increasing barriers to movement and reducing the effective area of habitat patches, whether through land conversions or patch contamination, will act upon the species' behavior trajectories, which often involve much larger habitat areas than commonly studied by biologists (Smallwood 1999).

Consider that the mass density of populations of species of mammalian Carnivora averages nine kg/km^2 when scaled to the spatial area encompassing the population, referred to as the threshold area (Smallwood 2001a). This scaling of the spatial extent of a population to achieve a common mass density suggests an ecological allometry that is more sensitive to the spatial extent at which the species' population operates than it is to density or body mass. Indeed, threshold area was proportional to average female brain mass among species of Carnivora, and it was more responsive to female brain mass than to body mass (Smallwood 2001a). Species of Carnivora aggregate in similar numbers, but these numbers spread across larger spatial extents in proportion to increasing average brain mass. I hypothesize that the habitat areas intervening populations also scale with female brain mass. I also hypothesize that the habitat areas at which the populations of other animal species operate also scale with brain mass. Female brain mass as the axis of similitude would implicate parental care and other life-history attributes, along with basal metabolic rate and sensory perception. Species are probably much more sensitive to habitat fragmentation than can be measured by extrapolations of average density to the remaining areas of habitat fragments.

Discussion of habitat fragmentation often centers on individual species, but also often it encompasses multiple species. This dual use of the term causes confusion because the original definition of the term *habitat* applies to single species, but *habitat* is increasingly used to characterize vegetation complexes associated with multiple species, i.e., *habitat types*. When the discussion refers to wildlife habitat or habitat types, the basis of habitat fragmentation is typically a map of vegetation cover types that are convenient for mapping but probably ill-suited for analysis of habitat fragmentation. Measuring species richness in response to habitat fragmentation implies the same concept of habitat type as the basis of the analysis. In reality, a map of a species' habitat, i.e., that part of the environment where the species lives, would rarely match a mapped habitat type intended to apply to multiple species. That is, the non-habitat matrix derived from such a map probably rarely applies to a given species. This source of confusion in discussions of habitat fragmentation will be discussed in the following discussion.

Corridors Defined

Beier and Loe (1992) defined the wildlife movement corridor as a linear habitat whose primary function is to connect habitat patches. Beier and Noss (1998) defined corridor as linear habitat connecting habitat patches of the same type and separated by a dissimilar matrix. Beier and Noss specifically excluded linear reaches of vegetation that were dissimilar to the habitat patches at issue; e.g., they excluded riparian corridors that might stretch between two patches of chaparral. In their meta-analysis of corridor effectiveness, Gilbert-Norton et al. (2010) defined corridor "as a narrow, linear (or near-linear) piece of habitat that connects two larger patches of habitat that are surrounded by a nonhabitat matrix." They also distinguished between natural and constructed corridors. Simberloff and Cox (1987) characterized what they termed *conservation corridors* as constructed corridors intended to connect habitat reserves to facilitate immigration and genetic exchange. Forman and Godron (1981) defined another two types of corridor as (1) strip corridor: a protected strip of landscape that is wide enough to include interior conditions typical of the interior conditions of habitat patches being connected by the strip, and (2) line corridor: a protected strip of landscape that is too narrow to include interior conditions typical of the interior conditions of habitat patches being connected by the strip but instead includes edge conditions that are typical of the edge conditions of the connected habitat patches. There are many other definitions of *corridor* in the literature, including movement corridor, habitat corridor, dispersal corridor, and landscape linkages, so it is clear that the corridor concept varies among investigators according to their application.

Beier and Noss (1998) concluded that corridors enhance population viability in connected habitat patches, but the papers they reviewed generally did not support their conclusion due to pseudoreplication and confounding caused by inadequate experimental design. In response to criticism of the evidence supporting the claim that habitat corridors improve population viability (Simberloff et al. 1992), Beier and Noss asserted that an experimental design was unnecessary for concluding that corridors are effective.

Synthesis of Habitat Fragmentation and Corridors

The concepts of habitat fragmentation and corridors are interrelated. These terms are part of the same discussion about a particular type of anthropogenic impact on wildlife species and whether and to what degree this type of impact can be moderated by the protection or creation of habitat or movement corridors. Habitat fragmentation can certainly happen naturally (Morrison et al. 1998), but naturally occurring habitat fragmentation has been rarely discussed and is of relatively low interest. Much greater focus has been on anthropogenic habitat fragmentation and whether and how to mitigate its adverse effects via corridor protection or construction. Discussions about corridors have been more confusing on the distinction of whether the context is about corridors as naturally occurring landscape or habitat features or as a conservation asset that needs to be protected or constructed to counter the effects of habitat fragmentation.

Corridor implies concentrated movement of one or more species, or disproportionate use of a linear portion of a landscape. Naturally occurring corridors are typically characterized as linear features of the landscape, such as stream basins and the vegetation that grows along streams. Linear features of the landscape

have been thought to be used by animals as convenient guides to migration, dispersal, home range patrol, and other types of long-distance movement. However, biologists often assume that this relationship exists, rather than actually measuring it. In fact, animals move across all types of landscape features and vegetation patterns, not just linear features. The capacity for animals and plants to move across the "matrix" landscape might be just as important as the capacity for movement along corridors.

Conservation corridors also imply concentrated movement of one or more species but only in the context of navigating between habitat patches in a changing environment. Whether protected or constructed, conservation corridors are intended to enable continued or improved movement of wildlife between habitat fragments. However, the effectiveness of conservation corridors is rarely tested, and many so-called corridors may themselves be nothing more than remnant fragments of habitat. Furthermore, many conservation corridors have been established to suit the convenience of the planners. I have rarely seen evidence that animal movement patterns were measured prior to the establishment of conservation corridors. The common assumption appears to have been that the animals surviving habitat fragmentation will use the corridors, perhaps because they have no other choice. Nevertheless, the purpose of conservation corridors has been to offset the adverse impacts of habitat fragmentation. The only overlap in the scientific dialogue between naturally occurring corridors and conservation corridors is in the protection of the former as the latter as habitat fragmentation proceeds.

Habitat Fragmentation and Corridors in Practice

The concepts of habitat fragmentation and corridors have been favorite topics of debate in the scientific literature, having inspired thousands of papers and books with one or both of these concepts being addressed. Scientists debate definitions related to these concepts, and they debate how to measure habitat fragmentation and how to identify corridors from within a landscape matrix. While these scientific debates have continued for nearly three decades, other human actions have transformed large tracts of wildlife habitat for human purposes. It is at the business end where the concepts of habitat fragmentation and corridors have been in most urgent need of general understanding and implementation.

It is in environmental reviews, management plans, environmental laws, and, ultimately, actions where the concepts of habitat fragmentation and corridors matter most (Salwasser 1990). The practitioners who need to understand these concepts include environmental consultants, biologists working for regulatory and natural resource agencies (i.e., permitting agencies), members of the public who participate with project planning, environmental lawyers, and decision-makers including city councils, county supervisors, state commissions, and judges. Unless the concepts of habitat fragmentation and corridors are understood clearly by practitioners, and unless these concepts influence management and conservation decisions in meaningful ways, then, to be frank, the scientific discussion of them has been of little value.

To ascertain the effectiveness of the concepts of habitat fragmentation and corridors, I examined environmental review documents with which I participated as an expert witness over the past twenty years (appendix 7.A). I examined how the concepts of habitat fragmentation and corridors were portrayed and implemented in the face of real proposed projects that, if approved, could potentially affect wildlife through the process of habitat fragmentation and destruction or protection of corridors. Some of these projects involved very small areas, but even a project on a small habitat area can potentially sever habitat linkages, thereby contributing to habitat fragmentation. Many of these projects were constructed, some were not, and others are waiting for permits or construction starting dates, but the actual outcomes of these proposed projects were irrelevant to my examination of how the concepts of habitat fragmentation and corridors were treated in the environmental reviews.

Under the California Environmental Quality Act (CEQA), which is the law that I most often work under as an expert witness, a project would have a significant impact on the environment if it were to "interfere substantially with the movement of any native resident or migratory fish or wildlife species or with established native resident or migratory wildlife corridors, or im-

pede the use of native wildlife nursery site." This language is the closest that CEQA comes to recognizing habitat fragmentation as an impact and corridors as the solution. Consultants and resource agency biologists often cite this CEQA standard before discussing wildlife movement corridors in the context of project impacts or mitigation. Because environmental reviews performed in California are supposed to address the project's impacts on wildlife movement, habitat fragmentation and corridors should be expected to be discussed in environmental review.

I assessed the environmental review documents of fifty-seven proposed projects involving 83,413 ha (app. 7.A). Wildlife movement corridors were mentioned in thirty-four reviews (60%) and analyzed in zero reviews. Habitat fragmentation was mentioned in twenty-six reviews (46%) and analyzed in two reviews (3.5%). Where the concepts of habitat fragmentation and corridors matter most—where they matter to environmental decision-making—about half of the environmental reviews made no mention of them.

In one of the two reviews that included an analysis of habitat fragmentation (app. 7.B, 5), the level of project impact was calculated in acres and was based on the starting habitat value and ending habitat value. Habitat values were 100% for parcels zoned ≥40 acres, 75% for parcels zoned ≥20 acres, and 25% for parcels zoned ≥10 acres. However, because >90% of the project area was composed of parcels zoned ≥40 acres, the project area was concluded to have 100% habitat value. In the other review including a quantitative analysis (app. 7.B, 36), habitat fragmentation was analyzed as the amount of habitat area that will be lost as a percentage of the cumulative impacts in the county.

An analysis that is sufficient for the California Environmental Quality Act or the National Environmental Policy Act is not the same type or level of analysis that is practiced by research scientists, so even the two crude analyses that I encountered in environmental review documents would not have been regarded as scientific by the research community. Essentially, there has been no scientific analysis performed and no scientific predictions made of the effects of habitat fragmentation or corridor loss or protection resulting from any of the environmental reviews with which I have been professionally involved. Consultants and agency biologists preparing the environmental reviews for about half of the projects have expressed familiarity with the terms *habitat fragmentation* and *corridors*, but neither knew how to perform impact assessments relevant to these concepts, or chose not to do so.

Descriptions of Habitat Fragmentation

Environmental reviews defined or described habitat fragmentation in various ways, expressing some overlap as well as considerable divergence in understanding of the concept. One review defined habitat fragmentation as habitat loss resulting in displacement of individuals from the developed area (app. 7.B, 1). This review equated habitat fragmentation with habitat loss characterized by displacement rather than by any overall population reduction. In another review, this notion of displacement was rolled into one of crowding within fragmented patches of Swainson's hawk (*Buteo swainsoni*) foraging habitat, "increasingly fragmented with more intensive uses (agricultural/residential or urban development) on smaller minimum parcel sizes" (app. 7.B, 5). In this version of habitat fragmentation, the affected species suffers the inconvenience of crowding but not necessarily any overall numerical reduction.

In contrast to the notion of simple displacement, or at worst crowding due to habitat fragmentation, another review explained that small fragments of habitat can only support small populations and are more vulnerable to extinction (app. 7.B, 6). Alluding to habitat fragmentation, an environmental review document included the statement, "The ability for wildlife to freely move about an area and not become isolated is considered connectivity and is important to allow dispersal of a species to maintain and exchange genetic characteristics, forage (food and water), and escape from predation" (app. 7.B, 8). Imposing barriers to movement qualified as habitat fragmentation in one review (app. 7.B, 41), and in two other reviews barriers were implied: "Fragmentation of open space areas by urbanization creates 'islands' of wildlife habitat that are more or less isolated from each other" (app. 7.B, 4), and "Habitat fragmentation involves the potential for dividing sensitive habitat and thereby lessening its biological value" (app. 7.B, 10). In another review, the public learned that "habitat fragmentation can result when development occurs within larger regions of natural habitat. The effects of habitat fragmentation

can extend beyond the boundaries of an area proposed for development" (app. 7.B, 13). This review went on to describe how fragmentation of a creek could adversely affect downstream wetlands by altering hydration periods. Thus, some environmental reviews recognized the potential for net greater impacts caused by movement barriers and habitat fragmentation as compared to simple habitat loss.

Descriptions of Corridors

Environmental reviews defined or described corridors more often than they defined habitat fragmentation. One review explained that corridors link blocks of habitat, where the habitat in corridors resembles habitat preferred by the target species (app. 7.B, 33), and another defined wildlife corridors as "pathways or habitat linkages that connect discrete areas of natural open space otherwise separated or fragmented by topography and changes in vegetation." (app. 7.B, 31).

Corridor definitions sometimes appeared tailored to serve the strategy of the impacts assessment or mitigation plan. Kern County (app. 7.B, 2, 4.3–20) wrote, "Wildlife movement corridors, also referred to as dispersal corridors or landscape linkages, are generally defined as linear features along which animals can travel from one habitat or resource area to another . . . drainages, ridgelines, and other natural and built linear features and barriers often serve as areas that wildlife routinely use to access essential natural resources. It is assumed that wildlife species would use such features for movement if they occurred within the survey area." This definition implies that wildlife will use the land left undeveloped as a corridor, thereby mitigating the project's impacts.

In the review of a project seeking to mitigate its impacts by leaving undeveloped land as corridors, PBS&J (app. 7.B, 4, 5.4–8) wrote,

> Terms such as habitat corridors, linkages, crossings, and travel routes are used to describe physical connections that allow wildlife to move between patches of suitable habitat in undisturbed landscapes as well as environments fragmented by urban development . . . Wildlife corridors are usually bounded by urban land areas or other areas unsuitable for wildlife. The corridor generally contains suitable cover, food, and/or water to

support species and facilitate movement while in the corridor. Wildlife corridors link areas of suitable habitat that are otherwise separated by areas of non-suitable habitat such as rugged terrain, changes in vegetation, or human disturbance. Wildlife corridors are essential to the regional ecology of a species because they provide avenues of genetic exchange and allow animals to access alternative territories as dictated by fluctuating population densities . . . Wildlife corridors are typically relatively small, linear habitats that connect two or more habitat patches that would otherwise be fragmented or isolated from one another.

This last part of the definition appeared to rationalize leaving only a narrow strip of land as the corridor.

The review of another large residential development—for which the principal mitigation measure was the protection of land to be left undeveloped as movement corridors (app. 7.B, 9)—defined wildlife movement corridors as the "gentlest topography and more open habitat." A subsequent review document for this project identified a wildlife corridor as a drainage or riparian vegetation in a canyon, which traverses no other topographic features and will not be surrounded by development in the future. Similarly, the consultants who prepared the planning for a multispecies habitat conservation plan across a large portion of San Diego County (app. 7.B, 33) defined a wildlife corridor as "a route used by one or more species to move between two areas of habitat. A corridor can be defined by topographical features such as ridges or valleys, habitat types such as bands of riparian vegetation, areas of natural open space passing between two man-made constraints, or even game trails used by many generations of animals" (page 6-1). The review defined two types of corridor—local and regional (page 6-1): "Regional corridors connect open space in a region and allow activities such as dispersal of young, genetic transfer between subpopulations, and seasonal migration. Local corridors are routes within a habitat used regularly by an animal to commute between resources such as denning sites, water sources, and hunting or foraging areas."

According to another review (app. 7.B, 10), "Wildlife corridors are areas of habitat used by wildlife for seasonal or daily migration." In another (app. 7.B, 14), "Wildlife corridors are routes frequently utilized by

wildlife that provide shelter and sufficient food supplies to support wildlife species during migration. Movement corridors generally consist of native or undeveloped matrix habitats (e.g., greenbelts, parks, or other open spaces) that span contiguous acres of unfragmented habitat. Wildlife movement corridors are an important element of resident species home ranges." Pacific Municipal Consultants (app. 7.B, 16, 4.4–20) defined wildlife movement corridors as "traditional routes used by wildlife to travel within their home range. Movement corridors typically provide wildlife with undisturbed cover and foraging habitat and are generally composed of several trails in contiguous spans of forested, riparian, riverine, and woodland habitats. The width of movement corridors varies depending on the topography. Movement corridors are an essential element of home ranges of a wide variety of wildlife including black bear (*Ursus americanus*), grey fox (*Urocyon cinereoargenteus*), mountain lion (*Felis concolor*), and other migratory wildlife."

According to EGI (app. 7.B, 34), "Wildlife movement corridors and habitat linkages are areas that connect suitable wildlife habitat areas in a region otherwise fragmented by rugged terrain, changes in vegetation, or human disturbance. Corridors are generally local pathways connecting short distances usually covering one or two main types of vegetation communities. Linkages are landscape level connections between very large core areas and generally span several thousand feet and cover multiple habitat types. Natural features such as canyon drainages, ridgelines, or areas with vegetation cover provide corridors and linkages for wildlife travel. The habitat connectivity provided by corridors and linkages is important in providing access to mates, food, and water, allowing the dispersal of individuals away from high-density areas, and facilitating the exchange of genetic traits between populations."

According to ECORP (app. 7.B. 36), corridors are where surrounding habitat concentrates wildlife movement, or which link large areas of undeveloped open space. Other reviews defined wildlife corridors as "areas that connect suitable wildlife habitat areas within a region, especially where species are known to migrate" (app. 7.B, 37), and "where resident and migratory animals freely move, generally within or between preferred habitat areas" (app. 7.B, 30).

Environmental reviews sometimes identified particular parts of the project area as wildlife movement corridors, which were then claimed to be unaffected by the project or proposed to be protected or enhanced as mitigation (app. 7.B, 2, 9, 18, 25, 41). Corridor designations were arbitrary in these reviews and may have been influenced by convenience. More often, environmental reviews claimed that the project site did not occur in an established or major wildlife movement corridor (app. 7.B, 2, 3, 5, 7, 11, 12, 14, 15, 17, 19, 20, 26, 30, 34, 36, 46). The premise implied by this strategy was that there exists a registry or inventory of wildlife movement corridors that was consulted and turned up a negative result. Even if there was an inventory of corridors, the inventory's scientific credibility would hinge on species-specificity and empirical foundation.

Summary of Habitat Fragmentation and Corridors in Practice

Environmental review documents rarely cited peer-reviewed papers on the topics of habitat fragmentation and corridors. Given their varied characterizations of habitat fragmentation and corridors, it was apparent that practitioners often only vaguely understood these concepts. The concept of habitat fragmentation was the most poorly understood of the two concepts, having been omitted from discussion in most reviews and defined erroneously in most reviews that addressed the topic. In some cases, the practitioners may have crafted definitions of these terms to minimize estimates of project impacts or to minimize the cost of mitigation. Among the environmental documents I reviewed, none of the alleged existing corridors were measured on any variables for corridor function. None of the corridors proposed for protection as project mitigation were accompanied by commitments to test for corridor functionality following project development. Instead of testing for corridor functionality, those who performed environmental reviews of the fifty-seven projects addressed herein relied on hopeful speculation that land designated as corridors would actually function as corridors.

Based on my review of environmental assessments on the impacts and mitigation planning of fifty-seven projects, I conclude that the concepts of habitat fragmentation and corridors have too often failed to matter where they most needed to be implemented. There

might exist nonscientific reasons for this poor translation of scientific concepts into conservation practice, but even so, it should be more difficult to abuse concepts that are clearly and consistently understood. If these concepts are going to matter in the future, if they are going to conserve wildlife in the face of massive anthropogenic land use changes, then barriers must be overcome to effective measurement and consistent understanding.

Barriers to Measuring Habitat Fragmentation and Corridors
The Dependent Variable

The effects of habitat fragmentation and corridors must be measureable if they are going to translate into meaningful theory or sound conservation planning and decision-making. It is not enough for conservation biologists to proclaim that "we know that habitat fragmentation is happening and corridors are effective at rectifying the effects of fragmentation." The effects need to be quantified and relevant hypotheses tested at appropriate spatial and temporal scales. It is therefore necessary to identify what it is that fragmentation and corridors are thought to be affecting.

Species richness and species diversity were early favorites as dependent variables responding to habitat fragmentation, due to their frequent use in island biogeography theory and the often-claimed likeness between habitat fragments and islands. Species richness remains in use, as Gilbert-Norton et al. (2010) used it as an indirect measure in their meta-analysis of corridor effectiveness. However, the comparability of terrestrial habitat fragments to islands is questionable because the fragments occur in very different ecological contexts than do islands. Furthermore, species richness and species diversity are community measures, which, by a proper definition of habitat (Hall et al. 1997), qualify as poor measures of habitat fragmentation. In other words, because habitat is a species-specific concept, its fragmentation is best measured for the species to which the habitat is associated. Species richness and species diversity might be more suitable dependent variables in analyses of landscape or ecosystem fragmentation.

Population persistence would probably generate the greatest agreement among ecologists and wildlife biologists as the ultimate expression of the effects of fragmentation and corridors, although there might not be agreement on what qualifies as a population or what time period is sufficient for measuring persistence. Abundance could be measured as a variable that indicates the population and could be expressed as the number of individuals, breeding pairs, or colonies or populations. The health of the population can also be measured, such as by genetic variation, productivity, and survival, but these variables are difficult to measure and are probably more proximal expressions of the effects of fragmentation and corridors.

Discussed more often than applied, a common measure of corridor effectiveness has been movement of individuals within the corridor, or corridor use. This measure goes to a hypothesized principal mechanism of population decline caused by habitat fragmentation and of population rescue provided by corridors. Organisms must be able to disperse, which habitat fragmentation is hypothesized to hamper and which corridors are hypothesized to facilitate. Indeed, Gilbert-Norton et al.'s (2010) meta-analysis of corridor effectiveness tested whether movement increased with the existence of natural or created corridors. However, a dependent variable measured as the rate of movement between habitat patches does not express the capacity of the remaining habitat. The investigators using this dependent variable only assume that improved movement translates into greater habitat capacity. They selected movement as their dependent variable because movement is critical to colonization and gene flow, which they believed increases population viability and decreases extinction probability. In reality, the movement that is measured could be of the same individual traveling back and forth between habitat patches, or it could be of individuals dispersing into an ecological sink. Whereas the measurement of movement rates along a corridor can be reassuring, it cannot alone express corridor effectiveness because it is in the connected habitat patches where the corridor's effects ultimately matter.

Investigators have often discussed or used indicators of habitat fragmentation, including percentage of habitat lost, perimeter-to-area ratio, mean patch size, and fractal dimension. The usefulness of these indicators depends on the degree of correspondence between the indicators and direct measures of fragmentation.

Measuring Habitat

The concept of habitat fragmentation has been poorly understood by the community in greatest need of understanding it. This poor understanding might be exacerbated by a generally poor understanding of habitat. The habitat concept itself has lacked consistent definition (Guthery and Strickland, this volume), leading to ample confusion among practitioners as well as theorists. The simplest and perhaps most appropriate definition of habitat would be that part of the environment within a species' geographic range where the species actually lives (Morrison and Hall 2002). However, Hall et al. (1997) found vague or imprecise characterizations of habitat in 82% of the fifty papers they reviewed. A popular characterization of habitat has been the classification of vegetation into "habitat types," to which species are then associated based on relative abundance of the species in each type. Even vaguer is the characterization of habitat as all types of natural vegetation bounded by lands converted for human use (Bolger et al. 1997). This characterization of habitat glosses over what specifically it is in the environment that species are selecting; it overlooks the resources in the environment that each species needs for metabolism and reproduction, and which limits the species' abundance.

The very fact that species are selecting resources from the environment indicates that the second law of thermodynamics applies to the distribution and abundance of wildlife (Smallwood 1993, 2002). The primary regulator of the distribution and abundance of any wildlife species is the degree to which individuals can metabolize sufficiently and for long enough to successfully reproduce. The primary resource must be energy for metabolism, but then other resources are needed to locate and process this energy, including atmosphere (water), nutrients, ecosystem processes, and the suite of other biological species in the environment. Because energy needed for metabolism is the primary resource that defines a species' habitat, these energy sources should be the core of any definition of a species' habitat, and the portions of the environment that facilitate the metabolism of these resources should round out the definition. Thus, habitat is defined as the suite of environmental elements—soils, plant species, and other animal species—that includes the species' limiting resources. Habitat is a vague, indicator-level representation of the parts of the environment that are essential to the species' metabolism and reproduction, and hence to its persistence (Morrison et al. 1998).

The complexities of measuring habitat fragmentation increase when considering the more fundamental challenges of measuring habitat selection. After all, measuring habitat fragmentation presupposes that the habitat undergoing fragmentation was accurately characterized as selected habitat. A fundamental problem with this presupposition is that the species performing the measurements is not the species whose habitat selection is being measured, so we should not assume that we perceive the species' environment in the same manner as the species we are studying. We draw inferences from patterns in the data, and tend to conclude that a species has selected a portion of the environment based on the species' disproportionate number of individuals or time spent in that selected area. Our inferences, however, could be confounded by multiple factors, including our measurement of the wrong environmental variables, the wrong spatiotemporal grain or extent, and our counting of the wrong study unit, e.g., individuals versus breeding pairs versus colonies versus populations (discussed in the following section). In some cases, disproportionate use of an area might represent an ecological sink (Smallwood 2002).

A second fundamental problem with measuring habitat selection is the epistemological context of the measurements, such as whether the investigator measures meaningful pattern as deviation from a uniform or random null pattern (Smallwood 2002). The starting point matters for measuring meaningful associations with places on the Earth where environmental resources might be selected. The null condition in which the investigator assumes that species' individuals would be uniformly spaced would result in greater deviations between the null condition and the measured condition, so long as the appropriate variable(s) is being measured at the appropriate spatiotemporal grain and extent. If the species' individuals space themselves out in a regular pattern due to social rules (home ranges), then the study unit might more appropriately consist of social groups rather than individuals, and measurements might be needed at much larger spatio-

temporal grain and extent. As for the null condition in which the investigator assumes that species' individuals would be randomly spaced, deviations between the null condition and the measured condition would be shorter, and these deviations might vary depending on what is meant by *random*.

A uniform pattern is easy to define. A uniform pattern results when the study units are maximally distant from each other. It is the pattern one would expect if there were no relationships between organisms and environmental resources. An interesting exception to this expectation would be when regular patterns of distribution result from social rules organized around the species' long-term experience with the resources, but even in cases of socially driven regularity, deviations from a uniform pattern should be evident at larger spatial extents. Not only is meaningful pattern relatively easy to measure as deviations from a null pattern of uniformity, but these deviations can also be interpreted from laws of thermodynamics and information theory (Shannon and Weaver 1949; Kullback 1959; Phipps 1981).

Many might regard a random pattern as also easy to define. A random pattern is the pattern of distribution resulting from lack of influence from the locations of other individuals (Fisher 1950; Taylor 1961). Because many observed or simulated patterns of distribution can be regarded as random, the starting point for measuring meaningful patterns of aggregation is less clear than it is for uniformity. Along a theoretical continuum from uniform to aggregated patterns, randomness occupies an unknown and perhaps varying proportion of this range somewhere between uniformity and completely aggregated (see Smallwood 2002, fig. 6.3). Furthermore, the definition of a random pattern highlights interactions among individuals, which serve as the most common study unit in habitat selection studies. It implies that interactions among individuals result in the meaningful patterns of aggregation that we measure, but in so doing it also obscures the influence of environmental resources. Of course, we know that individuals interact and that these interactions influence the patterns of distribution that we observe, but randomness as the null condition invokes these interactions as paramount and starts the analysis of habitat selection from a different worldview than starting it from uniformity as the null condition.

A third fundamental problem with measuring habitat selection—and this problem is related to the second when it comes to detecting meaningful patterns of habitat selection—is the spatiotemporal grain and extent of the measurements, whether the measurements are made during a snapshot of the organisms life (or over a season, a year, multiple years, multiple generations, or multiple periods of the species' multiannual cycle of abundance) and whether the measurements are made within an animal's home range, across multiple home ranges, or across an area capable of supporting an entire population or larger demographic unit. In other words, the sampling resolution and measurement sensitivity must be sufficient for testing hypotheses of habitat selection (Huston 2002) before any meaningful tests can be performed on the effects of habitat fragmentation or corridors. As an example, measuring habitat selection immediately after land conversions fragmented the habitat could be misleading due to temporary crowding of individuals in the remaining habitat fragments (Whitcomb et al. 1981; Lovejoy et al. 1986; Scott 1993).

The capacity of a species to utilize an environment (i.e., its habitat), or to populate its numbers within an environment, will be limited by the most limiting resource according to Liebig's Law of the Minimum. This limitation might vary across the species' environment or through time at given locations. This variation in where and when Liebig's Law of the Minimum will operate defines the ultimate source of instability in the environment with which a species must contend. Therefore, each species evolves strategies to contend with the instability in its most limiting resource(s), and the most effective strategies tend to be behavioral and social in nature, often resulting in time lags and feedback loops affecting numerical responses. Species of wildlife tend to buffer themselves against instability in the most limiting resources by developing stubborn social rules, such as home range tenure (home ranges often remain unchanged even in the face of superabundant food resources) and mating hierarchies. They also buffer themselves against resource instability by developing stubborn behaviors such as seasonal migrations, foraging self-limitations, and foraging by gestalt where resources are expected to be found rather than where they are known to be found (Hutto 1990). Therefore, not only is measuring habitat selection prone to con-

founding due to resource instability in past and even relic environments, but measuring the effects of habitat fragmentation is prone to further complications due to a new type of resource instability in the current environment.

A fourth fundamental problem with measuring habitat selection is achieving replication and interspersion of "treatments," which in the context of habitat selection studies are the categories of vegetation, landscape, or specific resources being associated with counts of the species under study. Pseudoreplication results from inadequate replication and interspersion of treatments, but these conditions are often more difficult to achieve in mensurative studies (Hurlbert 1984). Testing the effects of habitat fragmentation and corridors is likely even more prone to pseudoreplication than is testing for habitat selection.

A fifth fundamental problem is often not knowing why the animals that were counted were located where they were found by researchers. The aggregations we encounter can be one of four types (Smallwood 2002): resource, demographic, early stage, and constrained. A resource aggregation would likely be a temporary collection of individuals where a limiting resource has become available. A demographic aggregation might be a temporary collection of breeding pairs or a gathering of juveniles or subadults developing social relationships. An early-stage aggregation would be the collection of individuals corresponding with the level of colonization at the time of the habitat survey. A constrained aggregation would be the type of aggregation that is most prone to confounding in the context of fragmented habitat, because it is a forced aggregation, or even an expression of crowding. Either way, the constrained aggregation is likely temporary, so measurement of it can yield an overoptimistic interpretation of habitat selection (or corridor use).

Also related to habitat selection and how its measurement can influence interpretation of the effects of habitat fragmentation is the effect of location. Although there has been discussion of the influences of the extent of habitat loss and degree of isolation on habitat fragmentation, as well as the edge-to-interior ratio and the types of activities and species composition in the areas surrounding habitat fragments, there has been little, if any, discussion of the effect of location. The influence of location on corridor effectiveness has often been discussed, but its influence on habitat fragmentation has been neglected. Certain habitat areas are likely strategic to a species for one or more reasons, so their removal would have disproportionate consequences for the species. An example would be a landscape bottleneck, where animals must be able to negotiate for dispersal. Another example would be systematic loss of habitat in a particular landscape context, such as the upland grassland areas surrounding ponds. Many species forage or breed in ponds but spend the rest of their lives in the surrounding grasslands. Therefore the effects of habitat fragmentation might vary by location, all else being equal.

Permeable Movement Barriers

Just as difficult as it might be to characterize habitat fragments as places where a species would really choose to reside is to characterize the matrix areas surrounding the fragments as nonhabitat and as barriers to movement between fragments. For some species, the matrix might be easily shown to block movement, but for others the matrix might serve as a portion of the species' habitat and be completely or partially permeable to movement. La Polla and Barrett (1993) tested corridor width and presence on captive-reared meadow vole (*Microtus pennsylvanicus*) density in grassland fragments, but the matrix was simply mowed grass. The mowed grass may have been completely permeable to vole movement between unmowed fragments. Tests for habitat fragmentation and corridor effectiveness need to clearly characterize the ability of each species' ability to move across or reside within the matrix areas.

Conclusions

Habitat fragmentation likely qualifies as a major threat to the persistence of many plant and animal species, and corridors might help to mitigate this threat. However, conceptualizations of habitat fragmentation and corridors vary widely in the scientific literature and are either poorly understood or intentionally abused by many practitioners. The science supporting these concepts needs to improve. Appropriate dependent variables representing responses to habitat fragmentation and corridors need to be measured at appropriate spatial grain and extent and over appropriate time periods. Quantitative tests of habitat fragmentation and corridor effectiveness should be made to single species that are sufficiently understood to measure habitat variables rather than using overly simplistic maps of vegetation cover types that are applied to suites of wildlife species (e.g., Bolger et al. 1997). Much more care is needed on the measurement of habitat fragmentation and corridor effectiveness, and in the discussion of these concepts.

Definitions of habitat fragmentation and corridors need to be more explicit. Scientists need to more effectively convey their concepts of habitat fragmentation and corridors to biologists in natural resource agencies and environmental consulting firms, as well as to attorneys, political decision-makers, and the public. Where these concepts really matter is where decisions are being made about whether and how to further fragment habitat and how to mitigate the effects of this fragmentation. These decisions are being made now, and with the proliferation of habitat conservation plans these decisions increasingly involve very large areas of wildlife habitat. Vague and poorly understood concepts such as habitat fragmentation and corridors can cause more harm than good in wildlife conservation, as I have seen too many times in the aftermath of the environmental reviews I addressed in this chapter.

ACKNOWLEDGMENTS

I appreciate the review comments on this chapter from A. M. Long and an anonymous reviewer. I also appreciate M. L. Morrison for our many discussions of the habitat concept.

APPENDIX 7.A

Summary of environmental reviews with which I have had personal experience and whether the reviews mentioned or analyzed the potential project impacts on habitat fragmentation or movement corridors. Citations appear in Appendix 7.B by reference (Ref) number in the right column. MW refers to megawatts in energy projects.

Source	Project	Acres	Habitat fragmentation	Movement corridors	Ref
			Mention/analysis of potential impacts related to:		
BLM 2012a	104 MW Steens Mt Wind Energy Project	8,700	yes/no	no/no	1
Kern County 2012	250 MW Beacon Photovoltaic Project	2,301	no/no	yes/no	2
Yolo County 2008	Clark-Pacific Concrete Precast Factory	90	no/no	yes/no	3
PBS&J 2008	Delta Shores residential/commercial development	782	yes/no	yes/no	4
Sacramento LAFCo 2011	Expand City of Elk Grove's Sphere of Influence to accommodate future residential/commercial development	7,869	yes/yes	yes/no	5
BLM 2010	370 MW Ivanpah Solar Electric Generating System	4,073	yes/no	yes/no	6
City of Lancaster 2012	60 MW Summer Solar and Springtime Solar Projects	216	no/no	yes/no	7
HDR 2011	600 MW Mount Signal & Calexico Solar Farm Projects	4,228	no/no	yes/no	8
Impact Sciences 2002	Newhall Ranch residential development	12,000	yes/no	yes/no	9
Caltrans 2006, 2010	Route 84 Safety Improvement Project (Niles Canyon)	24	yes/no	yes/no	10
BLM 2012b	20 MW Ocotillo Sol Project	115	yes/no	yes/no	11
Foothill Associates 2006	Regional University development	1,282	yes/no	yes/no	12
EDAW 2008	Rio del Oro Specific Plan, housing development	3,828	yes/no	yes/no	13
Imperial County 2012	200 MW Solar Gen 2 Array Project	2,009	yes/no	yes/no	14
Design, Community & Environment 2010	St. John's Church expansion, requiring removal of mature forest canopy	3	no/no	yes/no	15
Pacific Municipal Consultants 2005	Yuba Highlands Specific Plan residential development	2,902	yes/no	yes/no	16
Ecology and Environment 2007a, b	67 MW Windy Flats West Wind Energy Project	616	no/no	no/no	17
	Antonio Mountain Ranch residential development	523	yes/no	yes/no	18
BLM 2012c	150 MW Desert Harvest Solar Project	1,464	yes/no	yes/no	19
Ecology and Environment 2012	49.9-MW Hudson Ranch Power II Geothermal Project and Simbol Calipatria Plant II	100	yes/no	yes/no	20
WRA Environmental Consultants 2009	J&J Ranch, 24 Adobe Lane residential development	20.5	no/no	yes/no	21
Harzoff 1992; Huffman & Associates 1992	Putah Creek Parkway Demonstration Project, adding bike path and fence along centerline of riparian corridor	10	no/no	no/no	22
	UC Davis Long-Range Development Plan		no/no	no/no	23
Yeast 1998	Sunset SkyRanch Airport expansion	108	no/no	no/no	24
Raney Planning & Management, Inc. 2004	Covell Village residential development	422	no/no	yes/no	25
Planning Consultants Research	West Bluff residential development, the last upland area bordering the Ballona Wetlands	44	no/no	yes/no	26
Thomas Reid Associates 1997	Natomas Basin HCP for residential development on wetlands and fields used for rice cultivation	21,300	yes/no	no/no	27
County of Yolo 2010	365-foot-tall Results Radio Tower with guy wires	1.7	no/no	no/no	28
Glazner Environmental Consulting 1999	Atwood Apartments	55.6	yes/no	no/no	29
City of Sierra Madre 2001; EIP Associates 1999	Maranatha High School	63	yes/no	yes/no	30
Grassetti Environmental Consulting 2004; Edelstein 2003	Creekside Highlands residential development along riparian zone of Ward Creek	7.1	yes/no	yes/no	31

Source	Project	Acres	Habitat fragmentation	Movement corridors	Ref
			Mention/analysis of potential impacts related to:		
Humboldt Bay Harbor & US Army Corps of Engineers 1995	Dredging and disposal of 5.6 million cubic yards of benthic marine sediment in Humboldt Bay Harbor	—	no/no	no/no	32
Ogden 1998	San Diego Multiple Habitat Conservation Program for incidental take permits of 85 species	121,023	yes/no	yes/no	33
EGI 2012	140 MW Campo Verde Solar Project	1,990	no/no	yes/no	34
ESA 2012	33 MW Casa Diablo IV Geothermal Development Project	52.5	no/no	yes/no	35
ECORP 2011; Kern County 2012	20 MW FRV Orion Solar Project	260	yes/yes	yes/no	36
Chambers Group 2012	30 MW Imperial Valley Solar Company 2 Project	158.5	no/no	yes/no	37
Glazner Environmental Consulting 1999	Atwood Ranch Unit III Subdivision	55.6	no/no	yes/no	38
City of Sacramento 2011; AES 2011	20 MW Sutter Landing Park Solar Photovoltaic Project	104	no/no	no/no	39
CEC 2007	660 MW Colusa Generating Station, gas-fired power plant	31	yes/no	no/no	40
City of Petaluma 2013	Residential development	59	yes/no	yes/no	41
Kern County & ESA 2012	125 MW Pioneer Green Solar Project	720	no/no	no/no	42
Swaim 2011	Oakland Zoo expansion	62	yes/no	no/no	43
Contra Costa County 2001	Expansion of Temple B'nai Tikyah, Walnut Creek, CA	0.3	no/no	no/no	44
Jones & Stokes Associates 1999	Valley elderberry longhorn beetle HCP for gravel mine in Yolo County, CA	9.9	no/no	no/no	45
City of Anderson 2003	Anderson Marketplace (Walmart) commercial development	26.6	no/no	no/no	46
Sacramento County 2002	The Promenade commercial development	8	no/no	no/no	47
North Fork Associates 2000	Silver Bend Apartments	6	no/no	no/no	48
Foothill Associates 2002	Winters Highlands residential development	102.6	no/no	no/no	49
EIP Associates 2001	UC Merced NCCP/HCCP	2133	yes/no	yes/no	50
URS 2001	UC Merced Long Range Development Plan EIR	2000	yes/no	no/no	51
Beck 2003	Table Mountain mining	50	no/no	no/no	52
Foothill Associates 2004; City of Winters 2004	Callahan Estates residential development	26.4	no/no	no/no	53
CEC 2000	Blythe Energy Project	76	yes/no	no/no	54
Kern County 2013	100 MW Rosamond Solar Project	400	yes/no	yes/no	55
Arnold 2013	7.5 MW LANDPRO Solar Project	80.6	no/no	yes/no	56
ESA 2012, Sage Institute, Inc. 2011	Expansion of Metropolitan Air Park, City of San Diego	331	no/no	yes/no	57

APPENDIX 7.B

References for environmental review documents reviewed for treatment of corridors and habitat fragmentation.

Ref	Reference
1	BLM (Bureau of Land Management). 2012. North Steens 230-kV Transmission Line Project Draft Environmental Impact Statement. DOI-BLM-OR-B060-2010-0035-EIS, Hines Oregon.
2	Kern County. 2012. Draft Environmental Impact Report, SCH# 2012011029, Beacon Photovoltaic Project by Beacon Solar LLC, Conditional Use Permit 11, Map 152, Kern County Planning and Community Development Department, Bakersfield, California.
3	Yolo County. 2008. Clark Precast, LLC's "Sugarland" Project, Initial Study/Mitigated Negative Declaration, Zone File # 2007-078, Yolo County Planning & Public Works Department, Woodland, CA.
4	PBS&J. 2008. Delta Shores Draft Environmental Impact Report. City of Sacramento. Sacramento, CA.

continued

Appendix 7.B continued

Ref	Reference
5	Sacramento LAFCo (Local Agency Formation Commission). 2011. Draft Environmental Impact Report: City of Elk Grove Proposed Sphere of Influence Amendment (LAFC # 09-10). State Clearinghouse No. 2010092076. City of Elk Grove, Sacramento County, California.
6	BLM (Bureau of Land Management). 2010. California Desert Conservation Area Plan Amendment / Final Environmental Impact Statement for Ivanpah Solar Electric Generating System, FEIS-10-31, Needles, CA.
7	City of Lancaster. 2012. Initial Study for Conditional Use Permits 12-08 and 12-09, Summer Solar and Springtime Solar Projects. Lancaster, CA.
8	HDR (Engineering, Inc.) 2011. Draft Environmental Impact Report Mount Signal and Calexico Solar Farm Projects, Imperial County, California. County of Imperial, El Centro, California.
9	Impact Sciences. 2002. Newhall Ranch November 2000 Administrative Draft EIR. Los Angeles County Department of Regional Planning, Los Angeles, California.
10	Caltrans (California Department of Transportation). 2006. Route 84 Safety Improvement Project Negative Declaration (CEQA). Dividing Border Between Union City and the City of Fremont in Alameda County, California, 04-ALA-84-KP 19.5/21.4, EA 174400. Caltrans, Oakland, California.
10	Caltrans (California Department of Transportation). 2010. Niles Canyon Safety Improvement Project Draft Environmental Impact Report/Environmental Assessment. Alameda County, California.
11	BLM (Bureau of Land Management). 2012a. El Centro Office Draft Environmental Impact Statement / Draft CDCA Plan Amendment Ocotillo Sol Project. US Department of the Interior, BLM/CA/ES-2012-009+1793, DES 20-12, DOI-BLM-CA-D000-2012-0005-EIS. El Centro, California.
12	Foothill Associates. 2006. Biological Resources Assessment: Regional University site and off-site improvements, Placer County, California. Prepared for KT Development. County of Placer, Auburn, California.
13	EDAW. 2008. Recirculated Draft Environmental Impact Report/Supplemental Draft Environmental Impact Statement, Rio del Oro Specific Plan Project, State Clearinghouse #2003122057, City of Rancho Cordova and US Army Corps of Engineers, Sacramento District.
14	Imperial County. 2012. Draft Environmental Impact Report: Solar Gen 2 Solar Array Power Project. Imperial County Planning & Development Services Department, El Centro, California.
15	Design, Community & Environment. 2010. St. John's Church Project Draft Environmental Impact Report, Oakland, California, State Clearing House Number 2008032031.
16	Pacific Municipal Consultants. 2005. Yuba Highlands Specific Plan Draft Environmental Impact Report, State Clearing House Number 2001032070, Yuba County, Marysville, CA.
17	Ecology and Environment, Inc. 2007a. Windy Flats Wind Energy Farm, Klickitat County, Washington: Environmental Report. Prepared for Windy Point Partners, LLC. Portland, Oregon.
17	Ecology and Environment, Inc. 2007b. Windy Flats Wind Farm, Klickitat County, Washington: State Environmental Policy Act Checklist WAC-197-11-960. Prepared for Windy Point Partners, LLC. Portland, Oregon.
18	Antonio Mountain Ranch Specific Plan Public Draft Environmental Impact Report. Placer County Planning and Redevelopment Department, Rocklin, California. State Clearing House Number 2000022053
19	BLM (US Bureau of Land Management). 2012c. Desert Harvest Solar Project Draft Environmental Impact Statement and Draft California Desert Conservation Area Plan Amendment, CACA-49491, DOI Control #: DES 12-17, Publication Index #: BLM/CA/ES-2012-006+1793, US Department of Interior, Bureau of Land Management, Palm Springs Field Office, Palm Springs, California.
20	Ecology and Environment, Inc. 2012. Hudson Ranch Power II CUP #G10-0002/Simbol II CUP #12-0005 Draft Environmental Impact Report, Volume I. County of Imperial, Department of Planning and Development Services, El Centro, CA.
21	WRA Environmental Consultants. 2009. Biological Resources Assessment, 24 Adobe Lane, Orinda, Contra Costa County, California. Prepared for City of Orinda, California.
22	Harzoff, D. 1992. Negative Declaration for the Putah Creek Parkway Demonstration Project. City of Davis, California.
22	Huffman, A. (Huffman & Associates, Inc.) 1992. Evaluation of North Fork of Putah Creek Channel, Davis, California. Report to City of Davis, California.
23	UC Davis. 2003. UC Davis Long-range Development Plan. Office of Resource Management & Planning, University of California, Davis, California
24	Yeast, D. E. 1998. Negative Declaration: Sunset SkyRanch Airport Use Permit. Control #: 97-UPP-0594. County of Sacramento, California.
25	Raney Planning & Management, Inc. 2004. Covell Village (SCH# 2004062089) Draft Program Level Environmental Impact Report. Prepared for City of Davis, California.
26	Planning Consultants Research. 1998. West Bluffs Project Subsequent Draft Environmental Impact Report. Prepared for City of Los Angeles, California.
27	Thomas Reid Associates. 1997. Natomas Basin Habitat Conservation Plan, Sacramento and Sutter Counties, California. City of Sacramento, California.

Ref	Reference
28	County of Yolo. 2010. Initial Study/Mitigated Negative Declaration for Results Radio Zone File #2009-001. Woodland, California.
29	Glazner Environmental Consulting. 1999. Biological Resources Assessment for the 55.6-acre Atwood No. 3 Property, Placer County, California. Report prepared for A. R. Associates, Auburn, California.
30	City of Sierra Madre. 2001. Negative Declaration and Initial Study for Proposed Maranatha High School Impact Reduction Plan. Sierra Madre, California.
31	Edelstein, D. 2003. Analysis of impacts of Creekside Highlands. App. A in Grassetti Environmental Consulting (2004). Initial Study and Draft Mitigated Negative Declaration, Vesting Tentative Map, Tract 7270, Creekside Highlands Project. Submitted to Alameda County Planning Department, Hayward, California.
31	Grassetti Environmental Consulting. 2004. Initial Study and Draft Mitigated Negative Declaration, Vesting Tentative Map, Tract 7270, Creekside Highlands Project. Submitted to Alameda County Planning Department, Hayward, California.
32	Humboldt Bay Harbor Recreation and Conservation District & US Army Corps of Engineers. 1995. Final Environmental Impact Statement / Environmental Impact Report: Humboldt Harbor and Bay Deepening Navigation Project. Eureka, California.
33	Ogden (Ogden Environmental and Energy Services Co., Inc.) 1998. Biological goals, standards, and guidelines for Multiple Habitat Preserve Design. Prepared for San Diego Association of Governments, San Diego, California.
34	EGI. 2012. Final Environmental Impact Report for the Campo Verde Solar Project, SCH. No. 2011111049. Prepared for Imperial County, El Centro, California.
35	ESA (Environmental Science Associates). 2012. Casa Diablo IV Geothermal Development Project, Draft Environmental Impact Statement and Draft Environmental Impact Report. Prepared for Bureau of Land Management, DOI Control #: DES 12-21, Publication Index #: BLM/CA-ES-2013-002+1793, State Clearinghouse No. 2011041008, Bishop Field Office, Bishop, California.
36	ECORP. 2011. Biological Evaluation and Impact Analysis Orion Solar Site, Kern County, CA. Report to SunEdison, San Francisco, California.
36	Kern County. 2012. Draft Environmental Impact Report, SCH# 2012031079, FRV Orion Solar Project (PP12232). Bakersfield, California.
37	Chambers Group, Inc. 2012. Draft Environmental Impact Report, Imperial Valley Solar Company 2 Project, Imperial County, California. County Of Imperial, Planning & Development Services Department, El Centro, California.
39	AES (Analytical Environmental Services). 2011. Biological Resources Assessment, City Of Sacramento 28th Street Solar Photovoltaic Farm. Report to City of Sacramento, California.
39	City of Sacramento. 2011. Community Development Department Solar Photovoltaic Park at 28th Street Landfill Project, Initial Study. Mitigated Negative Declaration, Sacramento, California.
40	CEC (California Energy Commission). 2007. Preliminary Staff Assessment: Colusa Generating Station, Application For Certification (06-AFC-9), Colusa County, CEC-700-2007-003-PSA, Sacramento, CA.
41	City of Petaluma. 2013. Davidon Homes Tentative Subdivision Map and Rezoning Project Draft Environmental Impact Report and Technical Appendices. State Clearinghouse No. 2004072137. City of Petaluma, California.
42	Kern County and ESA (Environmental Science Associates). 2012. Draft Environmental Impact Report, Pioneer Green Solar Project by Pioneer Green Solar II, LLC, SCH# 2012011025, Kern County Planning and Community Development Department, Bakersfield, California.
43	Swaim, K. (Swaim Biological, Inc.). 2011. Oakland Zoo California Project Biological Assessment. Report to East Bay Zoological Society, Oakland, California.
44	Contra Costa County. 2001. Initial Study for the Expansion of the Congregation Temple B'nai Tikyah, Walnut Creek, CA.
45	Jones & Stokes Associates. 1999. Habitat Conservation Plan for the Valley elderberry longhorn beetle for A. Teichert & Sons, Inc., Yolo County, California. Prepared for US Fish and Wildlife Service, Sacramento, California.
46	City of Anderson. 2003. Anderson Marketplace Project Draft Environmental Impact Report. City of Anderson Planning Department, Anderson, California.
47	Sacramento County. 2002. Revised Draft Environmental Impact Report: The Promenade Community Plan Amendment, Rezone, Use Permit, Variances, and Special Review of Parking. State Clearinghouse No. 1998042028. Sacramento County Department of Environmental Review and Assessment, Sacramento, California.
48	North Fork Associates. 2000. Biological Resources Assessment for the ffl6-acre Silver Bend Apartment Project, Placer County, California. Auburn, California.
49	Foothill Associates. 2002. Biological Resources Assessment: Winters Highlands Project Site. City of Winters, California.
50	EIP Associates. 2001. University Community Plan EIR. State Clearinghouse No. 2001021056. County of Merced, California.
51	URS. 2001. UC Merced Long Range Development Plan Draft Environmental Impact Report. Merced County, California.
52	Beck, A. J. 2003. Biological assessment of a portion of North Table Mountain, Central Butte County, California. Unpubl. report by Eco-Analysts, Chico, California.
53	City of Winters. 2004. Initial Study and Negative Declaration for Callahan Estates. City of Winters, California.
54	CEC (California Energy Commission). 2000. Preliminary Staff Assessment: Blythe Energy Power Plant Project, Application for Certification 99-AFC-8, Riverside County, Sacramento, CA.

continued

Appendix 7.B continued

Ref	Reference
55	Kern County. 2013. Addendum to the Environmental Impact Report for the Rosamond Solar Project, ROSAMOND SOLAR MODI-FICATION PROJECT (PP13292), SGS Antelope Valley Development, LLC, Specific Plan Amendment No. 18, Map No. 232, Specific Plan Amendment No. 19, Map 232, Zone Change Case No. 36, Map No. 232, Modification to Conditional Use Permit No. 27, Map No. 232, Streets and Highways—Nonsummary Vacation—Map No. 232. Kern County Planning and Community Development Department, Bakersfield, California.
56	Arnold, R. (RCA Associates). 2012a. General biological resources assessment, LANDPRO 7.5 MW Solar Project, APN0466-181-059, 060, 061 & 062, San Bernardino County, California. Unpublished report to Sunlight Partners, LLC, Mesa, Arizona.
57	ESA. 2012. Metropolitan Airpark Project Draft Environmental Impact Report, SCH No. 201007054, Project No. 208889, City of San Diego Development Services Department, San Diego, California.
	Sage Institute, Inc. (SII). 2011 (revised by ESA in 2012). Metropolitan Airpark Project Volume 1, 2011 Biology Survey Report, Project No. 208889. Prepared for Metropolitan Airpark, LLC.

LITERATURE CITED

Alberts, A. C., A. D. Richman, D. Tran, R. Sauvajot, C. McCalvin, and D. T. Bolger. 1993. "Effects of Habitat Fragmentation on Native and Exotic Plants in Southern California Coastal Scrub." In *Interface between Ecology and Land Development in California*, edited by J. E. Keeley, 103–10. Southern California Academy of Sciences, Los Angeles.

Beier, P., and S. Loe. 1992. "A Checklist for Evaluating Impacts to Wildlife Movement Corridors." *Wildlife Society Bulletin* 20:434–40.

Beier, P., and R. F. Noss. 1998. "Do Habitat Corridors Provide Connectivity?" *Conservation Biology* 12:1241–52.

Bender, D. J., T. A. Contreras, and L. Fahrig. 1998. "Habitat Loss and Population Decline: A Meta-analysis of the Patch Size Effect." *Ecology* 79:517–33.

Bolger, D. T., A. C. Alberts, R. M. Sauvajot, P. Potenza, C. McCalvin, D. Tran, S. Mazzoni, and M. E. Soule. 1997. "Response of Rodents to Habitat Fragmentation in Coastal Southern California." *Ecological Applications* 7:552–63.

Fisher R. A. 1950. "The Significance of Deviations from Expectation in a Poisson Series." *Biometrics* 6:17–24.

Forman, R. T. T., and M. Godron. 1981. "Patches and Structural Components for a Landscape Ecology." *Bioscience* 31:733–40.

Gilbert-Norton, L., R. Wilson, J. R. Stevens, and K. H. Beard. 2010. "A Meta-analytic Review of Corridor Effectiveness." *Conservation Biology* 24:660–68.

Hall, L. S., P. R. Krausman, and M. L. Morrison. 1997. "The Habitat Concept and a Plea for Standard Terminology." *Wildlife Society Bulletin* 25:173–82.

Hurlbert, S. H. 1984. "Pseudoreplication and the Design of Ecological Field Experiments." *Ecological Monographs* 54:187–211.

Huston, M. A. 2002. *Predicting Species Occurrences: Issues of Scale and Accuracy*, edited by J. M. Scott, P. J. Heglund, M. Morrison, M. Raphael, J. Haufler, and B. Wall. Island Press, Covelo, CA.

Hutto, R. L. 1990. "Measuring the Availability of Food Resources." *Studies in Avian Biology* 13:20–28.

Karr, J. R. 1994. "Landscapes and Management for Ecological Integrity." In *Biodiversity and Landscape: A Paradox for Humanity*, edited by K. C. Kim and R. D. Weaver, 229–51. Cambridge University Press, New York.

Kullback S. 1959. *Information Theory and Statistics*. John Wiley & Sons, New York, NY.

La Polla, V. N., and G. W. Barrett. 1993. "Effects of Corridor Width and Presence on the Population Dynamics of the Meadow Vole Microtus pennsylvanicus." *Landscape Ecology* 8:25–37.

Laurence, W. F., and E. Yensen. 1991. "Predicting the Impacts of Edge Effects in Fragmented Habitats." *Biological Conservation* 55:77–92.

Longcore, J. R., H. Boyd, R. T. Brooks, G. M. Haramis, D. K. McNicol, J. R. Newman, K. A. Smith, and F. Stearns. 1993. "Acidic Depositions: Effects on Wildlife and Habitats." *Wildlife Society Technical Review* 931.

Lovejoy, T. E., R. O. Bierregaard, A. B. Rylands, J. R. Malcom, C. E. Quintela, L. H. Harper, K. S. Brown Jr., A. H. Powell, G. V. N. Powell, N. O. R. Schubart, and M. B Hays. 1986. "Edge and Other Effects of Isolation on Amazonian Forest Fragments." In *Conservation Biology: The Science of Scarcity and Diversity*, edited by M. E. Soule, 257–85. Sinauer Associates, Sunderland, MA.

McCollin, D. 1993. "Avian Distribution Patterns in a Fragmented Wooded Landscape (North Humberside, U.K.): The Role of between-Patch and within-Patch Structure." *Global Ecology and Biogeography Letters* 3:48–62.

Morrison, M. L., and L. S. Hall. 2002. "Standard Terminology: Toward a Common Language to Advance Ecological Understanding and Application." In *Predicting Species Occurrences: Issues of Scale and Accuracy*, edited by J. M. Scott, P. J. Heglund, M. Morrison, M. Raphael, J. Haufler, and B. Wall, 43–52. Island Press, Covelo, CA.

Morrison, M. L., B. G. Marcot, and R. W. Mannan. 1998. *Wildlife-Habitat Relationships: Concepts and Applications*. 2nd edition. University of Wisconsin Press Madison, WI.

Phipps, M. 1981. "Entropy and Community Pattern Analysis." *Journal of Theoretical Biology* 93:253–73.

Porneluzi, P., J. C. Bednarz, L. R. Goodrich, N. Zawada, and J. Hoover. 1993. "Reproductive Performance of Territorial Ovenbirds Occupying Forest Fragments and a Continuous Forest in Pennsylvania." *Conservation Biology* 7:618–22.

Rich, C., and T. Longcore, eds. 2006. *Ecological Consequences of Artificial Night Lighting*. Island Press, Covelo, CA.

Roland, J. 1993. "Large-Scale Forest Fragmentation Increases the Duration of Tent Caterpillar Outbreak." *Oecologia* 93:25–30.

Salwasser, H. 1990. "Conserving Biological Diversity: A Perspective on Scope and Approaches." *Forest Ecology and Management* 35:79–90.

Saunders, D. A., R. J. Hobbs, and C. Margules. 1991. "Biological Consequences of Ecosystem Fragmentation: A Review." *Conservation Biology* 5:18–32.

Scott, T. A. 1993. "Initial Effects of Housing Construction on Woodland Birds along the Wildland Urban Interface." In *Interface between Ecology and Land Development in California*, edited by J. E. Keeley, 181–87. Southern California Academy of Sciences, Los Angeles, CA.

Shannon, C. E., and W. Weaver. 1949. *The Mathematical Theory of Communication*. University of Illinois Press, Urbana, IL.

Simberloff, D., and J. Cox. 1987. "Consequences and Costs of Conservation Corridors." *Conservation Biology* 1:63–71.

Simberloff, D., J. A. Farr, J. Cox, and D. W. Mehlman. 1992. "Movement Corridors: Conservation Bargains or Poor Investments?" *Conservation Biology* 6:493–504.

Smallwood, K. S. 1993. "Understanding Ecological Pattern and Process by Association and Order." *Acta Oecologica* 14:443–62.

———. 1999. "Scale Domains of Abundance among Species of Mammalian Carnivora." *Environmental Conservation* 26:102–11.

———. 2001a. "The Allometry of Density within the Space Used by Populations of Mammalian Carnivores." *Canadian Journal of Zoology* 79:1634–40.

———. 2001b. "Linking Habitat Restoration to Meaningful Units of Animal Demography." *Restoration Ecology* 9:253–61.

———. 2002. "Habitat Models Based on Numerical Comparisons." *Predicting Species Occurrences: Issues of Scale and Accuracy*, edited by J. M. Scott, P. J. Heglund, M. Morrison, M. Raphael, J. Haufler, and B. Wall, 83–95. Island Press, Covelo, CA.

Taylor, L. R. 1961. "Aggregation, Variance and the Mean." *Nature* 189:732–35.

Taylor, R. A. J., and L. R. Taylor. 1979. "A Behavioral Model for the Evolution of Spatial Dynamics." In *Population Dynamics*, edited by R. M. Anderson, B. D. Turner, and L. R. Taylor, 1–28. Blackwell Scientific Publications, Oxford.

Villard, M.-A., M. K. Trzcinski, and G. Merriam. 1999. "Fragmentation Effects on Forest Birds: Relative Influence of Woodland Cover and Configuration on Landscape Occupancy." *Conservation Biology* 13:774–83.

Whitcomb, R. F., C. S. Robbins, I. F. Lynch, B. L. Whitcomb, M. K. Klimkiewicz, and D. Bystrak. 1981. "Effects of Forest Fragmentation on Avifauna of the Eastern Deciduous Forest." In *Forest Island Dynamics in Man-Dominated Landscapes*, edited by R. L. Burgess and D. M. Sharp, 125–205. Springer-Verlag, New York, NY.

Wiens, J. A. 1997. "Metapopulation Dynamics and Landscape Ecology." In *Metapopulation Biology: Ecology, Genetics, and Evolution*, edited by I. A. Hanski and M. E. Gilpin, 43–62. Academic Press, San Diego, CA.

Wilcox, B. A., and D. D. Murphy. 1985. "Conservation Strategy: The Effects of Fragmentation on Extinction." *American Naturalist* 125:879–87.

Wilcox, B. A., K. S. Smallwood, and J. A. Kahn. 2002. "Toward a Forest Capital Index." In *Managing for Healthy Ecosystems*, edited by D. J. Rapport, W. L. Lasley, D. E. Rolston, N. O. Nielsen, C. O. Qualset, and A. B. Damania, 285–98. Lewis Publishers, Boca Raton, FL.

Woodroffe, R., and J. R. Ginsberg. 1998. "Edge Effects and the Extinction of Populations Inside Protected Areas." *Science* 280:2126–28.

Yahner, R. H. 1996. "Habitat Fragmentation and Habitat Loss." *Wildlife Society Bulletin* 24:592.

Zembal, R. 1993. "The Need for Corridors between Coastal Wetlands and Uplands in Southern California." In *Interface between Ecology and Land Development in California*, edited by J. E. Keeley, 205–8. Southern California Academy of Sciences, Los Angeles, CA.

8

Julie L. Lockwood and
J. Curtis Burkhalter

The Impact of Invasive Species on Wildlife Habitat

Nonnative species are an increasingly prevalent aspect of wildlife habitat. In today's world, it is nearly impossible to walk through a forest, meadow, or wetland and not readily encounter several nonnative plants and animals. Some of these nonnative species are so common that it is difficult to imagine them failing to have an effect on co-occurring native wildlife. Other nonnative species are so rare and localized in their distribution that they seem innocuous. What is a wildlife biologist to make of this range of nonnative species impacts? When should the presence of nonnative species become a concern in regard to their effects on native wildlife populations? The answers to these questions are, of course, quite complex. In this chapter, we tackle this complexity by summarizing currently accepted frameworks for defining nonnative species and measuring their impacts, and then reviewing a suite of mechanisms by which nonnative species can degrade the habitat quality of native wildlife. This review is designed to provide a scaffold onto which biologists can begin to place their own experiences in managing nonnative species for the benefit of wildlife, and with which to more readily access the expansive literature on biological invasions.

Definitions and Framework

The title of this chapter introduces three concepts (invasive, impact, and wildlife habitat) that we must define before moving to mechanisms of invasive species impacts. The first two terms are related in the sense that, by most definitions, nonnative species are labeled as "invasive" only if they are abundant and widespread enough to cause negative ecological impact (Lockwood et al. 2013). A negative impact occurs when the presence of an invasive species causes a reduction in the distribution, abundance, or individual performance (survival, reproduction) of one or more native species (Parker et al. 1999). Only a small fraction of the species that have the opportunity to be invasive (i.e., are transported out of their native range and released in a nonnative locale) are known to go on to become invasive (Williamson and Fitter 1996). Blackburn et al. (2011) provide a comprehensive and integrative framework for considering the process of species invasion (fig. 8.1). This framework defines the series of barriers and filters that a nonnative species must transit before it becomes invasive. These barriers are not trivial, and the proportion of species that transit each commonly hovers at around 10–20% (Williamson and Fitter 1996), although it can reach as high as 40–50% in some situations (Jeschke and Strayer 2006).

Morrison et al. (2006) define wildlife habitat as "an area with a combination of resources (like food, cover, and water) and environmental conditions (temperature, precipitation, presence or absence of predators or competitors) that promotes occupancy by individuals of a given species (or population) and allows those individuals to survive and reproduce." Given this definition, invasive species will negatively impact wildlife habitat by altering resources or conditions to the extent that survival and reproduction of a targeted species is affected. This impact can be imposed by (1) the invader maintaining a numerical dominance over native wild-

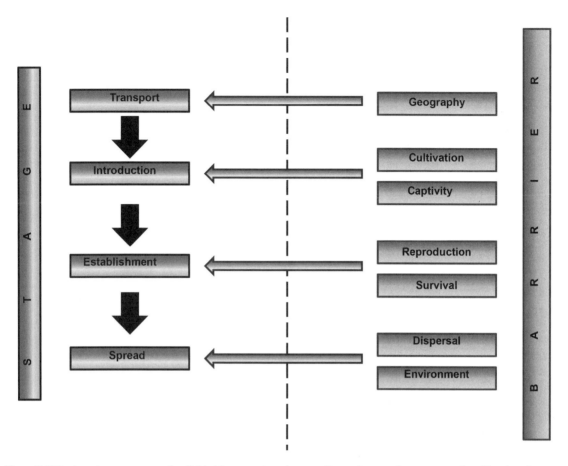

Figure 8.1 The invasion process can be divided into a series of stages. For each stage there are associated barriers that must be overcome before the species can proceed into the next stage. The number of nonnative species that successfully transit any one stage is generally thought to be quite small (<20%), thus dictating that most introduced nonnative species fail to reach an abundance and geographical range size that allow them to negatively impact native wildlife. Adapted from Blackburn et al. (2011)

life, (2) the invader being so common across space that very few individuals in a wildlife population will fail to come into contact with the invaders, (3) indirect effects mediated through one or more native species, (4) nonnative species changing ecosystem properties (e.g., changes in nutrient cycling, disturbance regimes), or (5) a high per capita negative effect of a single invader on a single native individual.

These definitions provide a clear way to link wildlife habitat and invasive species impacts. However, there is hidden complexity behind these concepts that stems from considering how impacts may vary through time and across space. Not all invasive species will show impacts on wildlife habitat over the duration of their tenure in an ecosystem, and the impact of the invader will not be felt evenly across all locations within that ecosystem.

Nonnative species typically will not impact native ecosystems until they reach some critical range size and abundance. This threshold is not well defined for any single invader and certainly varies according to the strength of the per capita effect of the invader on the native species (fig. 8.2). However defined, it will typically take several years between when a nonnative species is newly discovered in a habitat and its attainment of a population size and distribution that allow it to impact co-occurring native wildlife (fig. 8.2). This lag between arrival and becoming invasive may be relatively short or it can be quite prolonged (e.g., more than one hundred years), with the time dictated mostly by

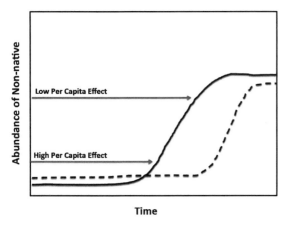

Figure 8.2 The impact of an invasive species is dependent on its abundance, geographical range size, and per capita effect. The solid line depicts standard logistic population growth in a nonnative population, which includes a period of time when the nonnative is inherently rare. The dashed line represents population growth with a prolonged lag, whereby abundance stays low for longer than logistic growth would dictate. Native wildlife will not suffer reduced survivorship or fecundity (impacts) until some threshold abundance (depicted here) or range size is achieved. With or without a prolonged lag, newly established nonnative populations will not reach this threshold until some time has passed. The time required in any one case is dependent on how long the population stays rare, as well as the per capita effects of nonnative on native individuals.

how fast the nonnative population can grow (Crooks 2005). The mechanisms that produce prolonged lags are numerous and include genetic impoverishment, Allee effects, or negative biotic interactions, although no one mechanism has yet explained the majority of lags (Crooks 2005; Lockwood et al. 2013). In addition to lags, there is the possibility that the impacts of invasive species will have an acute stage after which impacts will attenuate (Strayer et al. 2006; fig. 8.2). The attenuation of impacts can occur via behavioral or evolutionary adaptation of the local flora and fauna to the invader or through a drop in the abundance or geographical extent of the invasive population (e.g., overshooting a density limitation and then returning to a number of individuals that is better matched to local resource conditions). The element of time thus introduces the possibility that managers may dismiss a nonnative species as ecologically benign because it persists

at low numbers for a long period when in fact it is on its way to causing serious negative impacts (Crooks 2005). Additionally, unexpected events in time and space (e.g., disturbances) can also have unexpected consequences in regard to how seemingly benign nonnatives may rapidly proliferate into new areas following disturbance events (Keeley et al. 2003). Equally unwelcome is the possibility that the impacts of a particularly troublesome invader will lessen through time, thus leading a manager to expend great resources on a problem that will naturally dissipate (Strayer et al. 2006). In the future, climate change may also alter whether ecologically benign nonnative species become invasive (Hellmann et al. 2008).

It is relatively easy to extend this complexity in measuring the impact of invasive species across space. Not all locations within a particular habitat will sustain equal numbers of the invasive species due to relatively fine-grained differences in resources and environmental conditions. The ecological "rules" that govern how native species respond to such environmental conditions will also apply to any co-occurring nonnative species. Thus, some locations will provide optimal habitat for the invader and some only marginal. For a wildlife manager, a sound knowledge of a nonnative species' niche requirements will greatly inform where this species will impose strong impacts on co-occurring wildlife and where it will not. Management decisions can then proceed toward optimizing scarce human and economic capital to ameliorate invasive species impacts only where they are likely to occur.

With these definitions and frameworks in mind, we will devote the rest of this chapter to reviewing the various mechanisms by which an invasive species can affect wildlife habitat. In providing examples of these mechanisms, we will illustrate the complexity inherent in defining and measuring impact. Our review is organized so that we cover the direct impacts of invasive species on wildlife habitat first, followed by indirect mechanisms of impact, and finally synergisms between invasive species impacts and other known wildlife habitat stressors (e.g., fragmentation).

Direct Impacts

Invasive species can impact wildlife habitat by directly altering resource levels or environmental conditions such that they reduce the survival or fecundity of co-occurring wildlife. Invasive plants can impose such changes by (1) altering vegetation composition so that available cover is lost or degraded in usefulness, (2) replacing, or otherwise making inaccessible, plants or invertebrates that provide critical food, and (3) changing the chemistry, salinity, water flow, or other physical aspects of habitat so that suboptimal physiological conditions are created. Invasive animals can directly impact native wildlife either by eating or competing with native wildlife, or by producing toxins that harm native wildlife predators that consume them.

Invasive Plants

The structure that vegetation produces is a key component of wildlife habitat, providing suitable microclimates, nesting substrate, and protection from predation, and supporting critical food resources (Price and Waser 1984; Morrison et al. 2006). Vegetation structure also strongly influences the physical processes (e.g., fire and flood) and chemistry that characterize wildlife habitat (Brooks et al. 2004; Brown et al. 2006). Invasive plants can alter vegetation structure to varying degrees and concomitantly alter habitat features for wildlife that depend on these structures. Some invasive plants radically change vegetation structure, earning them the label of ecosystem engineer (Wang et al. 2006), while others seem to replicate the structure produced by their native plant counterparts. In the former case, managers should expect large impacts on native wildlife, while in the latter there may be no or only mild impacts. In table 8.1, we provide an overview of literature that addresses this continuum, including examples of invasive plants that alter cover, change food resources, and transform the physical environment.

An example of an invasive ecosystem engineer is the Amur honeysuckle (*Lonicera maackii*), which has established itself within the oak-hickory forests of the eastern and midwestern United States. This plant produces a dense shrub layer that is absent in the typically open understory of uninvaded forests (Collier et al. 2002). One of the many effects of this invasive honeysuckle is a reduction in soil surface temperature and an increase in soil-level humidity (Watling et al. 2011b). For species such as amphibians that have tight physiological requirements, such changes can drastically affect the

quality of that habitat. Watling et al. (2011b) tested for such an effect on a range of native amphibians that use oak-hickory forests in Missouri. They found that the higher the density of Amur honeysuckle, the fewer amphibian species recorded. In addition, the presence of honeysuckle tended to support the same suite of species across space, whereas native vegetation supported a wider variety of species. This change in spatial diversity patterns in native amphibians is perhaps due to the honeysuckle creating relatively homogeneous microclimates across space, whereas native vegetation created patches with differing conditions that support differing suites of amphibians (Watling et al. 2011b). There was also a clear shift in the set of amphibian species found across a gradient of honeysuckle density, which Watling et al. (2011b) attributed to the cooler microclimate produced by heavy honeysuckle cover.

Nonnative plants found in North American grasslands provide an example of more contextual and nuanced effects on wildlife. Grasslands have a particular structure characterized by low abundance of woody vegetation, few or no large trees, and a predominance of perennial grasses (White et al. 2000). Several native birds are considered grassland obligates, and thus the loss of grassland habitat directly results in the decline in these birds' populations (Johnson 2000). Some of these grassland losses are obvious (e.g., conversion to housing developments), while others are subtle, involving a switch from native to nonnative grasses, sometimes with the grassland being dominated by a single nonnative grass. However, the presence of nonnative grasses does not automatically transform a grassland into unsuitable habitat for native birds (e.g., Jones and Bock 2005; Kennedy et al. 2009). The extent to which invasion by nonnative grasses will influence native birds seems to be tied to the degree to which the nonnatives mimic the vegetation structures produced by natives (Kennedy et al. 2009). In this case, structure is important for nest construction, placement so that the nest is concealed from predators, and the support of invertebrates that provide high nutrition food items for breeding birds and their young. Whether a manager will need to shift grassland plant composition away from nonnatives to natives will depend crucially on the breeding biology of the birds he or she wishes to support and the degree to which the existing vegetation (native or nonnative) supports those needs (Kennedy et al. 2009).

Table 8.1 An overview of literature that addresses the continuum of impacts of invasive plants on wildlife habitat.

Impact	Example
Changes in physical properties	
Change in disturbance regime	*Andropogon gayanus* increases fire intensity in Australia (Rossiter et al. 2003)
Change in nutrient regimes	*Tamarix ramosissima* leaf litter decomposition changes dissolved organic matter dynamics in southwestern North America streams (Kennedy and Hobbie 2003)
Change in water quality	*Lythrum salicaria* increases tannin concentrations in freshwater ecosystems (Brown et al. 2006)
Change in nutrient regimes	Litter from suite of exotics decomposes and releases nitrogen significantly faster than natives in eastern US hardwood forests (Ashton et al. 2005)
Change in disturbance regime	Extensive review of various studies involving invasives and changes in disturbance regimes (Mack and D'Antonio 1998)
Replacing, or otherwise making inaccessible, food resources	
Changes in food availability	*Centaurea maculosa* reduces native plant diversity and, consequently, food availability for native birds in Montana (Ortega et al. 2006)
Seasonal changes in food availability	*Lonicera spp.* in eastern US forests provide more food for frugivorous avian species during winter (McCusker et al. 2010)
Changes in food abundance	Noxious weeds in western US rangelands shown to reduce food abundance for wildlife (DiTomaso 2009)
Decrease in food abundance	*Spartina alterniflora* invasions in China result in lower food abundances, thereby reducing arthropod abundance and community composition (Wu et al. 2009)
Changes in food web dynamics	*Elymus athericus*, a nonnative grass, invasion into European salt marshes induces a trophic cascade in which the dominant arthropod prey for most fish species shifts within invaded marshes, thereby disturbing trophic function (Laffaille et al. 2005)
Altering vegetation composition so that cover is lost or degraded	
Changes in vegetation composition	Invasion of *Lonicera spp.* and *Rosa multiflora* results in changes to plant cover and results in a doubling of avian nest depredation (Borgmann and Rodewald 2004)
Indirect impact of changes in plant composition	Invasion of *Lonicera maackii* changes water chemistry that results in greater levels of activity and/or increased surfacing behavior, exposing tadpoles to increased predation risk (Watling et al. 2011a)
Decreases in native plant cover	*Eragrostis lehmanniana* and *Cenchrus ciliaris* invade native grasslands in Texas and reduce ground-nesting avian community abundance (Flanders et al. 2006)
Changes in vegetation cover and structure	*Eragrostis lehmanniana* invasion into native grasslands in Arizona results in changes in vegetation structure and cover, resulting in reductions in small mammal abundance (Litt and Steidl 2011)
Changes in plant composition	Nest predation levels increase early in breeding season for birds utilizing the nonnative shrub *Lonicera maackii* (Rodewald et al. 2010)

In terms of physical processes, a common mechanism by which invasive plants alter wildlife habitat is by altering natural disturbance regimes such as fire. Wildlife that live in fire-prone ecosystems typically have evolved mechanisms to withstand fire events and perhaps profit from them by taking advantage of the temporary change in vegetation structure or the increase in available food resources (DiTomaso 2009; Litt and Steidl 2011). Thus, invasive plants that alter the frequency, intensity, areal extent, or timing of fire can create conditions that disfavor native wildlife (Brooks et al. 2004). In grasslands, the invasion of fire-tolerant plants can lead to a grass-fire cycle whereby the habitat is burned much more often, at higher intensities, and over larger areas than is common when native plants dominated (Brooks et al. 2004). The end result has been the transformation of these grasslands into habitats that cannot support a whole suite of native wildlife (Brooks et al. 2004).

Litt and Steidl (2011), for example, explored the influence of a fire-tolerant invasive grass, Lehmann lovegrass (*Eragrostis lehmanniana*), on the small mammals native to the semideserts of Arizona. They found that each species responded differently to the structural changes that lovegrass imposed (Litt and Steidl 2011). Those mammals that did well in areas with dense grass cover and high vegetation found lovegrass-dominated areas just as attractive as habitats that contained mostly native grasses. However, those mammals that preferred areas that are sparsely vegetated decreased in the presence of lovegrass. Litt and Steidl (2011) attribute some of this variation in use of the lovegrass to the strongly divergent food resources invaded versus uninvaded habitat offered small mammals. These results mirror those for grassland birds, providing another example of how the effects of invasive plants can vary according to the requirements of native wildlife.

Litt and Steidl (2011) also measured the response of the native small mammals to fire, again across a gradient of lovegrass dominance. When these grasslands burned, the higher vegetation biomass and differences in plant species composition across a gradient of invaded sites tended to alter the magnitude, persistence, and direction of fire effects on small mammals. For those mammalian species that preferred dense vegetation cover (i.e., dominated by lovegrass), fire dramatically reduced the suitability of a site. For species that preferred open, sparse vegetation, fire tended to improve habitat conditions.

In addition to changing disturbance regimes, invasive species have the ability to affect ecosystem properties such as nutrient cycling and water quality. Loss et al. (2012) demonstrated that the invasion of nonnative earthworms into North American hardwood forests has caused a multitude of changes across trophic levels. Nonnative earthworms increase the rate of leaf litter decomposition, resulting in faster rates of nutrient depletion by plant communities. This in turn resulted in changes in plant communities, which subsequently precipitated declines in the number of ground-nesting bird species. An additional example comes from tidal wetlands of the United States. Brown et al. (2006) demonstrated that invasion by purple loosestrife (*Lythrum salicaria*) increases tannin concentrations in the invaded wetlands, which in turn has negative impacts on aquatic food webs.

Animal Invasions

Animal invaders will directly affect co-occurring native wildlife by serving as their predators, their prey, or their competitors. The evidence for invasive animal impacts comes largely from invasive predators, although there are certainly recorded cases of impact entailing all of these mechanisms (Blackburn et al. 2004). Invasive predators will consume adults, young, or eggs of native wildlife. They will also have nonconsumptive effects whereby their presence adds costs to the energy budget of native prey via mechanisms such as the need for increased vigilance at the expense of foraging or using habitats that are safe from the invasive predator but suboptimal for foraging or breeding (Preisser et al. 2005).

Invasive predators are found in terrestrial island and continental regions, and in many freshwater and marine habitats (Cox and Lima 2006). The list of invasive predators that have profoundly affected native prey include some rather obvious examples of species that have the teeth and body size to eat most anything that they come across, a list that includes the brown tree snake (*Boiga irregularis*), Burmese python (*Python molurus bivittatus*), and red fox (*Vulpes vulpes*) (Harding et al. 2002; Wiles et al. 2003; Dove et al. 2011). Others may be a bit more surprising such as wild boar (*Sus scrofa*), red imported fire ant (*Solenopsis invicta*), and ship rat (*Rattus rattus*). This latter group includes species that are either not typically considered ravenous predators (boar) or are very small relative to their prey (ants and rats). However, it is not the size of the predator that matters as much as its evolutionary "fit" within the ecosystem it is invading (Cox and Lima 2006).

Some of the best-known examples of extinction due to invasive species impacts come from situations where a nonnative predator establishes in an ecosystem where the native prey are naïve to its presence (Cox and Lima 2006; Lockwood et al. 2013). More formally, these hunting behaviors (and other associated features) are termed predator "archetypes," and prey naiveté results from the prey having never evolved, or having lost, anti-predator defenses for the archetype of the invader (Cox and Lima 2006; Sih et al. 2010). In the context of wildlife conservation, a manager should expect large negative impacts of an invasive predator when there is a mismatch between how that predator pursues its prey and how the native prey respond to the presence and behavior of this predator. These relationships can be quite complex in detail, and there is a renewed focus in ecology on understanding the mechanisms by which prey respond to their predators and what level of mismatch is necessary for the prey to drop in numbers toward extinction (e.g., Banks and Dickman 2007; Salo et al. 2007). What is relatively well known, however, is that native prey species are more likely to become extinct if they are already rare or have naturally localized distributions (Cox and Lima 2006). A particularly vexing situation can arise when the invasive predators are able to maintain very high densities even when they have consumed a large percentage of the individuals of a native prey species (Courchamp et al. 1999; Roemer et al. 2001). Such invasive predators are able to eat the prey to extinction without themselves suffering an associated population decline because their numbers are subsidized by human food sources or by less behaviorally naïve (native or nonnative) alternative prey (e.g., hyperpredation, Courchamp et al. 2003).

If we reverse the role of the invader and make it the novel prey item that is consumed by the native wildlife predator, there is the potential for the invasive prey to drive predator population numbers down via its unique antipredation defenses. Here again it is the mismatch between what antipredator mechanisms a prey will demonstrate and what the prey has evolved to recognize and avoid. There are a variety of antipredator defenses that animals will employ (Caro and Girling 2005; Pease 2011), but in this context it is the production of toxins that seem to have the most clear negative effects on native predators. Perhaps the best-known example is the effect of invasive cane toads (*Bufo mariuns*) on native wildlife in Australia. Cane toads, which are native to South and Central America, exude onto their skin a powerful bufogenin that is highly toxic to their predators (Phillips et al. 2003). In Australia, where the cane toad was introduced to serve as a biocontrol agent in sugarcane plantations, the native wildlife are naïve to this toxin, and for many species if an individual ingests a single cane toad it will perish (Shine 1995). In locations where the cane toad is abundant, a large percentage of native predators will consume the toad with obvious effects on the consumed individual's survival probability. This added source of mortality has reduced the abundance of some native wildlife populations in Australia (Shine 2010).

Phillips et al. (2003) suggest that snakes may be particularly susceptible to the ill effects of cane toads since their feeding mechanism nearly ensures that they will come into contact with enough of the toxin to represent a lethal dose. The feeding trials and other data collected by Phillips et al. (2003) suggest that nearly 30% of terrestrial Australian snakes are threatened by cane toads. Cane toads have been purposefully introduced to a wide variety of other locations where sugarcane was (or is) a principle agricultural product, and there is growing evidence that in many of these places native wildlife populations may be at risk due to ingesting cane toads (e.g., Wilson et al. 2011).

Finally, there is the possibility that invasive animals will compete with native wildlife and thus reduce the native's survivorship or fecundity. Evidence for this effect is limited, as it is within invasions more broadly (Gurevitch and Padilla 2004; Lockwood et al. 2013). Competition, unlike predation, involves a more graduated reduction in the competing species' survivorship and reproduction. Thus, even though two species may compete for available resources, the effects of this competition on either population's abundance could be relatively small or fluctuate substantially across space and through time, especially as compared to other mechanisms reducing abundance such as disease or predation (Davis 2009). Nevertheless, there are a few examples of invasive competitors clearly impacting the populations of co-occurring wildlife. Perhaps the best example comes from Australia where the invasive common myna (*Acridotheres tristis*) aggressively excludes native parrots from using secondary tree cavities for nesting (Pell and Tidemann 1997). A similar competitive effect has been posited between European starlings (*Sturnus vulgaris*) and secondary cavity-nesting birds in North America, although the evidence for a strong effect is equivocal (Koenig 2003).

Indirect Impacts

Invasive species can impact wildlife habitat through a variety of indirect pathways. Such pathways link two or more species via a third species (Morin 2011; fig. 8.3). There are few fully worked examples of invasive species having indirect impacts on native wildlife habitat, no doubt due to the difficulty in fully documenting the complex interactions associated. Those that have been documented mostly involve the influence of invasive herbivores on island wildlife. Wagner and van Driesche (2010), for example, describe several cases where invasive herbivores such as feral pigs, cattle, goats, and sheep consume the host plants of specialist insect herbivores on islands, sometimes to the point where the associated insect became extinct. Similarly,

Spruce-Fir Forest, Southern Appalachia

Introduced Balsam Woolly Adelgid

Remnant Spruce-Fir Forest

Declines in Canopy and Subcanopy Birds

Figure 8.3 Invasive species can affect native wildlife via their influence on a third species, thus producing indirect impacts. For example, the nonnative balsam woolly adelgid has devastated spruce-fir forests in the southern Appalachian mountains of North America. The loss of spruce and fir trees substantially reduces the potential feeding and breeding habitat of birds that specialize in the use of canopy and subcanopy trees. The end result has been a noted reduction in the abundance of in these birds (see Rabenold et al. 1998).

Barrios-Garcia and Ballari (2012) review cases where wild boar have negatively impacted native wildlife, including causing a reduction in the abundance of native Hawaiian honeycreepers via the boar, reducing nectar-producing plants in forest understories through its rooting behavior. The impact of invasive herbivores can be less direct, but perhaps more widespread, through their sometimes profound influence on vegetation cover (Rabenold et al. 1998; Courchamp et al. 2003). For example, Donlan et al. (2007) provide strong evidence that grazing by feral goats on the Galapagos Islands led to a substantive reduction in suitable habitat for the endemic and highly threatened Galapagos rail (*Laterallus spilonotus*). One example of an invasive plant having

an indirect effect on wildlife is given by Rodewald et al. (2010), in which they find invasive shrubs (*Lonicera* spp.) creating habitat in which birds experience reduced survival early in the breeding season.

Invasive species can also indirectly have an impact by altering microclimates on which native wildlife depend. A particularly compelling example comes from Siderhurst et al. (2010), who documented how the loss of eastern and Carolina hemlock trees (*Tsuga canadensis* and *T. caroliniana*, respectively) due to the hemlock woolly adelgid (*Adelges tsugae*) affected the habitat of brook trout (*Salvelinus fontinalis*). Hemlock woolly adelgids, natives of East Asia, are invasive forest pests that suck the sap of native hemlock trees in North America. Their effects on North American hemlock trees are harmful enough that most hemlocks will die within four to fifteen years after initial adelgid infestation (Orwig 2002). Hemlock is a dominant canopy tree in many North American forests, especially within riparian zones, and dictates a suite of forest microclimatic conditions including the degree of shading and temperatures at ground level (Siderhurst et al. 2010). Given the hemlock's central role in these forests, their loss from the canopy should precipitate many indirect impacts on native wildlife. Siderhurst et al. (2010) investigated whether, and when, the loss of hemlock would increase solar radiation to streams and thus also increase water temperatures. They estimated that the loss of hemlock trees due to adelgid damage increased light levels by over 20% in some streams in western Virginia, resulting in an increase of 2°C in mean daily water temperatures. Such an increase will bring water temperatures to near the maximum thermal tolerance of brook trout (Butryn et al. 2013). Siderhurst et al. (2010) provide evidence that such changes may not come to fruition over the long term, but their results provide a clear connection between invasive forest insects and the abundance of recreationally important fish.

Invasive plants can increase their abundance or geographical range through forming mutualistic relationships with native or nonnative animals. When these invasive plants have impacts on wildlife through any of the mechanisms in table 8.1, there exists an indirect route of impact (fig. 8.3). The most common example in the literature is the association between birds and invasive plants where the bird provides pollination and seed dispersal services to the invader (e.g., Lafleur et al. 2007). Aslan and Rejmánek (2010) report that nearly half of all their citizen science observations of bird-invasive plant interactions involved birds eating the seeds or fruit of the plants and thus likely serving as a dispersal mutualist for the invader. Best and Arcese (2009) show that selective grazing by Canada geese (*Branta canadensis*) on a suite of nonnative grasses had a net positive effect on the grasses themselves. Grazing pressure created growing conditions that favored the production of nonnative grass stems (i.e., reduced litter accumulation), and the seeds ingested by grazing geese were dispersed into locations suitable for the grasses to establish (Best and Arcese 2009). Mammals can also form mutualistic interactions with nonnative plants, thus increasing their dispersal capabilities (Davis et al. 2010). Examples of mutualistic interactions between mammals and nonnative plants include the much-maligned rat that can also directly produce negative impacts on native wildlife (Shiels and Drake 2011).

Given the volume with which nonnative species are transported and released worldwide, there is the distinct possibility that an ecosystem will harbor several invasive species (Lockwood et al. 2013). In such situations, there is the possibility that a suite of nonnative species will form mutualistic networks that together impose higher levels of impact than each could achieve individually (Simberloff and Von Holle 1999). When such networks materialize, the end result can be a transformation of the native ecosystem, or an "invasional meltdown," where the end result is the loss of native species (Simberloff and Von Holle 1999; Simberloff 2006). The evidence for invasion meltdowns is limited, although there are several compelling case studies (Simberloff 2006). However, no example directly ties the strong facilitation between two nonnative species to a decrease in the survivorship or fecundity of co-occuring native wildlife (Simberloff 2006). Although not a direct example of invasional meltdown, the introduction of new parasites/diseases by introduced animals could have far-reaching effects on native biota (Prenter et al. 2004). A gripping example of the devastation that introduced parasites can have on native wildlife is seen in the marked reduction in endemic Hawaiian aviafauna caused by the introduction of avian malaria (van Riper III et al. 1986).

In fact, from our literature review, there are no studies of facilitation involving nonnative species that "close the loop" and unequivocally show how the establishment of a mutualism will indirectly negatively affect local wildlife. Thus, for example, although birds commonly eat the fleshy fruits of many invasive plants, it is not clear the extent to which the increased dispersal of such a plant is serving to decrease the abundance of any native wildlife species of concern.

Synergisms and Synthesis

Wildlife populations are principally driven toward extinction through the loss, fragmentation, and degradation of their habitat (Damschen et al. 2006). As we reviewed previously, invasive species can heap yet more negative influences on the survivorship and reproduction of wildlife populations. The degree to which invasive species are prominent threatening factors in wildlife declines is the subject of debate (e.g., Gurevitch and Padilla 2004 vs. Clavero and Garcia-Berthou 2005). However, there are many examples of wildlife having become extinct where the negative impacts of invasive species are clear and significant (Miller et al. 1989; Clavero and Garcia-Berthou 2005; Wagner and Van Driesche 2010). But to what extent do habitat loss and fragmentation interact with species invasions and lead to the loss of native species? Despite the two forces playing clear roles in biodiversity loss (Hobbs 2000), they are very often studied independently (Didham et al. 2007; Vilà and Ibanez 2011). In a comprehensive literature survey by Didham et al. (2007), they found that of the nearly 12,000 studies of land-use change (e.g., fragmentation) and over 3,500 studies of species invasions (published between 2002 and 2007), only 1% considered both simultaneously and 0.03% explicitly considered the interaction of the two.

What we know from studies that do consider both in tandem is that the land uses that surround habitat fragments play a large role in the number and types of invasive plants that exist within those fragments (e.g., Hobbs 2000; Lindenmayer and McCarthy 2001). In a review of empirical evidence linking landscape context with invasive plants, Vilà and Ibanez (2011) condense this literature into a few solid generalizations. Well supported is the positive relationship between the degree of habitat fragmentation and the presence of invasive plants. This result is driven to a large degree by the increased presence of invasive plants near the edge of fragments. Additionally, there is a positive relationship between the percentage of urbanized land around the habitat fragment and the incidence of invasive plants within that fragment. Road density, frequency of road use, and the existence of road improvements near a habitat fragment tend to increase the number of invasive plants species in the fragment. In short, the types and relative abundance of land uses around habitat fragments dictate to a great extent the degree to which that fragment harbors invasive plants (Vilà and Ibanez 2011). The degree to which we can generalize these specific results to the presence of invasive animals is uncertain, although several examples show that the land uses around a habitat fragment will influence the abundance and impact of invasive animals within that fragment (e.g., Donovan et al. 1997; Tewksbury et al. 2006).

The work of Vilà and Ibanez (2011) also highlights understudied elements of the interplay between invasive species and habitat loss or fragmentation. For example, it is not clear how often habitat corridors aid in the dispersal of invasive species between habitat patches. This statement is true even though the possibility that corridors would increase the prevalence of invasive species within connected habitat fragments was made twenty years ago (Simberloff et al. 1992). The few studies that have tested this effect have produced contrasting results (e.g., Damschen et al. 2006 vs. Thiele et al. 2008), making generalizations difficult. Vilà and Ibanez (2011) suggest that the confusing message from these limited studies is due to their failure to recognize the land uses surrounding the corridors. The habitats surrounding corridors matter just as they do for habitat fragments, and thus without a knowledge of these land uses we cannot predict whether a corridor will funnel more invasive plants into a fragment. Also somewhat surprising is the lack of research on the influence of fragment size, shape, and time since isolation on the number of invasive species present in a fragment patch (Vilà and Ibanez 2011). These are obvious places where research will highly repay the effort needed to produce clear answers.

Given the growing number of nonnative species transported worldwide, the importance of interactions between invasive species and other drivers of wildlife habitat loss or degradation will only increase (Didham

et al. 2007). Existing evidence provides some indication of how the two forces will combine to determine the fate of native wildlife, but clearly much more remains to be explored. Such explorations must do a better job of directly tying landscape changes (e.g., fragmentation) to the numerical abundance and geographical range size of the invasive species and documenting how such landscape changes influence the per capita effect of invasive species on native species (Didham et al. 2007). These three factors (the abundance, range size, and per capita effect of the invader) have been suggested to be the key components of measuring impact on native species (see previous discussion, Parker et al. 1999). In the context of wildlife management and conservation, this is the end measure that is vital for us to understand in responding to crises in wildlife populations. Making these connections explicit clarifies the impact of our research and provides detailed guidance to those tasked with managing habitats for our native wildlife.

LITERATURE CITED

Ashton, I. W., L. A. Hyatt, K. M. Howe, J. Gurevitch, and M. T. Lerdau. 2005. "Invasive Species Accelerate Decomposition and Litter Nitrogen Loss in a Mixed Deciduous Forest." *Ecological Applications* 15:1263–72.

Aslan, C. E., and M. Rejmánek. 2010. "Avian Use of Introduced Plants: Ornithologist Records Illuminate Interspecific Associations and Research Needs." *Ecological Applications* 20:1005–20.

Banks, P. B., and C. R. Dickman. 2007. "Alien Predation and the Effects of Multiple Levels of Prey Naiveté." *Trends in Ecology & Evolution* 22:229.

Barrios-Garcia, M. N., and S. A. Ballari. 2012. "Impact of Wild Boar (*Sus scrofa*) in Its Introduced and Native Range: A Review." *Biological Invasions* 14(11): 2283–2300.

Best, R. J., and P. Arcese. 2009. "Exotic Herbivores Directly Facilitate the Exotic Grasses they Graze: Mechanisms for an Unexpected Positive Feedback between Invaders." *Oecologia* 159:139–50.

Blackburn, T. M., P. Cassey, R. P. Duncan, K. L. Evans, and K. J. Gaston. 2004. "Avian Extinction and Mammalian Introductions on Oceanic Islands." *Science* 305:1955–58.

Blackburn, T. M., P. Pyšek, S. Bacher, J. T. Carlton, R. P. Duncan, V. Jarošík, J. R. U. Wilson, and D. M. Richardson. 2011. "A Proposed Unified Framework for Biological Invasions." *Trends in Ecology & Evolution* 26:333–39.

Borgmann, K. L., and A. D. Rodewald. 2004. "Nest Predation in an Urbanizing Landscape: The Role of Exotic Shrubs." *Ecological Applications* 14:1757–65.

Brooks, M. L., C. M. D'Antonio, D. M. Richardson, J. B. Grace,

J. E. Keeley, J. M. DiTomaso, R. J. Hobbs, M. Pellant, and D. Pyke. 2004. "Effects of Invasive Alien Plants on Fire Regimes." *BioScience* 54:677–88.

Brown, C. J., B. Blossey, J. C. Maerz, and S. J. Joule. 2006. "Invasive Plant and Experimental Venue Affect Tadpole Performance." *Biological Invasions* 8:327–38.

Butryn, R. S., D. L. Parrish, and D. M. Rizzo. 2013. "Summer Stream Temperature Metrics for Predicting Brook Trout (*Salvelinus fontinalis*) Distribution in Streams." *Hydrobiologia* 703:47–57.

Caro, T., and S. Girling. 2005. *Antipredator Defenses in Birds and Mammals.* University of Chicago Press, Chicago, IL.

Clavero, M., and E. Garcia-Berthou. 2005. "Invasive Species Are a Leading Cause of Animal Extinctions." *Trends in Ecology and Evolution* 20:110.

Collier, M. H., J. L. Vankat, and M. R. Hughes. 2002. "Diminished Plant Richness and Abundance Below *Lonicera maackii*, an Invasive Shrub." *American Midland Naturalist* 147:60–71.

Courchamp, F., J. L. Chapuis, and M. Pascal. 2003. "Mammal Invaders on Islands: Impact, Control and Control Impact." *Biological Reviews* 78:347–83.

Courchamp, F., M. Langlais, and G. Sugihara. 1999. "Control of Rabbits to Protect Island Birds from Cat Predation." *Biological Conservation* 89:219–25.

Cox, J. G., and S. L. Lima. 2006. "Naiveté and an Aquatic–Terrestrial Dichotomy in the Effects of Introduced Predators." *Trends in Ecology & Evolution* 21:674–80.

Crooks, J. A. 2005. "Lag Times and Exotic Species: The Ecology and Management of Biological Invasions in Slow-Motion." *Ecoscience* 12:316–29.

Damschen, E. I., N. M. Haddad, J. L. Orrock, J. J. Tewksbury, and D. J. Levey. 2006. "Corridors Increase Plant Species Richness at Large Scales." *Science* 313:1284–86.

Davis, M. A. 2009. *Invasion Biology.* Oxford University Press, New York, NY, USA.

Davis, N. E., D. M. Forsyth, and G. Coulson. 2010. "Facilitative Interactions between an Exotic Mammal and Native and Exotic Plants: Hog Deer (*Axis porcinus*) as Seed Dispersers in South-eastern Australia." *Biological Invasions* 12:1079–92.

Didham, R. K., J. M. Tylianakis, N. J. Gemmell, T. A. Rand, and R. M. Ewers. 2007. "Interactive Effects of Habitat Modification and Species Invasion on Native Species Decline." *Trends in Ecology & Evolution* 22:489–96.

DiTomaso, J. M. 2009. "Invasive Weeds in Rangelands: Species, Impacts, and Management." *Weed Science* 48:255–65.

Donlan, C. J., K. Campbell, W. Cabrera, C. Lavoie, V. Carrion, and F. Cruz. 2007. "Recovery of the Galápagos Rail Following the Removal of Invasive Mammals." *Biological Conservation* 138:520–24.

Donovan, T. M., P. W. Jones, E. M. Annand, and F. R. Thompson III. 1997. "Variation in Local-Scale Edge Effects: Mechanisms and Landscape Context." *Ecology* 78:2064–75.

Dove, C. J., R. W. Snow, M. R. Rochford, and F. J. Mazzotti. 2011. "Birds Consumed by the Invasive Burmese Python (*Python molurus bivittatus*) in Everglades National Park, Florida, USA." *The Wilson Journal of Ornithology* 123:126–31.

Flanders, A. A., W. P. Kuvlesky Jr., D. C. Ruthven III, R. E. Zaiglin, R. L. Bingham, T. E. Fulbright, F. Hernández, L. A. Brennan, and J. Vega Rivera. 2006. "Effects of Invasive Exotic Grasses on South Texas Rangeland Breeding Birds." *The Auk* 123:171–82.

Gurevitch, J., and D. K. Padilla. 2004. "Are Invasive Species a Major Cause of Extinctions?" *Trends in Ecology & Evolution* 19:470–74.

Harding, E. K., D. F. Doak, and J. D. Albertson. 2002. "Evaluating the Effectiveness of Predator Control: The Non-Native Red Fox as a Case Study." *Conservation Biology* 15:1114–22.

Hellmann, J. J., J. E. Byers, B. G. Bierwagen, and J. S. Dukes. 2008. "Five Potential Consequences of Climate Change for Invasive Species." *Conservation Biology* 22:534–43.

Hobbs, R. J. 2000. "Land-Use Changes and Invasions." In *Invasive Species in a Changing World*, edited by H. A. Mooney and R. J. Hobbs, 55–64. Island Press, Washington, D.C.

Jeschke, J. M., and D. L. Strayer. 2006. "Determinants of Vertebrate Invasion Success in Europe and North America." *Global Change Biology* 12:1608–19.

Johnson, D. H. 2000. "Grassland Bird Use of Conservation Reserve Program Fields in the Great Plains." In *A Comprehensive Review of Farm Bill Contributions to Wildlife Conservation 1985–2000*, edited by L. P. Heard, A. W. Allen, L. B. Best, S. J. Brady, W. Burger, A. J. Esser, E. Hackett, D. H. Johnson, R. L. Pederson, R. E. Reynolds, C. Rewa, M. R. Ryan, R. T. Molleur, and P. Buck, 19–34. United States Department of Agriculture, Natural Resources Conservation Service, Wildlife Habitat Management Institute Technical Report USDA/NRCS/WHMI-2000.

Jones, Z. F., and C. E. Bock. 2005. "The Botteri's Sparrow and Exotic Arizona Grasslands: An Ecological Trap or Habitat Regained?" *The Condor* 107:731–41.

Keeley, J. E., D. Lubin, and C. Fotheringham. 2003. "Fire and Grazing Impacts on Plant Diversity and Alien Plant Invasions in the Southern Sierra Nevada." *Ecological Applications* 13:1355–74.

Kennedy, P. L., S. J. DeBano, A. M. Bartuszevige, and A. S. Lueders. 2009. "Effects of Native and Non-Native Grassland Plant Communities on Breeding Passerine Birds: Implications for Restoration of Northwest Bunchgrass Prairie." *Restoration Ecology* 17:515–25.

Kennedy, P. L., and S. E. Hobbie. 2003. "Saltcedar (*Tamarix ramosissima*) Invasion Alters Organic Matter Dynamics in a Desert Stream." *Freshwater Biology* 49:65–76.

Koenig, W. D. 2003. "European Starlings and Their Effect on Native Cavity-Nesting Birds." *Conservation Biology* 17:1134–40.

Laffaille, P., J. Pétillon, E. Parlier, L. Valéry, F. Ysnel, A. Radureau, E. Feunteun, and J. C. Lefeuvre. 2005. "Does the Invasive Plant, *Elymus athericus*, Modify Fish Diet in Tidal Salt Marshes?" *Estuarine, Coastal and Shelf Science* 65:739–46.

Lafleur, N. E., M. A. Rubega, and C. S. Elphick. 2007. "Invasive

Fruits, Novel Foods, and Choice: An Investigation of European Starling and American Robin Frugivory." *The Wilson Journal of Ornithology* 119:429–38.

Lindenmayer, D. B., and M. A. McCarthy. 2001. "The Spatial Distribution of Non-native Plant Invaders in a Pine–Eucalypt Landscape Mosaic in South-eastern Australia." *Biological Conservation* 102:77–87.

Litt, A. R., and R. J. Steidl. 2011. "Interactive Effects of Fire and Nonnative Plants on Small Mammals in Grasslands." *Wildlife Monographs* 176:1–31.

Lockwood, J. L., M. F. Hoopes, and M. P. Marchetti. 2013. *Invasion Ecology*. 2nd edition. Blackwell Publishing, Oxford, UK.

Loss, S. R., G. J. Niemi, and R. B. Blair. 2012. "Invasions of Non-native Earthworms Related to Population Declines of Ground-Nesting Songbirds across a Regional Extent in Northern Hardwood Forests of North America." *Landscape Ecology* 27:683–96.

Mack, M. C., and C. M. D'Antonio. 1998. "Impacts of Biological Invasions on Disturbance Regimes." *Trends in Ecology & Evolution* 13:195–98.

McCusker, C. E., M. P. Ward, and J. D. Brawn. 2010. "Seasonal Responses of Avian Communities to Invasive Bush Honeysuckles (*Lonicera* spp.)." *Biological Invasions* 12:2459–70.

Miller, R. R., J. D. Williams, and J. E. Williams. 1989. "Extinctions of North American Fishes during the Past Century." *Fisheries* 14:22–38.

Morin, P. J. 2011. *Community Ecology*. 2nd edition. Wiley-Blackwell, Oxford, UK.

Morrison, M. L., B. G. Marcot, and R. W. Mannan. 2006. *Wildlife-Habitat Relationships*. 3rd edition. Island Press, Washington D.C.

Ortega, Y. K., K. S. McKelvey, and D. L. Six. 2006. "Invasion of an Exotic Forb Impacts Reproductive Success and Site Fidelity of a Migratory Songbird." *Oecologia* 149:340–51.

Orwig, D. A. 2002. "Stand Dynamics Associated with Chronic Hemlock Woolly Adelgid Infestations in Southern New England." Proceedings, Hemlock Woolly Adelgid in the Eastern United States Symposium, 5–7.

Parker, I. M., D. Simberloff, W. Lonsdale, K. Goodell, M. Wonham, P. Kareiva, M. Williamson, B. Von Holle, P. Moyle, and J. Byers. 1999. "Impact: Toward a Framework for Understanding the Ecological Effects of Invaders." *Biological Invasions* 1:3–19.

Pease, K. M. 2011. "Rapid Evolution of Anti-predator Defenses in Pacific Tree Frog Tadpoles Exposed to Invasive Predatory Crayfish." PhD dissertation. University of California–Los Angeles, California.

Pell, A., and C. Tidemann. 1997. "The Impact of Two Exotic Hollow-Nesting Birds on Two Native Parrots in Savannah and Woodland in Eastern Australia." *Biological Conservation* 79:145–53.

Phillips, B. L., G. P. Brown, and R. Shine. 2003. "Assessing the Potential Impact of Cane Toads on Australian Snakes." *Conservation Biology* 17:1738–47.

Preisser, E. L., D. I. Bolnick, and M. F. Benard. 2005. "Scared to Death? The Effects of Intimidation and Consumption in Predator-Prey Interactions." *Ecology* 86:501–9.

Prenter, J., C. MacNeil, J. T. A. Dick, and A. M. Dunn. 2004. "Roles of Parasites in Animal Invasions." *Trends in Ecology & Evolution* 19:385–90.

Price, M. V., and N. M. Waser. 1984. "On the Relative Abundance of Species: Postfire Changes in a Coastal Sage Scrub Rodent Community." *Ecology* 65:1161–69.

Rabenold, K. N., P. T. Fauth, B. W. Goodner, J. A. Sadowski, and P. G. Parker. 1998. "Response of Avian Communities to Disturbance by an Exotic Insect in Spruce-Fir Forests of the Southern Appalachians." *Conservation Biology* 12:177–89.

Rodewald, A. D., D. P. Shustack, and L. E. Hitchcock. 2010. "Exotic Shrubs as Ephemeral Ecological Traps for Nesting Birds." *Biological Invasions* 12:33–39.

Roemer, G. W., T. J. Coonan, D. K. Garcelon, J. Bascompte, and L. Laughrin. 2001. "Feral Pigs Facilitate Hyperpredation by Golden Eagles and Indirectly Cause the Decline of the Island Fox." *Animal Conservation* 4:307–18.

Rossiter, N. A., S. A. Setterfield, M. M. Douglas, and L. B. Hutley. 2003. "Testing the Grass-Fire Cycle: Alien Grass Invasion in the Tropical Savannas of Northern Australia." *Diversity and Distributions* 9:169–76.

Salo, P., E. Korpimäki, P. B. Banks, M. Nordström, C. R. Dickman, E. Korpimäki, P. B. Banks, M. Nordström, and C. R. Dickman. 2007. "Alien Predators Are More Dangerous Than Native Predators to Prey Populations." *Proceedings of the Royal Society B: Biological Sciences* 274:1237–43.

Shiels, A. B., and D. R. Drake. 2011. "Are Introduced Rats (*Rattus rattus*) Both Seed Predators and Dispersers in Hawaii?" *Biological Invasions* 13:883–94.

Shine, R. 1995. *Australian Snakes: A Natural History*. Cornell University Press, Ithaca, NY.

———. 2010. "The Ecological Impact of Invasive Cane Toads (*Bufo marinus*) in Australia." *The Quarterly Review of Biology* 85:253–91.

Siderhurst, L. A., H. P. Griscom, M. Hudy, and Z. J. Bortolot. 2010. "Changes in Light Levels and Stream Temperatures with Loss of Eastern Hemlock at a Southern Appalachian Stream: Implications for Brook Trout." *Forest Ecology and Management* 260:1677–88.

Sih, A., D. I. Bolnick, B. Luttbeg, J. L. Orrock, S. D. Peacor, L. M. Pintor, E. Preisser, J. S. Rehage, and J. R. Vonesh. 2010. "Predator–Prey Naïveté, Antipredator Behavior, and the Ecology of Predator Invasions." *Oikos* 119:610–21.

Simberloff, D. 2006. "Invasional Meltdown 6 Years Later: Important Phenomenon, Unfortunate Metaphor, or Both?" *Ecology Letters* 9:912–19.

Simberloff, D., J. A. Farr, J. Cox, and D. W. Mehlman. 1992. "Movement Corridors: Conservation Bargains or Poor Investments?" *Conservation Biology* 6:493–504.

Simberloff, D., and B. Von Holle. 1999. "Positive Interactions of Nonindigenous Species: Invasional Meltdown?" *Biological Invasions* 1:21–32.

Strayer, D. L., V. T. Eviner, J. M. Jeschke, and M. L. Pace. 2006.

"Understanding the Long-Term Effects of Species Invasions." *Trends in Ecology & Evolution* 21:645–51.

Tewksbury, J. J., L. Garner, S. Garner, J. D. Lloyd, V. Saab, and T. E. Martin. 2006. "Tests of Landscape Influence: Nest Predation and Brood Parasitism in Fragmented Ecosystems." *Ecology* 87:759–68.

Thiele, J., U. Schuckert, and A. Otte. 2008. "Cultural Landscapes of Germany are Patch-Corridor-Matrix Mosaics for an Invasive Megaforb." *Landscape Ecology* 23:453–65.

van Riper III, C., S. G. van Riper, M. L. Goff, and M. Laird. 1986. "The Epizootiology and Ecological Significance of Malaria in Hawaiian Land Birds." *Ecological Monographs* 56:327–44.

Vilà, M., and I. Ibanez. 2011. "Plant Invasions in the Landscape." *Landscape Ecology* 26:461–72.

Wagner, D. L., and R. G. Van Driesche. 2010. "Threats Posed to Rare or Endangered Insects by Invasions of Nonnative Species." *Annual Review of Entomology* 55:547–68.

Wang, Q., S. Q. An, Z. J. Ma, B. Zhao, J. K. Chen, and B. Li. 2006. "Invasive *Spartina alterniflora*: Biology, Ecology and Management." *Acta Phytotaxon Sin* 44:559–88.

Watling, J., C. Hickman, E. Lee, K. Wang, and J. Orrock. 2011. "Extracts of the Invasive Shrub *Lonicera maackii* Increase Mortality and Alter Behavior of Amphibian Larvae." *Oecologia* 165:153–59.

Watling, J. I., C. R. Hickman, and J. L. Orrock. 2011. "Invasive Shrub Alters Native Forest Amphibian Communities." *Biological Conservation* 144:2597–2601.

White, R. P., S. Murray, M. Rohweder, S. D. Prince, and K. M. J. Thompson. 2000. *Grassland Ecosystems*. World Resources Institute, Washington, D.C.

Wiles, G. J., J. Bart, R. E. Beck Jr., and C. F. Aguon. 2003. "Impacts of the Brown Tree Snake: Patterns of Decline and Species Persistence in Guam's Avifauna." *Conservation Biology* 17:1350–60.

Williamson, M., and A. Fitter. 1996. "The Varying Success of Invaders." *Ecology* 71:1661–66.

Wilson, B. S., S. E. Koenig, R. van Veen, E. Miersma, and D. Craig Rudolph. 2011. "Cane Toads a Threat to West Indian Wildlife: Mortality of Jamaican Boas Attributable to Toad Ingestion." *Biological Invasions* 13:55–60.

Wu, Y. T., C. H. Wang, X. D. Zhang, B. Zhao, L. F. Jiang, J. K. Chen, and B. Li. 2009. "Effects of Saltmarsh Invasion by *Spartina alterniflora* on Arthropod Community Structure and Diets." *Biological Invasions* 11:635–49.

PART III • RESEARCH AND CONSERVATION

9

Thoughts on Models and Prediction

Bret A. Collier and
Douglas H. Johnson

Wildlife managers and conservationists face many challenges, such as convincing the public that wildlife professionals occasionally do know more about wildlife populations than the average citizen, encouraging youth to appreciate some things that are not electronic, and persuading elected officials to think beyond the immediate and consider long-term sustainability. These challenges would exist even in a stable world.

However, the world is very dynamic. The weather is fickle, the climate changes in long-term cycles, and humans now influence climate in unprecedented manners. The human population grows at exponential rates, its impact magnified by increasing per capita consumption of resources. And the world shrinks; a disease that once might have been confined to a small rural village in Asia now may become a worldwide epidemic in a matter of weeks. Well-intentioned legislation can effect enormous but unanticipated changes in land use over millions of hectares across the nation, and changes in energy prices can potentially have significant environmental consequences. How is a wildlife manager to deal with so many major changes well beyond his or her control?

This chapter will not resolve these difficulties. Instead, we contemplate how wildlife managers and conservationists might think about future planning and become more proactive relative to wildlife populations and their habitats.

Thoughts on Models and Prediction

Knowledge of where and when animals occur, and their success in surviving and reproducing, is crucial to un-derstanding their status and predicting changes. That knowledge is gained by studying the habitats they use, as well as when and how they use them. One of the greatest challenges wildlife managers must address is recognizing how the status and sustainability of wildlife populations will change as the Earth's ecosystems continue in flux. It is a simple model that predicts complete habitat loss will be deleterious to species that are restricted to that habitat. Assessing responses to other habitat changes requires more complex models and invokes greater uncertainty. Numerous models have been developed for a variety of objectives and a wide range of species, but their record for accurate prediction is spotty. As ecologists strive to assess current and future impacts of environmental change on wildlife resources with only limited knowledge of the system, what is the manager to do? In this chapter, there will be no discussion of model creation or application, as a vast array of literature is available on that topic (Shenk and Franklin 2001; Williams et al. 2002; Millspaugh and Thompson 2009). Rather, this chapter will discuss issues outside the approach of "collect data, analyze, and explain" that typifies the standard approach to prediction in wildlife ecology and focus on the interplay among modeling, monitoring, and careful thinking.

Known Unknowns

There are things that we think we know. We think we know how to measure habitat, or at least the aspects of habitat that we believe are relevant to an animal. We think we know what affects survival, or at least how to identify the primary biotic and abiotic factors that affect survival. There are many things that we know are unknowns, and wildlife researchers use models to turn known unknowns into known knowns, such that we know the extent to which we know.

The Role of Modeling

The need for accurate predictions of ecological outcomes in response to environmental changes has heightened interest in the application of various models. One useful definition of a model is that an object A* is a model of an object A if one can use A* to address relevant questions about A (Minsky 1968). Thus, models are used to reduce the infinite-dimensional set of ecological components and their interactions to manageable pieces relevant to the specific topic of interest (Muller et al. 2011). Components in a model are linked, typically via mathematical relationships, to describe characteristics of the system in an effort to mimic or predict the behavior of the system being modeled. Thus, models in wildlife ecology help us to reduce the complexity of biological systems by focusing on only components pertinent to the immediate need (Shenk and Franklin 2001). Note that we are not discussing purely *statistical models* (e.g., those used in linear regression) per se but rather a more overarching framework focusing on some system. For example, population models help us understand how the processes of births, deaths, immigration, and emigration influence population size, and models of an ecosystem can be developed to mimic the dynamic interactions among its plants, animals, and abiotic components.

It is convenient to view a *system* as receiving *input* from some source and responding with an *output*. A model of that system can be constructed for one of three basic objectives, one each aligned with input, model, and output. *Prediction* involves determining the output from the system based on some prescribed input and the assumed model. *Understanding* the system is based on inferring relationships between observed inputs and outputs. *Control* involves manipulating inputs to the system to achieve desired outputs. Clearly, however, a model built for understanding a system, if it adequately mimics the relationships in the real system, will be useful for both prediction and control.

Consider some wildlife examples. If a manager wanted to maintain harvest of a species at some level, a model for *control* would be appropriate; the output (desired harvest) is known and the input (harvest regulations) need to be determined; the necessary model would relate regulations to actual harvest. If wildlife managers were concerned about how grassland birds might be affected by the elimination of the Conservation Reserve Program (CRP) and a return of CRP lands to cultivation, they would find useful a model for the *prediction* of densities of grassland birds in various habitats (CRP and cultivated lands); inputs would be areas of land in each land-use category before and after the curtailment of CRP, and outputs would be numbers of grassland birds before and after. A scientist interested in the effects of an ingested chemical on egg production by a bird species would seek to develop a model for *understanding* that relates measurements of egg production to the amount of the chemical ingested for a (large) number of birds.

Models have many uses in wildlife-habitat ecology; models can describe functional relationships among ecological variables, quantify those relationships, and change as additional data are collected, and models can be used to predict future scenarios. Models represent a way for us to synthesize knowledge and explain a system. Further, models provide scientists with the ability evaluate available knowledge, identify areas where information is lacking, formulate hypotheses, design research focused on filling knowledge gaps, integrate information from a variety of sources, and serve as a means of communicating research results (Johnson 2001).

Designing a Model

What type of modeling approach is best? Wildlife-habitat models range widely, from models for species distributions (Elith et al. 2006), models of fine-scale habitat suitability (US Fish and Wildlife Service 1981), to models focused on response to hypothetical or predicted perturbations (Boyce 1992; Hanski 1998; DeAngelis and Mooij 2005). Beck and Suring (2009) identified forty frameworks for modeling wildlife-habitat relationships, which vary in model structure, input requirements, and output variables. Thus, we refer the interested reader to Johnson (2001), who states, "A model has value if it provides better insight, predictions, or control than would have been available without the model" (Johnson 2001, 113). Relatively few models, however, have been confronted with a truly independent set of data to determine their actual value for their intended purpose.

Not surprisingly, a wide variety of recommendations for wildlife-habitat modeling have been proffered. Most of these suggestions have been discussed elsewhere in significant detail, and all of these suggestions are somehow linked in model development; however, here we attempt to distill those discussions down to the following set of general guidelines for model development.

First, models must have precise, well-defined questions that focus on key conservation-relevant outputs, while limiting the number of assumptions used. As models represent a construct of a system that we are trying to understand, they should clearly define what ecological variables are relevant to the modeling exercise (Müller et al. 2011). Models can rarely answer *why* questions, typically focusing on *how* questions, but often asking *what if*, as most questions in ecology are focused on resolving how a system operates relative to system inputs and our preconceptions about the structure of the system. Thus, approaching development of any model should be (somewhat obviously) based on a well-defined question that relies on reasonable expectations regarding how the system under study works and the question(s) that the researcher wants to evaluate.

Ecological systems are inherently complex, thus we cannot possibly define all of the interacting components (although given the proliferation of model sizes, some have tried), nor can we realistically parameterize those components or identify the relationships of those components. Ecological and statistical complexity in models should be treated differently in that a model could have thousands of interacting values (ecologically complex), yet relationships between model parameters could be deterministic in nature (statistically simple). It is important to realize that models are only as good as the data and assumptions on which they are based. All models rely on data and have assumptions about data—how it was collected, how it is structured, and so forth. We often lack adequate estimates of popu-

lation parameters encompassing the range of the data (both temporally and spatially) needed for modeling most systems of interest. In these cases, approaches such as simulation can be used to identify the relative influence of model elements and identify where future data collection should occur, given that the posited relationships are accurate. Thus, if a model is simple, slightly biased, but useful, that is better than a model that is complex, unbiased, but not useful. Models that are slightly biased for most species yet useful for all species represent a viable modeling alternative versus process- and species-specific models, as data can be acquired and shared across system. Models with a significant number of integrated parts often provide good predictions when focused within sample periods (periods in which the range of data was collected) but can trend toward unreliability as time and space increase (Bunnell 1989; Araujo and Rahbek 2006; Beutel et al. 2006). We note that, while rarely discussed, data often used for comparisons to modeled predictions are frequently model based (e.g., density/abundance data corrected for imperfect detection, capture-recapture estimates of population growth) and are thus uncertain. Yet, estimates are regularly used to evaluate models, often without addressing the inherent uncertainty in these estimates, which may impact the accuracy of model evaluations. Thus, there exists a modeling loop, where model evaluation is dependent on modeled predictions.

Models and their metrics do not ever represent truth in and of themselves. Model output that is purported to track the observed fluctuation in a population is no more representative of truth than one whose predictions are uncoupled from observed fluctuations. Models don't give answers; rather, they provide estimates of the magnitude of difference in model predictions based on the structure and variation in a model. Thus, models can and should be evaluated at different levels of the system. As most ecological systems are hierarchical in nature, the level of aggregation at which a model is evaluated will impact model veracity and usefulness. Model-based aggregated parameters (e.g., population growth rates) are more likely to be robust to moderate variation in inputs; intermediate parameters (e.g., per-capita recruitment) are more likely to be impacted by the vagaries of internal model parameters and the mechanistic relationships between those parameters; whereas input parameters (annual survival) will be driven by both the posited statistical model and the amount of data used in the estimation routine. Thus, model robustness, in many cases, may rely more on structure and less on accuracy of input values.

Most relationships among variables in real systems are nonlinear; however relationships frequently posited between model parameters are regularly linear. How can this incompatibility be reconciled, beyond concluding that all models are wrong and giving up on them? The answer is that most nonlinear relationships are "close enough" to linear to be useful, as long as input variables do not vary "too much." Of course, "close enough" and "too much" are vague terms, the first depending on the objectives of the modeling, and the second providing a cautionary note about extrapolating beyond the range of the data used to develop the model. Space is curved, while Euclidean geometry assumes it is not. Therefore Euclidean geometry is invalid (wrong), but it still is useful as a very close approximation to the truth. The nonlinearity is so slight at distances usually measured on Earth that it can be ignored, but that is not so true with cosmological distances. Nonlinearities among other relationships may not be so obvious. Consider a hypothetical species whose abundance responds to some habitat feature, denoted by X, as shown in figure 9.1. A reasonable example might be the abundance of grasshopper sparrows (*Ammodramus savannarum*) in relation to biomass of live vegetation. A study conducted in a semiarid region (A in the figure) likely would find that the species is more common when vegetation is thicker. Someone studying grasshopper sparrows in a more mesic region (C in the figure) would reach the opposite conclusion, whereas a study in the intermediate region (B in the figure) might detect no effect at all of vegetation biomass on grasshopper sparrow abundance. Extrapolating results from any of the three studies to regions with different vegetation would be misleading. Viewing figure 9.1 makes this last conclusion obvious, but no researcher conducting a single study would have the benefit of such an opportunity. The bottom line is that using models for prediction, which always involves extrapolation over time, will require extrapolation beyond the range of data on which the models were based, so it is especially important to determine if and when the models are no longer useful.

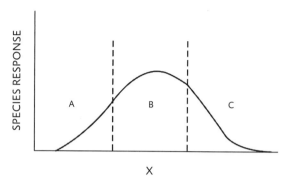

Figure 9.1 The nonlinear response of a species to an environmental gradient X.

Doing so will require careful monitoring of actual events in comparison to predictions.

Unknown Unknowns

Modeling is useful if something is known about the system and the relevant variables. For example, X affects Y, but you don't have complete knowledge, such as exactly how X affects Y. You know what you don't know. In this situation, one can propose and formulate various plausible relationships between X and Y and gather data, use extant data, or seek expert opinion about details of the relationship and the amount of uncertainty we should expect in the relationship. What, however, if you have no idea about the form of the relationship or, even worse, what the relevant variables are? That is, you don't know what you don't know—the unknown unknowns situation. Modeling now is not so straightforward. What *can* be done?

Possibly one could look backward in time. Imagine the world, say, fifty years ago, but knowing what is known now about what has transpired during that half century. Is it possible to identify events that we now realize have had a huge impact but were unappreciated a half-century ago?

For example, the world population was roughly three billion humans—a lot of people, but the exponential nature of its growth in developing countries was only beginning to become apparent. Some authorities, notably Paul Ehrlich and Ann Ehrlich (*The Population Bomb*) identified the patterns and described scenarios of calamities occurring within a decade or so. Such alarmists were widely ignored, especially when the anticipated catastrophes failed to occur as soon as predicted. Nonetheless, the problem of overpopulation continues today.

We suspect that virtually every major event that has occurred had been anticipated by someone much earlier. The difficulty for managers and those making decisions arises in part because numerous other events have been predicted but had not actually occurred. Who are the true prophets, who are the false prophets, and how can we tell them apart? Instead of seeking specific predictions about the future, perhaps we should be seeking environmental *leading indicators*, rescuing the term from the purely economics realm. What might useful environmental leading indicators be? Certainly, most organisms respond directly to various weather features, so climate should be assessed. And humans cause enormous changes to wildlife and their habitats, so human numbers, their geographical distribution, consumption patterns, and other behaviors somehow should be monitored. From a food web viewpoint, the top predators in a system rely on other species in the system and might well reflect the condition of those species. Because water flows downhill, it should be straightforward to identify locations where conditions reflect conditions throughout the entirety of the upstream area.

These ideas lead to the notion of monitoring without specific hypotheses in mind, or what has been disparaged by some as "surveillance monitoring." Nichols and colleagues (Yoccoz et al. 2001; Nichols and Williams 2006) have argued that monitoring is useful only when there is a specific hypothesis to test or a specific management program to inform. They criticize surveillance monitoring, which they defined as monitoring not guided by *a priori* hypotheses and their corresponding models (Nichols and Williams 2006). Certainly, if any management action is planned or a hypothesis needs to be tested, one should see what happens after the action or during the experiment. But many important events are neither planned nor anticipated. One can think of the Deepwater Horizon oil spill and its consequences to natural resources or the Exxon Valdez, Chernobyl, or the Japanese earthquake and subsequent tsunami. One cannot go to a disaster site and learn what it was like *before* the disaster; you need to have had a monitoring or analogous program in place well before the event. Thus, while not hypothesis driven, surveillance monitoring can be a critical for informing management, regardless of whether or not *a priori* hypotheses are posited.

Obviously, not everything can be monitored. However, we think that decisions about what to monitor should be based on at least the following considerations:

1. Importance of the resource: This criterion seems trivial as all resources are important to something, but it covers a multitude of attributes. Any resource that has major influence on a number of other resources qualifies here—such as krill, representing one of the lowest trophic levels in oceans, or freshwater, a resource regularly overlooked but on which both humans and wildlife directly depend.

2. Susceptibility: Which resources are deemed most susceptible? Certain forests are more susceptible to fire than others. Some wetlands are more readily drained than others. Some grasslands are more subject to encroachment by invasive species than others. Habitat specialists likely are more susceptible than generalists because they tolerate a more narrow range of conditions and would be more likely to be vulnerable when conditions change.

3. Potential rate of change: Prairie wetlands, as one trivial example, are much more dynamic than rock formations. Locations with a slow rate of change—such as mature forests, unless significant perturbations occur—likely require less intensive monitoring than, say, a coastal wetland.

4. Ease, objectivity, and repeatability: Monitoring is defined as information collected over time and space. Thus, methodological approaches must be objective (unbiased sampling, for instance) and allow for repeatability of the analytical methods such that accurate results can be obtained. For instance, one reason why singing birds are the object of so many monitoring programs is because they are much more readily amenable to current sampling and surveying methods than, say, crepuscular mammals or karst salamanders.

Control over Events

The basic approach used to further understanding of wildlife-habitat relationships is to identify locations where individuals are found, collect descriptive data on both the individuals and locations, analyze those data and develop an appropriate model, make predictions either for places other than the study area or for a later time period, and corroborate those results against the new data as evidence for a model's utility. The process of observation and inference represents the workhorse of wildlife science, where virtually every research output is somehow dependent on past data collected. Induction, or generalizing the specific, is then used to make predictions of future conditions. This process of evidence- (data) based science has served us well, and this reliance on previously gathered data comes with a cost, in that wildlife sciences are consistently reactive, rather than proactive, which limits the actual ability we have to manage or manipulate ecological systems.

Speculating what ecological calamities may occur is often met with significant skepticism and resistance, regardless of the potential benefits of looking ahead (McCann et al. 2006). Cynicism about such speculation has several foundations. One is the issue of false positives; many disasters that were anticipated did not occur. A second basis for cynicism is the fact that speculation is not *evidence based* and hence may not represent strong inference (Platt 1964). A third basis is innate conservatism and inertia; it is easier to continue doing what we have been doing rather than to change because of something that might—but might not—happen. Nonetheless, the benefit of strategically identifying and addressing imminent, yet perhaps unrecognized, issues could be immeasurable to both current and future management of wildlife and wildlife habitats (Sutherland et al. 2008). Horizon scanning is the process of identifying how a system may be at risk in the near or intermediate future relative to a suite of potential issues (disturbance, succession, climate variation, etc.) and prioritizing the issues, whether or not they currently occur, that deserve increased attention through research, monitoring, modeling, or evaluation (Sutherland et al. 2008). Regular application of horizon scanning would allow for identification of issues currently affecting an ecological system but also allow consideration of future conditions, not yet occurring, that may also impact the system. Regular use by wildlife ecologists would allow for both surveillance and hypothesis-driven monitoring strategies to be applied or implemented. Additionally, both species- and

ecosystem-level concerns could be elevated or lowered proactively as increased data is made available or as conditions change over time.

What's in the Future?

So, how do wildlife scientists and managers become more proactive? How should they respond to anticipated or predicted changes in the environment when often we can only speculate as to the causal relationships or appropriate response metrics? Despite the need, truly proactive wildlife management activities are rare; wildlife managers are better categorized as first responders than as actual managers, as they have very little control or even the ability to predict major events that will influence wildlife populations at large scales. This is regretful as we spend an inordinate amount of time trying to use models to predict what is going to happen after we respond to an issue; we perhaps do not spend enough time trying to construct and drive what should happen. Regardless, in some cases, researchers have had the foresight to synthesize previous research to identify potential overarching issues based on past patterns, such as the reduction of the ozone atmospheric layer due to human chemical use (Crutzen and Ehhalt 1977) or the plight of grassland birds in the United States (Droege and Sauer 1994; Knopf 1994). Consider two examples of the difficulties of proactive responses to management issues: the Conservation Reserve Program and the Deepwater Horizon oil spill, both of which have had impacts on wildlife and wildlife habitats.

The Conservation Reserve Program

First, the Conservation Reserve Program (CRP) nicely illustrates how the wildlife community responded fairly effectively to a major change in the landscape. CRP was a component of the 1985 Food Security Act (also known as the farm bill) that allowed owners of highly erodible cropland to plant it to grasses and forbs, leave it idle except for necessary maintenance, and receive annual payments for the period of the contract, usually ten or fifteen years. At that time, the United States had surpluses of major crops, and the Department of Agriculture was seeking a way to reduce those surpluses. The primary purpose of the CRP was to achieve that goal and thereby maintain adequately high prices for commodities. By focusing the program on highly erodible lands, the program also addressed its secondary objective of reducing soil erosion from wind and water. A third objective, seemingly almost an afterthought, was to restore habitat for fish and wildlife. The program was immensely popular with producers; the area enrolled in CRP climbed from 1,929,064 acres in 1986 to 32,522,380 acres in 1990 and remained above 30 million acres until 2012, when it dipped slightly (http://www.fsa.usda.gov/Internet/FSA_File /historystate8612.xls, accessed March 6, 2013).

Much of the area enrolled in CRP was in the Great Plains and Midwest, and the extensive conversion of cropland to perennial herbaceous cover resulted in a massive habitat alteration at landscape levels. Expectations within the wildlife community were that these grasslands would support far more birds and other wildlife than did the croplands they replaced (reviewed by Higgins et al. 1988). Wildlife researchers responded to this landscape change by initiating a number of studies to determine how wildlife, especially birds, responded. Studies of waterfowl nesting in CRP fields began in 1989 in the Dakotas and Minnesota (Kantrud 1993; Luttschwager et al. 1994), and CRP turned out to be a boon for nesting waterfowl, adding about two million ducks per year to the fall flight. Studies on grassland birds were initiated in 1990 in the northern prairies (Johnson and Schwartz 1993) and in several Midwestern states in 1991 (Best et al. 1997). To date, numerous grassland species have benefited from the conversion of cropland to perennial cover (e.g., Johnson and Igl 1995; Johnson 2005).

Because of the attention focused on wildlife benefits of CRP, wildlife received greater priority in subsequent farm bills. Modifications based on the obvious benefits were made in amendments to the farm bill in 1990 and 1995 to enhance consideration of wildlife (Heard 2000). In particular, environmental benefits were given consideration equal to soil and water conservation. A new practice was adopted by the Farm Service Agency specifically to benefit nesting waterfowl; the Duck Nesting Habitat Initiative prioritized lands in areas with abundant waterfowl and allowed greater areas of upland to be enrolled (http://www .ducksunlimited.org/conservation/habitat/new-crp -practice-emphasizes-ducks, accessed March 21, 2013).

Although the program's future is in doubt, as enrollment is expected to plummet with the currently high commodity prices, for decades CRP did provide habitat for an abundance of game and nongame wildlife, in part because wildlife scientists were able to identify its promise early, gather supportive information, and make informed recommendations to policy makers.

The Deepwater Horizon Oil Spill

A recent large oil spill, the Deepwater Horizon offshore drilling rig in the Gulf of Mexico, offers a compelling example of seriously deficient responses to events with huge potential consequences to wildlife populations and habitats. Although the event resulted from human error and could have been avoided, it would be wildly optimistic to assume that similar events would never occur. Here, we focus on the assessment of ecological damage caused by the Deepwater Horizon blowout. President Barack Obama formed the National Commission on the BP Deepwater Horizon Oil Spill and Offshore Drilling to assess the overall response to the event; the following information was derived from the Commission's report (National Commission 2011).

The Clean Water Act mandates the development of a National Contingency Plan, which prescribes the nationwide response structure for oil spills (265–66). Regional Response Teams, co-chaired by the US Coast Guard and EPA, include representatives from state and other federal agencies and are tasked with developing regional contingency plans as well as preauthorization protocols for certain response strategies. Area committees, which develop area contingency plans, similarly include federal and state representatives but are led by the Coast Guard. The area contingency plans are the most specific and relied-upon during the response to a spill. In addition, the industry had developed its own spill response plans (266). The Commission noted that industry plans were sent only to the Minerals Management Service, where few of them received environmental review. BP's response plan was "embarrassing" (133). It listed as a wildlife expert on whom BP would rely an individual who had died several years *before* BP submitted its plan. Among species of concern in case of a Gulf oil spill, the plan listed seals and walruses, which never occur in the Gulf.

Because of poor planning before the spill, the response to it was chaotic. Disagreement among federal agencies, between federal and state agencies, and between industry and public agencies contributed to this problem, as did political posturing and bureaucratic inertia. Further, independent scientific research related to the spill was hindered by the lack of timely access to the site and delayed funding (174). Specifically relevant to the topic of this chapter is the lack of "pretreatment" data. Regarding marine mammals, Tim Ragen, executive director of the federal Marine Mammal Commission, testified before a House of Representatives subcommittee and noted, "Unfortunately, the scientific foundation for evaluating the potential effects of the Deepwater Horizon spill on many marine mammals inhabiting the Gulf is weak" (181). The Commission also stated:

- "Unfortunately, comprehensive data on conditions before the spill—the natural 'status quo ante' from the shoreline to the deepwater Gulf—were generally lacking" (174).
- "A typical damage assessment can take years. Two sets of determinations—one concerning the baseline conditions against which damages to each species or habitat will be assessed and another concerning the quantification of those damages—are particularly difficult and consequential in terms of the overall results" (183–84).
- "Without well-established baseline conditions, there can be inaccurate quantification of damages or required restoration. Given that the ecological baseline can vary seasonally, annually, and over much longer time scales, it can be difficult to pinpoint the exact condition of an ecosystem prior to a spill. Because long-term historical data are often nonexistent or discontinuous, natural resource trustees are likely to be disadvantaged by a lack of sufficient information to fully characterize the condition of relevant ecosystems prior to the incident in question" (184).
- "[F]unding for academic and other scientists in the days and weeks immediately after the spill was limited. As a result, the nation lost a fleeting opportunity to maximize scientific understanding of how oil spills—particularly in the deep ocean—adversely affect individual organisms and the marine ecosystem. Such research depends on sampling, measurements, and investigations that can be accomplished only during and right after the spill" (184).

The bottom line is that while this specific incident could not have been predicted, it is reasonable to assume that some major oil spill would occur at some time in the future. With that anticipation, it would have been prudent to first develop a systematic monitoring program to establish an understanding of the condition of wildlife (and other ecological components) under natural (pre-"treatment") conditions and, second, develop a plan to rapidly implement additional research after an incident where effects from the incident are likely and at comparison sites unaffected by the incident. Such a plan should have a direct and short chain of command so that it can be effectively implemented on very short notice.

Conclusion

Unlike many of the so-called "hard sciences," wildlife ecology rarely can address major questions with designed experiments. The discipline largely relies on observational studies with severely limited opportunities for control, randomization, and replication (Shaffer and Johnson 2008). Accordingly, inferring causation is a much more daunting task. Moreover, major ecological treatments are rarely applied by scientists. They may be natural events, such as a hurricane or earthquake. They may be natural events with effects that were exacerbated by human actions, such as flooding. They may be purely results of human error, such as oil spills or wildfires. Or they may be consequences of actions taken largely for other purposes, such as the farm bill or reservoir dam building. The question is: "How can we capitalize on such events to learn about the affected system so that we may better predict, respond to, or mitigate similar events in the future?" Three scenarios are proposed depending on whether events can be predicted, are partly predictable, or are unpredictable.

A predictable event is one for which the action as well as its location and timing are known. Such events could include the construction or removal of a dam on a river, restoration of a particular plot of land, or predator removal. These situations permit the development of a solid experimental design, including a before-after, control-impact (BACI) design if certain variables are specified externally (e.g., the location of the event). These situations are conceptually straightforward, but details can often be messy. A partly predictable event is one that is "bound to occur" sometime, someplace, but the time and place are not known beforehand. As an example, an individual oil tanker may have only a 0.0001 probability of a major accident occurring while traversing a strait, but if enough ships pass through it, sooner or later one ship is "bound to" have a mishap. In this example, the approximate site of an event is known, but the timing is not. In anticipation of the event, it is feasible to gather appropriate "pretreatment" data in order to compare with measurements taken after the event. Additionally, of course, plans should be developed to respond to the event and manage the consequences as best able. Unpredictable events are, of course, the least amenable to planning. Here is where "surveillance monitoring" makes its mark. Such monitoring will provide the earliest warning possible about potentially calamitous events.

LITERATURE CITED

Araujo, M. B., and C. Rahbek. 2006. "How Does Climate Change Affect Biodiversity?" *Science* 313:1396–97.

Baccante, D. 2012. "Hydraulic Fracturing: A Fisheries Biologist Perspective." *Fisheries* 37:40–41.

Beck, J. L., and L. H. Suring. 2009. "Wildlife-Habitat Relationships Models: Description and Evaluation of Existing Frameworks." In *Models for Planning Wildlife Conservation in Large Landscapes*, edited by J. J. Millspaugh and F. R. Thompson III, 251–85. Academic Press, Burlington, MA, USA.

Best, L. B., H. Campa III, K. E. Kemp, R. J. Robel, M. R. Ryan, J. A. Savidge, H. P. Weeks Jr., and S. R. Winterstein. 1997. "Bird Abundance and Nesting in CRP Fields and Cropland in the Midwest: A Regional Approach." *Wildlife Society Bulletin* 25:864–77.

Beutel, T. S., R. J. S. Beeton, and G. S. Baxter. 2006. "Building Better Wildlife-Habitat Models." *Ecography* 22:219–23.

Boyce, M. S. 1992. "Population Viability Analysis." *Annual Review of Ecology and Systematics* 23:481–506.

Bunnell, F. L. 1989. "Alchemy and Uncertainty: What Good Are Models?" USDA Forest Service General Technical Report. PNW-GTR-232. United States Department of Agriculture, Forest Service, Washington D.C., USA.

Copeland, H. E., K. E. Doherty, D. E. Naugle, A. Pocewicz, and J. M. Kiesecker. 2009. "Mapping Oil and Gas Development Potential in the US Intermountain West and Estimating Impacts to Species." *PLOS One* 4:e7400.

Crutzen, P. J., and D. H. Ehhalt. 1977. "Effects of Nitrogen Fertilizers and Combustion on the Stratospheric Ozone Layer." *Ambio* 6:112–17.

DeAngelis, D. L., and W. M. Mooij. 2005. "Individual-Based Modeling of Ecological and Evolutionary Processes." *Annual Review of Ecology and Systematics* 36:147–68.

Department of Energy. 2013. "Monthly Energy Review, March 2013." US Energy Information Administration, Office of Energy Statistics, US Department of Energy, Washington, D.C. http://www.eia.gov/totalenergy/data/monthly/pdf/mer.pdf (accessed April 4, 2013).

Droege, S., and J. R. Sauer. 1994. "Are More North American Species Decreasing than Increasing?" In *Bird Numbers, 1992: Distribution, Monitoring and Ecological Aspects*, edited by J. M. Hahemeijer and T. J. Verstrael, 297–306. Statistic Netherlands, Voorsburg/Heerlen.

Ehrlich, P. 1968. *The Population Bomb*. Ballantine Books, New York.

Elith, J., C. H. Graham, R. P. Anderson, M. Dudik, S. Ferrier, A. Guisan, R. J. Hijmans, F. Huettmann, J. R. Leathwick, A. Lehmann, J. Li, L. G. Lohmann, B. A. Loiselle, G. Manion, C. Moritz, M. Nakamura, Y. Nakazawa, J. McC. Overton, A. T. Peterson, S. J. Phillips, K. Richardson, R. Scachette-Pereira, R. E. Schapire, J. Soberon, S. Williams, M. S. Wisz, and E. Zimmermann. 2006. "Novel Methods Improve Prediction of Species' Distributions from Occurrence Data." *Ecography* 29:129–51.

Entrekin, S., M. Evans-White, B. Johnson, and E. Hagenbuch. 2011. "Rapid Expansion of Natural Gas Development Poses a Threat to Surface Waters." *Frontiers in Ecology and the Environment* 9:503–11.

Hanski, I. 1998. "Metapopulation Dynamics." *Nature* 396: 41–49.

Heard, L. P. 2000. "Introduction." In *A Comprehensive Review of Farm Bill Contributions to Wildlife Conservation, 1985–2000*, edited by W. L. Hohman and D. J. Halloum, 1–4. US Department of Agriculture, Natural Resources Conservation Service, Wildlife Habitat Management Institute, Technical Report, USDA/NRCS/WHMI-2000.

Higgins, K. F., D. E. Nomsen, and W. A. Wentz. 1988. "The Role of the Conservation Reserve Program in Relation to Wildlife Enhancement, Wetlands, and Adjacent Habitats in the Northern Great Plains." In *Impacts of the Conservation Reserve Program in the Great Plains*, edited by J. E. Mitchell, 99–104. USDA Forest Service General Technical Report RM-158. Rocky Mountain Forest and Range Experiment Station, Fort Collins, CO.

Ingelfinger, F., and S. Anderson. 2004. "Passerine Responses to Roads Associated with Natural Gas Extraction in a Sagebrush Steppe Habitat." *Western North American Naturalist* 64:385–95.

Johnson, D. H. 1981. "The Use and Misuse of Statistics in Wildlife Habitat Studies." In *The Use of Multivariate Statistics in Studies of Wildlife Habitat*, edited by D. E. Capen, 11–19. US Forest Service General Technical Report RM-87.

———. 2001. "Validating and Evaluating Models." In *Modeling in Natural Resource Management: Development, Interpretation, and Application*, edited by T. M. Shenk and A. B. Franklin, 105–19. Island Press, Washington, D.C.

———. 2005. "Grassland Bird Use of Conservation Reserve Program Fields in the Great Plains." In *Fish and Wildlife Benefits of Farm Bill Conservation Programs: 2000–2005 Update. The Wildlife Society Technical Review 05–2*, edited by J. B. Haufler, 17–32. The Wildlife Society, Bethesda, Maryland.

Johnson, D. H., and L. D. Igl. 1995. "Contributions of the Conservation Reserve Program to Populations of Breeding Birds in North Dakota." *Wilson Bulletin* 107:709–18.

Johnson, D. H., and M. D. Schwartz. 1993. "The Conservation Reserve Program and Grassland Birds." *Conservation Biology* 7:934–37.

Kantrud, H. A. 1993. "Duck Nest Success on Conservation Reserve Program Land in the Prairie Pothole Region." *Journal of Soil and Water Conservation* 48:238–42.

Knopf, F. L. 1994. "Avian Assemblages on Altered Grasslands." *Studies in Avian Biology* 15:247–57.

Kuvlesky Jr., W. P., L. A. Brennan, M. L. Morrison, K. K. Boydston, B. M. Ballard, and F. C. Bryant. "Wind Energy Development and Wildlife Conservation: Challenges and Opportunities." *Journal of Wildlife Management* 71:2487–98.

Lange, H. J. D., J. Lahr, J. J. C. Van der Pol, Y. Wessels, and J. H. Faber. 2009. "Ecological Vulnerability in Wildlife: An expert Judgment and Multicriteria Analysis Tool Using Ecological Traits to Assess Relative Impact of Pollutants." *Environmental Toxicology and Chemistry* 28: 2233–40.

Luttschwager, K. S., K. F. Higgins, and J. A. Jenks. 1994. "Effects of Emergency Haying on Duck Nesting in Conservation Reserve Program Fields." *Wildlife Society Bulletin* 22:403–8.

McCann, R. K., B. G Marcot, and R. Ellis. 2006. "Bayesian Belief Networks: Applications in Ecology and Natural Resource Management." *Canadian Journal of Forest Research* 36:3053–62.

Millspaugh, J. J., and F. R. Thompson III. 2009. *Models for Planning Wildlife Conservation in Large Landscapes*. Academic Press, Burlington, MA, USA.

Minsky, M. 1968. "Matter, Mind and Models." In *Semantic Information Processing*, edited by M. Minsky, 425–32. MIT Press, Cambridge, MA, USA.

Morrison, M. L., B. G. Marcot, and R.W. Mannan. 2006. *Wildlife Habitat Relationships: Concepts and Applications*. University of Wisconsin Press, Madison, WI, USA.

Müller, F., B. Breckling, F. Jopp, and H. Reuter. 2011. "What Are the General Conditions Under which Ecological Models Can Be Applied?" In *Modeling Complex Ecological Dynamics*, edited by F. Jopp et al., 13–28. Springer-Verlag, Berlin.

National Commission on the BP Deepwater Horizon Oil Spill and Offshore Drilling (National Commission). 2011. "Deep Water: The Gulf Oil Disaster and the Future of Offshore Drilling." http://cybercemetery.unt.edu/archive/oilspill/20121210200431/http://www.oilspillcommission.gov/final-report (accessed September 18, 2013).

Naugle, D. E. 2011. *Energy Development and Wildlife Conservation in Western North America*. Island Press, Washington D.C., USA.

Nichols, J. D., M. C. Runge, F. A. Johnson, and B. K. Williams. 2007. "Adaptive Harvest Management of North American Waterfowl Populations: A Brief History and Future Prospects." *Journal of Ornithology* 148 (Supplement 2): S343–49.

Nichols, J. D., and B. K. Williams. 2006. "Monitoring for Conservation." *Trends in Ecology and Evolution* 21:668–73.

Platt, J. R. 1964. "Strong Inference." *Science* 146:347–53.

Sawyer, H., M. J. Kauffman, and R. M. Nielson. 2006. "Winter Habitat Selection of Mule Deer Before and during Development of a Natural Gas Field." *Journal of Wildlife Management* 70:396–403.

———. 2009. "Influence of Well Pad Activity on Winter Habitat Selection Patterns of Mule Deer." *Journal of Wildlife Management* 73:1052–61.

Shaffer, T. L., and D. H. Johnson. 2008. "Ways of Learning: Observational Studies versus Experiments." *Journal of Wildlife Management* 72:4–13.

Shenk, T. M., and A. B. Franklin. 2001. "Models in Natural Resource Management: An Introduction." In *Modeling in Natural Resource Management: Development, Interpretation, and Application*, edited by T. M. Shenk and A. B. Franklin, 1–8. Island Press, Washington, D.C.

Sutherland, W. J., R. Aveling, L. Bennum, E. Chapman, M. Clout, I. M. Côte, M. H. Depledge, L. V. Dicks, A. P. Dobson, L. Fellman, E. Fleishman, D. W. Gibbons, B. Keim, F. Lickorish, D. B. Lindenmayer, K. A. Monk, K. Norris, L. S. Peck, S. V. Prior, J. P. W. Scharlemann, M. Spalding, A. R. Watkinson. 2011. "A Horizon Scan of Global Conservation Issues for 2012." *Trends in Ecology and Evolution* 27:12–18.

Sutherland, W. J., M. J. Bailey, I. P. Bainbridge, T. Breeton, J. T. A. Dick, J. Drewitt, N. K. Dulvy, N. R. Dusic, R. P. Freckleton, K. J. Gaston, P. M. Gilder, R. E. Green, A. L. Heathwaite, S. M. Johnson, D. W. Macdonald, R. Mitchell, D. Osborn, R. P. Owen, J. Pretty, S. V. Prior, H. Prosser, A. S. Pullin, P. Rose, A. Stott, T. Tew, C. D. Thomas, D. B. A. Thompson, J. A. Vickery, M. Walker, C. Walmsley, S. Warrington, A. R. Watkinson, R. J. Williams, R. Woodroffe, and H. J. Woodroof. 2008. "Future Novel Threats and Opportunities Facing UK Biodiversity Identified by Horizon Scanning." *Journal of Applied Ecology* 45: 821–33.

Traill, L. W., C. J. A. Bradshaw, and B. W. Brook. 2007. "Minimum Viable Population Size: A Meta-analysis of 30 Years of Published Estimates." *Biological Conservation* 139:159–66.

Trombulak, S. C., and C. A. Frissell. 2000. "Review of Ecological Effects of Roads on Terrestrial and Aquatic Communities." *Conservation Biology* 14:18–30.

US Fish and Wildlife Service. 1981. "Standard for the Development of Habitat Suitability Index Models." 103 ESM, US Fish and Wildlife Service, Division of Ecological Services, Washington, D.C.

Walker, B. L., D. E. Naugle, K. E. Doherty. 2007. "Greater Sage-Grouse Population Response to Energy Development and Habitat Loss." *Journal of Wildlife Management* 71:2644–54.

Williams, B. K., J. D. Nichols, and M. J. Conroy. 2002. *Analysis and Management of Animal Populations*. Academic Press, San Diego, CA, USA.

Yoccuz, N. G., J. D. Nichols, and T. Boulinier. 2001. "Monitoring of Biological Diversity in Space and Time." *Trends in Ecology and Evolution* 16: 446–53.

— **10** —

Michael K. Schwartz,
Jamie S. Sanderlin, and
William M. Block

Manage Habitat, Monitor Species

Monitoring is the collection of data over time. We monitor many things: temperatures at local weather stations, daily changes in sea level along the coastline, annual prevalence of specific diseases, sunspot cycles, unemployment rates, inflation, commodity futures—the list is virtually endless. In wildlife biology, we also conduct a lot of monitoring, most commonly measuring state variables that either directly or indirectly relate to the size and areal extent of wildlife populations. Most of these activities, both in wildlife biology and more generally, are not strongly targeted at answering specific questions. Rather, they reflect measurements of things we consider to be of general interest, but which, as data streams temporally extend, often produce a variety of novel insights. For instance, no one could have predicted the benefits of long-term, fixed local weather stations or atmospheric CO_2 measurements in evaluating climate change, a purpose for which neither data stream was initially intended (Lovett et al. 2007; Harris 2010).

The general value of monitoring is not debated: without a time series of previously collected data we are limited in our ability to evaluate current conditions or make projections into the future. However, the collection and maintenance of data across extensive time and space that characterize an effective monitoring program are both expensive and difficult. Maintaining both resources and infrastructure to regularly collect and archive information with established methods and quality controls is challenging. Thus, it is reasonable to ask not whether these data are useful (they invariably are) but rather how useful they are when compared with the associated opportunity costs and whether monitoring data could be collected more efficiently.

Choosing exactly what to monitor in wildlife biology is no trivial matter. For example, should we monitor populations, habitats, or both? If populations are monitored, what is the appropriate state variable(s) to measure: abundance, density, presence, age-class structure, genetic variation or structure, survival, reproduction, movement? At what scale and resolution should managers monitor these variables? Can we use surrogate measures for monitoring variables of interest that are either impossible or extremely expensive to obtain on their own?

In this review, we focus on recent thinking concerning the role of monitoring in wildlife science, choice of state variables, and the degree to which monitoring can and should be designed and targeted to answer specific questions of management significance. Lastly, we discuss the paradigm of linking monitoring to risk assessment, decision analysis, and adaptive management.

Why Monitor?

We start this review by asking a simple question: why monitor at all? Although this may seem like a simple question to address, it is not. First, monitoring must be differentiated from assessments and research. We define monitoring as the measuring of a parameter of management interest (e.g., abundance of a species), or a surrogate of this parameter, over time (Schwartz et al. 2007). The critical part of this definition is the temporal element. Without considering time, we are not monitoring but rather are conducting an assessment. In essence by our definition, when multiple assessments are combined it becomes a monitoring effort. Assessments can be very valuable in setting management objectives. In fact, an assessment is one of the first steps for initiating a monitoring program. Assessments can be particularly useful if we ask if a state variable of interest is meeting a certain threshold. For example, instead of asking if estimates of elk (*Cervus canadensis*) abundance (abundance being the state variable of interest) are changing over time, we can ask: do we have a minimum of number of elk in a particular area after a given management action? The latter question is much simpler and less expensive to accomplish. We may even tie these assessments to triggers to make the situation more powerful. For instance, we could state (a priori to the assessment) that if we do not have a minimum number of elk in a particular area, post a management action, then we will take particular and precise follow-up actions. These assessments often are not representative measures of the state variable, nor do they need to be. For example, surveys can concentrate on those areas and populations most likely to change or where detecting change is of the greatest interest. The point here is that although monitoring is very useful, it can be expensive and may not be needed in all circumstances. An assessment can help us decide if monitoring is needed, provide some baseline data for the monitoring effort, and can help us choose which variables may be most precise and powerful for monitoring.

Once we decide that we need to accomplish more than a simple assessment and instead need trend, we are in the realm of monitoring. If we are interested in how the abundance of elk changes over time, we need to consider a suite of technical questions including those relating to statistical power, such as the length of time that we need to monitor a species before we would reasonably expect to see an increase or decline, or how we plan to estimate abundance given budget constraints. Yet, before we consider technical details of a monitoring effort, we first need to clearly define our study objectives and ensure that the product of our monitoring will satisfy our original goals. That is, monitoring abundance of elk over a ten-year time horizon and having 80% certainty that the population has declined by greater than 25% still does not provide any

explanation of the mechanism(s) behind the decline. If our initial goal was to identify what causes elk declines in an area, this monitoring effort alone, even if successful in detecting a decline, will have failed as our effort didn't collect data in a framework that could determine why the decline occurred (see adaptive management in the following discussion). Thus, monitoring may be necessary to track trends in wildlife populations of interest, but it may not be sufficient for us to have the information necessary to change a population trajectory (e.g., a decline).

We could argue that all "important" species should be monitored, and if we see declines, we could subsequently address the cause. However, this is much like the monitoring approach that the US Forest Service (USFS), one of the world's largest land owners, adopted from their 1982 planning regulations. Some of the biggest problems associated with the monitoring requirements from the 1982 planning regulations were that monitoring a large number of species was too expensive and thus often not conducted, leading to litigation. Furthermore, these monitoring efforts were seldom followed by any attempt to understand mechanisms behind the changes that were observed. Lastly, the USFS often relied on monitoring vegetation (habitat) as a proxy for presence or abundance of wildlife. Overall, this approach was untenable and was replaced with a new planning rule as of 2012 (see the following discussion), where monitoring is conducted at broad scales, focused on ecological conditions, and used only sparingly on focal species.

Should Habitat Surrogates Be Used?

Monitoring some state variables can be very expensive. One potential alternative to monitoring the state variable of interest directly is monitoring a surrogate. There is much debate in the literature on the appropriateness of using surrogates for monitoring (Carignan and Villard 2002). However, we need to differentiate and clearly define what we mean by surrogates. Surrogates are meant to indicate a variable of interest but are either less expensive or less complicated to measure. In science, surrogates should have a strong statistical link or causal connection to the state variable of interest (e.g., as one variable increases, the other does by a predictable amount). Many surrogates are used for monitoring, with different levels of success, depending on what is being monitored. For example, there is evidence that species accumulation indices, an index relating the percentage of an area surveyed to the percentage of the target species represented scaled by random and maximum possible curves, can act as weak indicators of biodiversity (Rodrigues and Brooks 2007). In wildlife management, we are often interested in surrogate measures for abundance, although there are other state variables that may be more meaningful than abundance.

Unfortunately, too often the surrogate for abundance that is used in practice is vegetation or habitat (Hunter et al. 1988). While the idea behind using habitat or vegetation communities as a coarse filter is very tempting, and has been used extensively in the conservation planning literature, the justification behind it is extremely weak (Oster et al. 2008; Seddon and Leech 2008; Cushman et al. 2010; Lindenmayer and Likens 2011). For example, Cushman et al. (2008, 2010) examined abundance patterns of birds in the Oregon Coast Range and found little support for the idea that vegetation measures could predict abundance patterns at both a plot and subbasin level. They noted that for this concept to work, several criteria must be met. The first is that habitat is a proxy or surrogate for population abundance. This requires that abundance of a species is tightly linked with the current and future environmental conditions such that changes in the environment are predictably reflected in abundance (Cushman et al. 2008). Second, vegetation maps at multiple scales must be reliable as a surrogate for habitat. One problem is that even if there are high correlations between vegetation and habitat, and between habitat and abundance, the compounding of relationships makes the use of vegetation maps a poor proxy of abundance. Furthermore, many wildlife-habitat models that relate species occurrence to broad vegetation types suffer from large rates of commission and omission errors (Block et al. 1994). This is not to say that there are not cases where there are tight statistical links between vegetation and abundance, but this may be the exception, not the rule.

Thus, unless a clear statistical or causal relationship between a surrogate measure such as habitat and species abundance, occurrence, or another state variable is established, one should not assume it exists (Lindenmayer and Likens 2011). Based on theoretical

and empirical concerns, we recommend that if one decides to monitor, it needs to be done on focal species, unless adequate data has been collected to determine the causal or statistical relationship between a surrogate and population metric of interest. Thus, we recommend that agencies should manage habitat and monitor species in order to meet wildlife monitoring mandates.

Adaptive Management

Once a decision has been made to monitor a focal species, there is still the issue that monitoring only provides information about trend or trajectory but does not provide any mechanism behind the trend (Nichols and Williams 2006). In recognition of this fact, the study of mechanism behind change in population trajectory has been combined with monitoring to create an "adaptive management" framework (Walters 1986; Johnson et al. 1993; Williams 1996; Block et al. 2001; Yoccoz et al. 2001; Westgate et al. 2013). Adaptive management is the structured, iterative process of making a management decision when there is considerable uncertainty. The adaptive cycle relies on monitoring the system after a management action by collecting additional data and revising models to reduce uncertainty and ultimately inform future decisions. Ideally, studies are structured as experiments to assess cause and effect; in reality, studies are typically pseudoexperiments or observational (Block et al. 2001). In essence, systematic, monitoring data are collected in a hypothesis framework (Nichols and Williams 2006).

Resource managers are often quick to assert that they practice adaptive management. In a general sense, adaptive management is a structured process to learn from what was done in the past. The key here is that the process is structured with a series of steps and feedback loops that facilitate learning. Moir and Block (2001) provided a conceptual model for adaptive management on public lands. It consisted of a seven-step process with the steps linked by a flow of information and feedback loops to inform whether or not management actions were meeting their intent. Information provided in feedback loops is largely the result of monitoring, and without monitoring, adaptive management fails (Biber 2011). These feedback loops can evaluate whether: (1) the monitoring design is effective and the correct state variable(s) are being measured; (2) monitoring is being done as designed; and (3) whether or not the management action is achieving its goal. Clearly, adaptive management hinges on successful monitoring. The failure to conduct valid monitoring is considered the "Achilles' heel" to adaptive management (Moir and Block 2001).

While the adaptive management approach can work well for species in which data are relatively easy to collect and abundant, such as game species, many monitoring efforts are focused on rare, endangered, and sensitive species. Threatened, endangered, and sensitive species often occur in low numbers and can be secretive and difficult to sample (Neel et al. 2013). Data will inherently be limited compared to more abundant game species. Thus, the adaptive management framework is likely to fail for rare species where experiments are unethical and data is sparse, unless adequate sampling effort, sometimes approaching a census, is allotted.

Articulating the Monitoring Goal

One failure of many monitoring programs is the lack of articulation of clear and measurable objectives. Clearly, one of the advantages of the adaptive management framework is that objectives must be stated if data are to inform competing management hypotheses of how the system functions (Nichols and Williams 2006). However, adaptive management also has been criticized for not articulating and including triggers to change management actions, leading to a lack of accountability (Nie and Schultz 2012). Outside of the adaptive management framework, there are hundreds of ongoing monitoring efforts that occur on public lands that have no clearly stated objectives or thresholds and are not tied back into decision-making processes.

The first monitoring goal that needs to be clearly stated is the purpose of the monitoring effort. Are the data being collected for general context monitoring (*sensu* Holthausen et al. 2005), which is defined as monitoring a broad array of ecosystem components at multiple scales? Breeding Bird Survey data (e.g., Boulinier et al. 1998) are an example of these types of data. Inherent in context monitoring is the idea that monitoring will detect emerging threats to wildlife populations and unanticipated changes in species or multispecies metrics (e.g., abundance, distribution, species richness, changes in local extinction and local colonization). At the other end of the spectrum from context monitoring is targeted monitoring. This is the monitoring of how a particular species is responding to a management action or general threat (Holthausen et al. 2005). Targeted monitoring also can be the monitoring of sensitive, threatened, or endangered species that may not be suffering from ongoing management actions but are threatened by small population processes (Caughley 1994). Context and targeted monitoring have very different objectives and costs, and likely involve collection of different types of data. Therefore, the first step in designing any monitoring program will be stating this broad objective.

For the remainder of this review, we will largely focus on targeted monitoring. We do not mean to imply that large, broad-scale context monitoring efforts have been unsuccessful or are not useful. Clearly, efforts like the Breeding Bird Survey, which relies on volunteers across the United States to observe birds during the breeding season, has led to many useful insights into changes in bird distributions and abundances (Link and Sauer 1998; Sauer et al. 2003; Sauer and Link 2011; Stegen et al. 2013). Whereas we see the value in context monitoring and advocate the need for national-level monitoring programs to accomplish this, success requires well-coordinated, well-funded, broad-scale data collection on many species or assemblages. It therefore represents the minority of monitoring efforts designed each year worldwide (e.g., Greenwood 2003; Kéry and Schmid 2006; Kéry and Plattner 2007; Biodiversity Monitoring Switzerland [http://www.biodiversitymonitoring.ch]). Thus, in the subsequent sections, we discuss issues of targeted monitoring related to metrics to monitor, sampling area, scale, cost, power, effect size, and duration.

Metrics to Measure and Monitor

An investigator must first decide *what* it is that he or she wants to measure (see table 10.1 adapted from Holthausen et al. 2005). As noted earlier, the ultimate piece of information sought is how management affects the population(s) under consideration. The most common metric to monitor is abundance. Yet, abundance may not be the most biologically meaningful metric

Table 10.1 Population and community measures that may be employed in a monitoring program. This table is adapted from Holthausen et al.'s (2005) table 4.

Population Measures	
Abundance/density	Abundance and density are the most commonly used monitoring metrics. Abundance estimates require capturing, marking, and recapturing individuals multiple times. Note that individuals do not need to be physically captured (e.g., noninvasive DNA capture-recapture) to provide an estimate with reasonable variance. Because these measures are often expensive, indices of abundance are frequently used. Density is abundance per area, although the estimation of the area being considered is not always clear. Abundance change over a unit of time (λ, finite rate of increase) also can be monitored.
Occupancy	Occupancy assesses presence of an organism at sample locations. Showing presence at a site necessitates only proof of the species' existence. Demonstrating absence is more complex and requires information on the probability of detecting the organism, given it is present. These data are often placed in an occupancy framework where multiple site visits at multiple stations can be used to infer changes in occupancy, which can be linked to abundance (Royle and Nichols 2003) or monitored as its own metric (MacKenzie et al. 2006). Occupancy models (MacKenzie et al. 2006) also can be used to detect changes in local colonization and local extinction of a species.
Vital rates (birth/death; immigration/emigration)	Vital rates are age-specific birth and death rates, and emigration and immigration rates. Monitoring vital rates sometimes provides a better measure of trend than measures of abundance. Collecting vital rate data allows the use of population viability analyses, which can estimate probability of persistence and allow modeling of the impact of management actions on persistence.
Range distribution measures	Geographic range is the range of a given species at a particular point in time. Monitoring range over time provides information on if a species is expanding, contracting, or remaining constant. Groups like the IUCN often use range distribution to monitor change in species status (Mace et al. 2008).
Genetic measures	Genetic monitoring has been defined as quantifying temporal changes in population genetic metrics or other population data generated using molecular markers. In general, three categories of genetic monitoring are recognized. Category I is simply the use of diagnostic molecular markers for traditional monitoring such as estimating changes in population abundance, vital rates, hybridization, or geographic range. Category II used the population genetic data itself to monitor changes in genetic variation, effective population size, dispersal, or population substructure. Category III addresses questions of evolutionary potential, such as change in frequencies of adaptive genes (see Hansen et al. 2012).

Community Measures	
Diversity measures	Species diversity measures used in community and ecosystem ecology include species richness and evenness. Species richness can be based on repeated estimates of the presence or absence of taxa, whereas species evenness requires abundance data. With recent advances in analytical techniques (Dorazio et al. 2006; Royle and Kéry 2007; Kéry and Royle 2008), we can estimate species richness and community dynamics from local colonization and local extinction (e.g., Sauer et al. 2013) while accounting for detection.
Integrity measures	Biological integrity is a comparative approach that examines if conditions (which can be measured using other approaches mentioned previously) are similar today to historical times or to other locations that are considered "intact."

to choose (Van Horne 1983) because it only provides one piece of the puzzle, and other state variables may be the driving factors for population change. For example, imagine a stream that has two adult fish in the early spring that produces between five hundred to one thousand fry in the summer, which are subject to heavy mortality. Imagine this dynamic continuing for a decade. An estimate of abundance in the fall would be less useful than a measure of connectivity or effective population size. Even if abundance is the desired goal, as it is easily articulated and fits well into management models (Fancy et al. 2009), estimates of abundance may not be obtainable, especially across an entire geographic range. In lieu of collecting data to estimate population abundance, investigators consider various proxy measures (table 10.1) and often consider these at small spatial scales, sometimes distributed across a species geographic range.

It is important here to note that while we did not advocate the use of habitat as a surrogate measure, there are often many surrogates that can be used for formal estimates of abundance (Slade and Blair 2000; McKelvey and Pearson 2001; Tallmon et al. 2010; Tallmon et al. 2012; Noon et al. 2012). The key difference is that modeling efforts that have tested the use of these surrogates for abundance often find them to be good predictors of abundance and, in some cases, more robust for detecting trends than estimates of abundance. In other words, the causal connection has been studied and validated. For example, Tallmon et al. (2010) showed that declines in effective population size can be a more sensitive measure of declining populations than monitoring estimates of abundance under many conditions. McKelvey and Pearson (2001) demonstrated that M_{t+1} (otherwise known as the number of unique individuals captured at time t+1) showed lower variance and less sensitivity to model violations than formal abundance estimates when sample sizes were small (but characteristic of most field studies). Similarly, there has been a recent explosion of the use of occupancy estimates as a way to monitor trends in wildlife populations (Gaston et al. 2000; MacKenzie and Nichols 2004; MacKenzie et al. 2006; Estrada and Arroyo 2012).

When monitoring populations, it is beneficial to determine which parameters are most sensitive to change and focus on those. Typically, focus is on population abundance or density of breeding individuals, which

could be misleading, if the population includes a large number of non-territorial animals (e.g., nonbreeding individuals) not easily sampled using traditional methods. In a situation where there is high mortality of territorial animals that are immediately replaced by surplus, the population of territorial animals may appear stable while the overall population is declining.

We recommend conducting sensitivity analyses to determine which state variables contribute most to population change (e.g., Caswell 1978; Williams et al. 2002). If demographic data are not available, the investigator will need to conduct literature reviews, consult with established experts (useful for Bayesian priors [Gelman et al. 2004]), or use results of similar studies to establish the basis for using specific state variables. All of these sources can be used in a sensitivity analysis, although some sources will be more valuable than others. For example, there will be more variation in parameter estimates from a literature review of studies in different geographic locations or resulting from different study methods compared to parameter estimates from a study conducted in the same area (Mills 2012).

Overall, we recommend that selecting the variables to measure should be based on strong statistically based information. One must understand the biology and population dynamics of the species being monitored to make better decisions on exactly what to monitor.

Selection of Sampling Areas

Once the study objective is established, the scale of resolution chosen by ecologists is perhaps the most important decision in monitoring because it predetermines procedures, observations, costs, and results (Green 1979; Hurlbert 1984). A major step in designing a monitoring program is to define the target population. Defining the target population essentially defines the area to be sampled. This first step establishes the sampling universe from which samples can be drawn and the extent to which inferences can be extrapolated.

Although this seems rather straightforward, the mobility of wildlife can muddle the inferences drawn from the established area. The basis for establishing the sampling frame includes information on habitat-use patterns, range sizes, movement patterns, and logistics of sampling a large area. Ideally, the bounds of the sampling area would correspond to some natural barrier that would keep a population closed, but this is often not the case. Without such bounds, open forms of estimators will need to be used (Marucco et al. 2011; Wilson et al. 2011; Ivan et al. 2013), or a combination of open and closed estimators, where sampling occurs during a short interval when the population is closed and over multiple seasons or years when the population is open (i.e., robust design [Pollock 1982; Kendall and Nichols 1995; Kendall et al. 1995]).

To reduce this issue associated with population closure, there have been developments in using an approach to estimating density (abundance per area) instead of abundance, using spatially explicit capture-recapture models (e.g., Efford 2004; Royle and Young 2008; Gardner et al. 2009; Russell et al. 2012). The distribution of individuals and their movements relative to a trapping array are explicitly modeled using this framework. For instance, the use of telemetry data can be used to evaluate geographic closure (White and Shenk 2001; Ivan et al. 2012) and to explicitly model animal movement in spatially explicit capture-recapture models (Sollmann et al. 2013). To evaluate geographic closure, animals are marked and time spent in the study area is estimated and used to scale abundance estimates and provide density (Ivan et al. 2013). Similarly, genetic data can be used to evaluate violations of closure assumptions by using approaches such as spatial autocorrelation among genotypes (Peakall and Smouse 2006). These methods provide indices of average animal gene flow in an area. Unfortunately, there has been little attention paid to the integration of genetic approaches to gene flow and abundance estimation, although this is beginning to change (Luikart et al. 2010).

When to Collect Data and for How Long?

A key aspect of monitoring is to know when to collect data, including the timing of data collection and the length of time over which data should be collected. Both are influenced by the biology of the organism, the objectives of the study, intrinsic and extrinsic factors that influence the parameter(s) to be estimated, and resources available to conduct the study.

Obviously, studies of breeding animals should be conducted during the breeding season, studies of non-breeding animals should be conducted during the nonbreeding period, and so on. Within a season, timing can be critically important because detectability of individuals can change for different activities or during different phenological phases (Bailey et al. 2004; MacKenzie 2005). Male passerine birds, for example, are generally more conspicuous during the early part of the breeding season when they are displaying as part of courtship and territorial defense activities. Detection probabilities for many species will be greater during this period than at other times. Another consideration is that the very population under study can change within a season. For example, age-class structures and numbers of individuals change during the course of the breeding season as juveniles fledge from nests and become a more entrenched part of the population.

Population estimates for a species therefore may differ depending on the timing of data collection. Once the timing decision is made, data must be collected during the same time in subsequent years to control for some of the within-season variation. Objectives of a monitoring effort also dictate when data should be collected. If monitoring is conducted to evaluate population trends of a species based on a demographic model, sampling should be done at the appropriate time to ensure precise and unbiased estimates of the relevant population parameters. Demographic models typically require fecundity and survival data to estimate the finite rate of population increase. Sampling for each of these parameters may be necessary during distinct times to ensure accurate estimates for the respective measures. Timing also is essential in the field of genetic monitoring (Goudet et al. 2002; Schwartz et al. 2007). For example, different measures of gene flow and genetic variation have been found to be more sensitive to the timing (e.g., pre- or postdispersal) of sample collection (Goudet et al. 2002; Hammond et al. 2006). This makes intuitive sense as populations should be most differentiated if sampling occurs postbreeding, but predispersal, and most similar after a pulse of dispersal. Consistency in timing and methods is essential.

Length of the study refers to how long a study should be conducted for reliable and accurate parameter estimates. A number of factors including study objectives, field methodology, ecosystem processes, biology of the species, budget, feasibility, and statistical power issues influence how long a monitoring study should occur. Monitoring studies also must consider temporal qualities of the ecological process or state being measured (e.g., population cycles, successional patterns including their frequency, magnitude, and regularity; Franklin 1989). Furthermore, a study should engage in data collection over a sufficiently long period to allow the population(s) under study to be subjected to a reasonable range of extrinsic environmental conditions. Consider two hypothetical wildlife populations that exhibit cyclic behaviors: one that cycles on average ten times per fifty years and the other exhibiting a complete cycle just once every twenty-five years. The population cycles are the results of various intrinsic and extrinsic factors that influence population growth and decline. A monitoring program established to sample both populations over a twenty-five-year period may be adequate to understand population trends in the species with frequent cycles but may be misleading for the species with the long population cycle. A longer timeframe would be needed to monitor the population of species with lower frequency cycles.

Considering only the frequency of population cycles may be inadequate, as the amplitude or magnitude of population shifts also may influence the length of a study. Sampling the population of a species exhibiting greater amplitude in population increases and declines would require a longer period to detect a population trend or effect size than a species exhibiting lower amplitude in population shifts.

Power to Detect Change

A major question that arises in any study is: "Am I sampling enough to detect a trend, should one exist?" "Enough" can be enough area, enough individuals per area, enough time, and so forth. The best way to address this issue is through simulations or calculations of statistical power (Zielinski and Stauffer 1996). That is, it is important to ask a question like: "If I am interested in changes in abundance and I mark one hundred individuals per year for ten years, will I have a 95% chance of detecting a 20% decline?" Or ask: "How many samples do I need to have 80% certainty of detecting a 20% decline given a ten-year monitoring program?" Too many monitoring efforts fail to quantify

what effect size they can detect with a specific effort—only to find at the end of the monitoring period that they never had power to detect change.

There are several computer programs available to examine interactions between sample size, effect size, type I error, and statistical power to detect trends in abundance, count, or occupancy under different sampling schemes (Gerrodette 1993; Gibbs 1995; Bailey et al. 2007). For example, program MONITOR estimates statistical power of a population-monitoring program when supplied with the number of plots, counts per plot and their variation, duration of the monitoring program, interval of monitoring, magnitude and direction of ongoing population trends, and significance level associated with trend detection. Similarly, program TRENDS conducts a power analysis of linear regression for monitoring wildlife populations (Gerrodette 1993). TRENDS estimates statistical power of a monitoring program after providing data related to the rate of change, duration of study, precision of the estimates, and the type I error rate. Alternatively, power can be set to a desired level and program TRENDS estimates one of the other input parameters. For example, if 80% power is desired, one can ask how many years a monitoring effort must be conducted given a precision of the estimate and type I error rate.

Recently, there has been an increase in reliance on occupancy-based approaches to estimate population trends. A new spatially based simulation program called SPACE has been created to evaluate power for detecting population trends over time based on species detections (Ellis et al. 2013). SPACE has been used to evaluate the power of monitoring trend occupancy for territorial animals while accounting for natural history (e.g., territorial overlap, different space use of different life stages), habitat use, and sampling scheme (e.g., size of the grid; Ellis et al. 2013).

We suggest that an analysis of power to detect trend is initiated before any monitoring effort is conducted. If specific programs do not exist to evaluate statistical power for a given method, consulting papers that have conducted simulations on power (and type I error rates) with that method will be insightful. Overall, it would be very unrewarding, if not embarrassing, to come to the end of a long-term monitoring program where no trend was found, only to realize that power to detect trend with the given protocol was no greater than a coin flip.

Resources to Detect Change

All of the components of monitoring that we have discussed so far are important but rely on sufficient monetary resources to complete these efforts. Agencies tasked with developing monitoring programs often work with limited budgets; therefore, cost-efficient sampling designs are paramount. Thus, we advocate using cost as an additional constraint for study design optimization (e.g., Sanderlin et al. 2009; Sanderin et al. 2014). We can determine the power to detect an effect size and the amount of sampling needed for accurate parameter estimates, but if there are not enough resources available to complete a study with these design specifications, there is no point in continuing with the monitoring effort.

Case Study: Lessons from USFS Monitoring under the Planning Rule

We can learn a lot about monitoring programs from the successes and failures of the USFS's efforts to monitor species under the 1982 "planning rule." The US Forest Service is a United States land management agency responsible for facilitating multiple-use activities on seventy-eight million hectares of land. These activities include the maintenance of healthy wildlife populations, clean water, recreational opportunities, and timber production. Guidance on how to balance competing demands on the landscape is given in the National Forest Management Act (NFMA) of 1976. One of the most controversial aspects of the NFMA has been the tiered planning approach and how planning rules accomplish NFMA's requirement to "provide for a diversity of plant and animal communities based on the suitability and capability of the specific land area in order to meet overall multiple-use objectives." Direction on how to achieve NFMA's mandate to provide for biodiversity was given in promulgated planning rules, which until 2012 called for the maintenance of viable populations of existing native and desired non-native vertebrate species in a planning area. Part of that maintenance of viable populations included identifying management indicator species and noting that "popu-

lation trends of the management indicator species will be monitored" (section 219.19 of the 1982 planning regulations). Interestingly, these management indicator species, which were proxies for other species, were often monitored through the use of habitat availability. This was considered a "proxy-by-proxy" approach and was deemed legally sound (*Inland Empire Public Lands Council v. USFS* 1995; *Lands Council v. McNair* 2008), although biologically questionable.

Monitoring under the 1982 planning regulations was seen as essential to the success of the multiple-use mandate and maintenance of viability clause. Unfortunately, the regulations relied on focused monitoring of a multitude of management indicator species, which proved too expensive and difficult to accomplish (Schultz 2010; Schultz et al. 2013). Thus, monitoring under the 1982 planning regulations often devolved to assessing the status and trend of the amount of habitat the species required without strong statistical links between the habitat and the species. We are currently in a new era where as of April 2012 the USFS is acting under a new planning rule (which has been in preparation for roughly fifteen years) with extensive public and scientific involvement. This new direction provides a great opportunity to improve management of wildlife on public lands (Schultz et al. 2013). Our goals in the remainder of this section are to review the new rule as it relates to monitoring and then provide advice as to how to choose species to monitor and points to consider.

Monitoring under the 2012 planning rule requires monitoring indicators that relate to the status of ecological conditions and focal species, which are defined as a narrow group of species whose status allows inference to the integrity of the larger ecosystem. These focal species are to "provide meaningful information regarding the effectiveness of the plan in maintaining or restoring the ecological conditions to maintain the diversity of plant and animals communities" (USFS 2012). Interestingly, the new monitoring rule has no special monitoring requirement for "species of conservation concern" (e.g., Threatened and Endangered Species), although it does have a viability requirement for this category. What remains to be tested in this new planning rule is how individual forests choose focal species to monitor and then implement a monitoring program.

In the following list, we discuss the characteristics of a focal species that we think will be effective for making inference to the larger ecosystem. An ideal focal species would have many of the following characteristics:

1. The species must index something greater than itself: A species that only reflects its own success in an ecosystem will likely be too narrow for capturing ecological integrity. While we would like to have data on all species in a planning area, common sense and the rule itself dictate that focal species are only limited subsets of species in the ecosystem. Species such as ecosystem engineers (e.g., beavers [*Castor canadensis*]) or keystone species (e.g., pileated woodpeckers [*Dryocopus pileatus*]) are good examples of animal species that index something greater than themselves. Ovenbirds (*Seiurus aurocapillus*) are another example, as they appear to be sensitive to fragmentation of mature forests and may index other species and processes of these habitats.

2. Low range of natural variability: If a species has large natural fluctuations that are not linked to climate or other environmental processes, statistical trend will be difficult to obtain. In this case, the temporal variation swamps out process variation, hence obviating the ability to identify trend. Similarly, a species known to cycle, like snowshoe hare (*Lepus americanus*), would make a poor choice as a focal species, as statistically detecting trend will be complicated. However, there are times when a species with a large range of natural variability will be exquisitely sensitive to environmental change, and this relationship (the signal) will override concerns about the large range of natural variability (the noise).

3. Species need to be statistically linked to habitat at the appropriate scale: If the purpose of the focal species is to provide inference to or monitor the integrity of the ecosystem, then there needs to be a strong statistical link to the vegetation. Likewise, selected focal species must be sensitive to particular land management actions. As many ecological processes are scale dependent, this statistical evaluation should be conducted at a multiple array of scales from plot to landscape levels.

4. Species can be monitored: A species may have a low range of variability and be a great reflection of land-use changes, but if it is too difficult or expensive to monitor then it should not be chosen. Some species' life history traits make them easier to monitor than others (e.g., small mammals that have smaller home ranges, are less vagile, and are abundant may be preferable to a carnivore that is secretive, rare, and vagile).

5. Background knowledge about national/global trend is needed for context: It is important to tie the local dynamics of a species into a broader monitoring program. Imagine picking boreal toads (*Bufo boreas*) or other amphibians for monitoring thirty years ago. The population's numbers would have likely declined in a forest, and this may have been attributed to habitat modifications or other management actions. This would be the wrong conclusion in the face of global amphibian decline. Knowing the magnitude of global decline compared to local declines would have provided relevant context.

Conclusions, Creative Approaches to Monitoring, and a Need for Large Scale National Efforts

Wildlife monitoring means many things to many different people. For some it is the broad-scale, long-term efforts to provide background information on species, while for others monitoring is very focused on a specific species for a specific reason (e.g., to detect a change in abundance after a wildfire or other natural event), and still for others monitoring is meant to provide data to better understand mechanisms underlying ecosystem change (i.e., adaptive management). Regardless of the reason to enter a monitoring effort, effective monitoring is a challenging endeavor as it can be prohibitively expensive, take a long time frame to achieve results, and may require coordination among multiple groups, especially if the effort is broad in scope. This will clearly require creative approaches to monitoring, which include the development of new field approaches and statistical techniques that can make monitoring more feasible. We have already seen many of these developments make monitoring efforts that were once inconceivable, such as the monitoring of abundance of wolves (*Canis lupus*) across the Alps (Marucco et al. 2009), now possible with the advent of DNA technology. Similarly, other technologies, such as trail cameras, have greatly improved our ability to obtain reliable data in an inexpensive manner. Developments in occupancy modeling and abundance estimation also are improving our ability to monitor. And combining the two approaches—where we use occupancy estimation with environmental DNA sampling (eDNA), which detects presence/absence of species by sampling the environment (e.g., water or air samples)—will certainly reduce costs and allow for more monitoring data to be collected.

Another recent trend to improve monitoring is to involve local citizens in monitoring efforts as part of "citizen-science" projects (e.g., Schmeller et al. 2008; Silvertown 2009; Miller-Rushing et al. 2012). This can reduce costs, allow for more monitoring data to be collected, and foster communications between the public and governmental agencies. These are unique opportunities to educate the public on research that is conducted on public lands and give a local community a sense of ownership with these public lands.

We wish to revisit several critical points that we covered earlier in this chapter, which are essential when designing more effective and efficient monitoring programs. First is the concept of the use of habitat as a proxy or surrogate measure for wildlife populations. Although it is very tempting to use habitat as a proxy for abundance of a species, there is very little evidence that this is a defensible position. The burden of proof is therefore on the group wishing to use habitat as a proxy to statistically show the tight relationship between the two variables before using them in a monitoring program. This does not mean that we are against surrogates or indices in general, but we urge caution in their use without first establishing a causal or statistical relationship, which is often not established when monitoring habitat as a surrogate for species' abundance. In fact, there is strong statistical evidence that under some circumstances indices of abundance are as good as collecting abundance, and some of these indices and surrogates may reflect change better than estimated abundance (McKelvey and Pearson 2001; Engemann 2003).

Our next summary point is that once a decision is made to monitor, we implore groups to be very specific in their objectives. This starts with the statement of intent of whether this is a broad-scale, context monitoring effort or a focal/targeted monitoring effort. The goal should be stated in context of an overarching objective. For instance, it is not enough to simply claim that we want the USFS to do focal monitoring of a species, but it is essential to understand that the intent of this focal monitoring may be to measure a broader ecosystem process. After these goals are established, we urge those responsible for monitoring efforts to consider all methodological approaches and conduct rigorous power analyses that include cost constraints to ensure the data collected will ultimately be able to answer the question posed. We also urge consistency in methods and timing of data collection. The world is filled with too many partial datasets, where someone changed methods without proper calibration linking the approaches.

Penultimately, we encourage the use of the adaptive management framework when appropriate. While this approach may not always work, we think that, when possible, setting monitoring in a framework that helps us also understand mechanisms behind trends is essential. We hardly have enough fiscal resources available now to conduct the initial monitoring, let alone funds to go back and try to conduct studies to understand trends once they are detected. Adaptive management and structured decision making allow us to kill the proverbial "two birds with one stone." However, we caution that adaptive management is best accomplished when thresholds or triggers are in place to force certain conservation actions, should these triggers be tripped.

Lastly, in this chapter we focused on the monitoring of focal species, which included species that reflect broader ecosystem processes and species that are of special management or conservation concern. While we believe that most monitoring efforts will be targeted, this does not mean that we do not see the value in context monitoring approaches. We can only imagine the valuable data that could be produced if we had a national wildlife monitoring program that was on the scale of our Forest Inventory and Analysis program (cf. Manley et al. 2004, 2005). Perhaps such an effort would have allowed the detection of emerging threats, such as invasive species, faster than our *ad hoc* detection methods currently in place could, ultimately saving billions of dollars. A commitment to large-scale monitoring also might enable us to detect patterns of faunal change resulting from overarching effects of climate change. Furthermore, the background information such a program would provide would help us interpret local trends in local monitoring datasets.

Overall, monitoring when done well provides important data on changes occurring on the landscape, yet when done poorly or in an *ad hoc* manner is simply a waste of valuable resources. We hope our framework here helps those intent on designing new monitoring efforts in the future.

LITERATURE CITED

Bailey, L., J. E. Hines, J. D. Nichols, and D. I. MacKenzie. 2007. "Sampling Design Trade-offs in Occupancy Studies with Imperfect Detection: Examples and Software." *Ecological Applications* 17: 281–90.

Bailey, L. L., T. R. Simons, and K. H. Pollock. 2004. "Estimating Site Occupancy and Species Detection Probability Parameters for Terrestrial Salamanders." *Ecological Applications* 14:692–702.

Biber, E. 2011. "The Problem of Environmental Monitoring." *University of Colorado Law Review* 83:1–82.

Block, W. M., A. B. Franklin, J. P. Ward Jr., J. L. Ganey, and G. C. White. 2001. "Design and Implementation of Monitoring Studies to Evaluate the Success of Ecological Restoration on Wildlife." *Restoration Ecology* 9:293–303.

Block, W. M., M. L. Morrison, J. Verner, and P. N. Manley. 1994. "Assessing Wildlife-Habitat-Relationships Models: A Case Study with California Oak Woodlands." *Wildlife Society Bulletin* 22:549–61.

Boulinier, T., J. D. Nichols, J. R. Sauer, J. E. Hines, and K. H. Pollock. 1998. "Estimating Species Richness: The Importance of Heterogeneity in Species Detectability." *Ecology* 79:1018–28.

Carignan, V., and M. A. Villard. 2002. "Selecting Indicator Species to Monitor Ecological Integrity: A Review." *Environmental Monitoring and Assessment* 78:45–61.

Caswell, H. 1978. "A General Formula for the Sensitivity of Population Growth Rate to Changes in Life History Parameters." *Theoretical Population Biology* 14:215–30.

Caughley, G. 1994. "Directions in Conservation Biology." *Journal of Animal Ecology* 63:215–44.

Cushman, S. A., K. S. McKelvey, D. H. Flather, and K. McGarigal. 2008. "Do Forest Community Types Provide a Sufficient Basis to Evaluate Biological Diversity?" *Frontiers in Ecology and the Environment* 6:13–17.

Cushman, S. A., K. S. McKelvey, B. R. Noon, and K. McGarigal. 2010. "Use of Abundance of One Species as a Surrogate for Abundance of Others." *Conservation Biology* 24:830–40.

Dorazio, R. M., J. A. Royle, B. Söderström, and A. Glimskär. 2006. "Estimating Species Richness and Accumulation by Modeling Species Occurrence and Detectability." *Ecology* 87:842–54.

Efford, M. G. 2004. "Density Estimation in Live-Trapping Studies." *Oikos* 106:598–610.

Ellis, M., J. Ivan, and M. K. Schwartz. 2013. "Spatially Explicit Power Analyses for Occupancy-Based Monitoring of Wolverines in the U.S. Rocky Mountains." *Conservation Biology*. doi: 10.1111/cobi.12139.

Engemann, R. M. 2003. "More on the Need to Get the Basics Right: Population Indices." *Wildlife Society Bulletin* 31:286–87.

Estrada, A., and B. Arroyo. 2012. "Occurrence vs Abundance Models: Differences between Species with Varying Aggregation Patterns." *Biological Conservation* 152:37–45.

Fancy, S. G., J. E. Gross, and S. L. Carter. 2009. "Monitoring the Condition of Natural Resources in US National Parks." *Environmental Monitoring and Assessment* 151:161–74.

Franklin, J. F. 1989. "Importance and Justification of Long-Term Studies in Ecology." In *Long-term Studies in Ecology: Approaches and Alternatives*, edited by G. E. Likens, 3–19. Springer-Verlag, New York.

Gardner, B., J. A. Royle, and M. T. Wegan. 2009. "Hierarchical Models for Estimating Density from DNA Mark-Recapture Studies." *Ecology* 90:1106–15.

Gaston, K. J., T. M. Blackburn, J. J. Greenwood, R. D. Gregory, R. M. Quinn, and J. H. Lawton. 2000. "Abundance–Occupancy Relationships." *Journal of Applied Ecology* 37:39–59.

Gerrodette, T. 1993. "Trends: Software for a Power Analysis of Linear Regression." *Wildlife Society Bulletin* 21:515–16.

Gelman A., J. B. Carlin, H. S. Stern, and D. B. Rubin. 2004. *Bayesian Data Analysis.* 2nd edition. Chapman and Hall/CRC, New York.

Gibbs, J. P. 1995. "Monitor 7.0: Software for Estimating the Statistical Power of Population Monitoring Programs." USGS-Patuxent Wildlife Research Center, Laurel, MD.

Goudet J., N. Perrin, and P. Waser. 2002. "Tests for Sex-Biased Dispersal Using Bi-parentally Inherited Genetic Markers." *Molecular Ecology* 11(6): 1103–14.

Green, R. H. 1979. *Sampling and Statistical Methods for Environmental Biologists.* John Wiley and Sons, New York.

Greenwood, J. J. D. 2003. "The Monitoring of British Breeding Birds: A Success Story for Conservation Science?" *Science of the Total Environment* 310:221–30.

Hammond, R. L., L. J. L. Handley, B. J. Winney, M. W. Bruford, and N. Perrin. 2006. "Genetic Evidence for Female-Biased Dispersal and Gene Flow in a Polygonous Primate." *Proceedings of the Royal Society B* 273:479–84.

Hansen, M. M., I. Olivieri, D. M. Waller, and E. E. Nielsen. 2012. "Monitoring Adaptive Genetic Responses to Environmental Change." *Molecular Ecology* 21:1311–29.

Harris, D. C. 2010. "Charles David Keeling and the Story of Atmospheric CO_2 Measurements." *Analytical Chemistry* 82:7865–70.

Holthausen R., R. L. Czaplewski, D. DeLorenzo, G. Hayward, W. B. Kessler, P. Manley, K. S. McKelvey, D. S. Powell, L. F.

Ruggiero, M. K. Schwartz, B. Van Horne, and C. D. Vojta. 2005. "Strategies for Monitoring Terrestrial Animals and Habitats." Gen. Tech. Rep. RMRS-GTR-161. Fort Collins, CO: US Department of Agriculture, Forest Service, Rocky Mountain Research Station. 34p.

Hunter, M. L., G. L. Jacobson, and T. Webb. 1988. "Paleoecology and the Coarse-Filter Approach to Maintaining Biological Diversity." *Conservation Biology* 2:375–85.

Hurlbert, S. A. 1984. "Pseudoreplication and the Design of Ecological Field Experiments." *Ecological Monographs* 54:187–211.

Inland Empire Public Lands Council v. USFS, 88 F.3d 754 (9th Cir. 1996).

Ivan, J. S., G. C. White, and T. M. Shenk. 2012. "Using Auxiliary Telemetry Information to Estimate Animal Density from Capture-Recapture Data." *Ecology* 94:809–16.

———. 2013. "Using Simulation to Compare Methods for Estimating Density from Capture-Recapture Data." *Ecology* 94:817–26.

Johnson, F. A., B. K. Williams, J. D. Nichols, J. E. Hines, W. L. Kendall, G. W. Smith, and D. F. Caithamer. 1993. "Developing an Adaptive Management Strategy for Harvesting Waterfowl in North America." *Transactions of the North American Wildlife and Natural Resources Conference* 58:565–83.

Kendall, W. L., and J. D. Nichols. 1995. "On the Use of Secondary Capture-Recapture Samples to Estimate Temporary Emigration and Breeding Proportions." *Journal of Applied Statistics* 22:751–62.

Kendall, W. L., K. H. Pollock, and C. Brownie. 1995. "A Likelihood-Based Approach to Capture-Recapture Estimation of Demographic Parameters Under the Robust Design." *Biometrics* 51:293–308.

Kéry, M., and H. Schmid. 2006. "Estimating Species Richness: Calibrating a Large Avian Monitoring Programme." *Journal of Applied Ecology* 43:101–10.

Kéry, M., and M. Plattner. 2007. "Species Richness Estimation and Determinants of Species Detectability in Butterfly Monitoring Programmes." *Ecological Entomology* 32: 53–61.

Kéry, M., and J. A. Royle. 2008. "Hierarchical Bayes Estimation of Species Richness and Occupancy in Spatially Replicated Surveys." *Journal of Applied Ecology* 45:589–98.

Lands Council v. McNair, 537 F.3d 981 (9th Cir. 2008).

Lindenmayer, D. B., and G. E. Likens. 2011. "Direct Measurement versus Surrogate Indicator Species for Evaluating Environmental Change and Biodiversity Loss." *Ecosystems* 14:47–59.

Link, W. A., and J. R. Sauer. 1998. "Estimating Population Change from Count Data: Application to the North American Breeding Bird Survey." *Ecological Applications* 8:258–68.

Lovett, G. M., et al. 2007. "Who Needs Environmental Monitoring?" *Frontiers in Ecology and the Environment* 5(5): 253–60.

Luikart, G., N. Ryman, D. A. Tallmon, M. K. Schwartz, and F. W. Allendorf. 2010. "Estimation of Census and Effective Population Sizes: The Increasing Usefulness of DNA-Based Approaches." *Conservation Genetics* 11:355–73.

Mace, G. M., N. J. Collar, K. J. Gaston, C. Hilton-Taylor, H. R. Akcakaya, N. Leader-Williams, et al. 2008. "Quantification of Extinction Risk: IUCN's System for Classifying Threatened Species." *Conservation Biology* 22:1424–42.

MacKenzie, D. I. 2005. "What Are the Issues with Presence-Absence Data for Wildlife Managers?" *Journal of Wildlife Management* 69:849–60.

MacKenzie, D. I., and J. D. Nichols. 2004. "Occupancy as a Surrogate for Abundance Estimation." *Animal Biodiversity and Conservation* 27:461–67.

MacKenzie, D. I, J. D. Nichols, J. A. Royle, K. P. Pollock, L. L. Bailey, and J. E. Hines. 2006. *Occupancy Estimation and Modeling: Inferring Patterns and Dynamics of Species Occurrence*. Academic Press, San Diego, California.

Manley, P. N., and B. Van Horne. 2005. "The Multiple Species Inventory and Monitoring Protocol: A Population, Community, and Biodiversity Monitoring Solution for National Forest System Lands." Proceedings of the International Monitoring Science and Technology Symposium. General Technical Report GTR-RMRS-in press. US Forest Service, Rocky Mountain Research Station, Fort Collins, Colorado, USA.

Manley, P. N., W. J. Zielinski, M. D. Schlesinger, and S. R. Mori. 2004. "Evaluation of a Multiple-Species Approach to Monitoring Species at the Ecoregional Scale." *Ecological Applications* 14:296–310.

Marucco, F., L. Boitani, D. H. Pletscher, and M. K. Schwartz. 2011. "Bridging the Gaps between Non-invasive Genetic Sampling and Population Parameter Estimation." *European Journal of Wildlife Research* 57:1–13.

Marucco, F., D. H. Pletscher, L. Boitani, M. K. Schwartz, K. L. Pilgrim, and J. D. Lebreton. 2009. "Wolf Survival and Population Trend Using Non-invasive Capture-Recapture Techniques in the Western Alps." *Journal of Applied Ecology* 46:1003–10.

McKelvey, K. S., and D. E. Pearson. 2001. "Population Estimation with Sparse Data: The Role of Estimators versus Indices Revisited." *Canadian Journal of Zoology* 79:1754–65.

Miller-Rushing, A., R. Primack, and R. Bonney. 2012. "The History of Public Participation in Ecological Research." *Frontiers in Ecology and the Environment* 10:285–90.

Mills, L. S. 2012. *Conservation of Wildlife Populations: Demography, Genetics, and Management*. 2nd edition. Wiley-Blackwell, New York.

Moir, W. H., and W. M. Block. 2001. "Adaptive Management on Public Lands in the United States: Commitment or Rhetoric? *Environmental Management* 28:141–48.

Neel, M. C., and J. P. Che-Castaldo. 2013. "Predicting Recovery Criteria for Threatened and Endangered Plant Species on the Basis of Past Abundances and Biological Traits." *Conservation Biology* 27:385–97.

Nichols, J. D., and B. K. Williams. 2006. "Monitoring for Conservation." *Trends in Ecology and Evolution* 21:668–73.

Nie, M. A., and C. A. Schultz. 2012. "Decision-Making Triggers in Adaptive Management." *Conservation Biology* 26:1137–44.

Noon, B. R., L. L. Bailey, T. D. Sisk, and K. S. McKelvey. 2012. "Efficient Species-Level Monitoring at the Landscape Scale." *Conservation Biology* 26:432–41.

Öster, M., K. Persson, and O. Eriksson. 2008. "Validation of Plant Diversity Indicators in Semi-natural Grasslands." *Agriculture, Ecosystems & Environment* 125:65–72.

Peakall, R., and P. E. Smouse. 2006. "GENALEX 6: Genetic Analysis in Excel. Population Genetic Software for Teaching and Research." *Molecular Ecology Notes* 6:288–95.

Pollock, K. H. 1982. "A Capture-Recapture Design Robust to Unequal Probability of Capture." *Journal of Wildlife Management* 46:757–60.

Rodrigues, A. S. L., and T. M. Brooks. 2007. "Shortcuts for Biodiversity Conservation Planning: The Effectiveness of Surrogates." *Annual Review of Ecology, Evolution, and Systematics* 38:713–37.

Royle, J. A., and M. Kéry. 2007. "A Bayesian State-Space Formulation of Dynamic Occupancy Models." *Ecology* 88:1813–23.

Royle, J. A., and J. D. Nichols. 2003. "Estimating Abundance from Repeated Presence-Absence Data or Point Counts." *Ecology* 84:777–90.

Royle, J. A., and K. Young. 2008. "A Hierarchical Model for Spatial Capture-Recapture Data." *Ecology* 89:2281–89.

Russell, R. E., J. A. Royle, R. Desimone, M. K. Schwartz, V. L. Edwards, K. P. Pilgrim, and K. S. Mckelvey. 2012. "Estimating Abundance of Mountain Lions from Unstructured Spatial Sampling." *Journal of Wildlife Management* 76:1551–61.

Sanderlin, J. S., W. M. Block, and J. L. Ganey. 2014. "Optimizing Study Design for Multi-species Avian Monitoring Programs." *Journal of Applied Ecology*. DOI: 10.1111/1365-2664.12252.

Sanderlin, J. S., N. Lazar, M. J. Conroy, and J. Reeves. 2012. "Cost-Efficient Selection of a Marker Panel in Genetic Studies." *Journal of Wildlife Management* 76:88–94.

Sauer, J. R., P. J. Blank, E. F. Zipkin, J. E. Fallon, and F. W. Fallon. 2013. "Using Multi-species Occupancy Models in Structured Decision Making on Managed Lands." *Journal of Wildlife Management* 77:117–27.

Sauer, J. R., J. E. Fallon, and R. Johnson. 2003. "Use of North American Breeding Bird Survey Data to Estimate Population Change for Bird Conservation Regions." *Journal of Wildlife Management* 67:372–89.

Sauer, J. R., and W. A. Link. 2011. "Analysis of the North American Breeding Bird Survey Using Hierarchical Models." *The Auk* 128:87–98.

Schmeller, D. S., P.-Y. Henry, R. Julliard, B. Bruber, J. Clobert, F. Dziock, S. Lengyel, P. Nowicki, E. Deri, E. Budrys, T. Kull, K. Tali, B. Bauch, J. Settele, C. Van Swaay, A. Kobler, V. Babij, E. Papastergiadou, and K. Henle. 2008. "Advantages of Volunteer-Based Biodiversity Monitoring in Europe." *Conservation Biology* 23:307–16.

Schultz, C. 2010. "Challenges in Connecting Cumulative Effects Analysis to Effective Wildlife Conservation Planning." *BioScience* 60:545–51.

Schultz, C. A., T. D. Sisk, B. R. Noon, and M. A. Nie. 2013. "Wildlife Conservation Planning Under the United States Forest Service's 2012 Planning Rule." *Journal of Wildlife Management* 77:428–44.

Schwartz, M. K., G. Luikart, and R. S. Waples. 2007. "Genetic Monitoring as a Promising Tool for Conservation and Management." *Trends in Ecology and Evolution* 22:25–33.

Seddon, P. J., and T. Leech. 2008. "Conservation Short Cut, or Long and Winding Road? A Critique of Umbrella Species Criteria." *Oryx* 42:240–45.

Silvertown, J. 2009. "A New Dawn for Citizen Science." *TRENDS in Ecology and Evolution* 24:467–71.

Slade, N. A., and S. M. Blair. 2000. "An Empirical Test of Using Counts of Individuals Captured as Indices of Population Size." *Journal of Mammalogy* 81:1035–45.

Sollmann, R., B. Gardner, A. W. Parsons, J. J. Stocking, B. T. McClintock, T. R. Simons, K. H. Pollock, and A. F. O'Connell. 2013. "A Spatial Mark-Resight Model Augmented with Telemetry Data." *Ecology* 94:553–59.

Stegen, J. C., A. L. Freestone, T. O. Crist, M. J. Anderson, J. M. Chase, L. S. Comita, H. V. Cornell, K. F. Davies, S. P. Harrison, A. H. Hurlbert, B. D. Inouye, N. J. B. Kraft, J. A. Myers, N. J. Sanders, N. G. Swenson, and M. Vellend. 2013. "Stochastic and Deterministic Drivers of Spatial and Temporal Turnover in Breeding Bird Communities." *Global Ecology and Biogeography* 22:202–12.

Tallmon, D. A., D. Gregovich, R. S. Waples, C. S. Baker, J. Jackson, B. L. Taylor, E. Archer, K. K. Martien, F. W. Allendorf, and M. K. Schwartz. 2010. "When Are Genetic Methods Useful for Estimating Contemporary Abundance and Detecting Population Trends?" *Molecular Ecology Resources* 10: 684–92.

Tallmon, D. A., R. S. Waples, D. Gregovich, and M. K. Schwartz. 2012. "Detecting Population Recovery Using Gametic Disequilibrium-Based Effective Population Size Estimates." *Conservation Genetics Resources* 4:987–89.

US Forest Service (USFS). 2012. "Final Programmatic Environmental Impact Statement: National Forest System Land Management Planning." USFS, Washington, D.C., USA.

Van Horne, B. 1983. "Density as a Misleading Indicator of Habitat Quality." *Journal of Wildlife Management* 47:893–901.

Walters, C. J. 1986. *Adaptive Management of Renewable Resources*. MacMillan, New York, New York.

Westgate, M. J., G. E. Likens, and D. B. Lindenmayer. 2013. "Adaptive Management of Biological Systems: A Review." *Biological Conservation* 158:128–39.

White, G. C., and T. M. Shenk. 2001. "Population Estimation with Radio-Marked Animals." In *Radio Tracking and Animal Populations*, edited J. J. Millspaugh and J. M. Marzluff, 329–50. Academic Press, San Diego, California.

Williams, B. K. 1996. "Adaptive Optimization and the Harvest of Biological Populations." *Mathematical Bioscience* 136:1–20.

Williams, B. K., J. D. Nichols, and M. J. Conroy. 2002. *Analysis and Management of Animal Populations*. Academic Press, New York.

Willson, J. D., C. T. Winne, and B. D. Todd. 2011. "Ecological and Methodological Factors Affecting Detectability and Population Estimation in Elusive Species." *Journal of Wildlife Management* 75:36–45.

Yoccoz, N. G., et al. 2001. "Monitoring of Biological Diversity in Space and Time." *Trends in Ecology and Evolution* 16:446–53.

Zielinski, W. J., and H. B. Stauffer. 1996. "Monitoring Martes Populations in California: Survey Design and Power Analysis." *Ecological Applications* 6:1254–67.

11

KEVIN S. MCKELVEY

The Effects of Disturbance and Succession on Wildlife Habitat and Animal Communities

This chapter discusses the study of disturbance and succession as they relate to wildlife. As such, the discussion is confined to those disturbance processes that change the physical attributes of habitat, leading to a postdisturbance trajectory. However, even with this narrowing of the scope of disturbances discussed, there remain formidable obstacles prior to any coherent discussion of disturbance. The first, and most fundamental, is definitional: what constitutes disturbance and succession, and what is habitat? The concepts of disturbance, habitat, and succession are highly scale-dependent; disturbances at one scale become part of continuous processes at a larger scale, and ideas associated with succession require assumptions of constancy, which become highly problematic as spatiotemporal scales increase. Literature on the effects of disturbance and succession on wildlife, however, focuses on a narrow range of spatial scales, primarily occurs within a narrow temporal window immediately following disturbance, and seldom includes interactions between areas within the disturbed patch and the landscape that surrounds it. While these largely descriptive studies undoubtedly have great local value, more general information about the relationships between organisms and environments shaped by disturbance and succession is remarkably limited. Multiple small-scale descriptive studies of the immediate postdisturbance environment do not appear to coherently aggregate into larger understandings of the effects of disturbance and succession on wildlife. Context is important: the conditions at the time of the disturbance, in adjacent undisturbed patches, and within the broader landscape all affect both the postdisturbance wildlife community and, more importantly, the trajectory of the postdisturbance community. Even for well-studied species, coherent understandings of their relationships to disturbance across time and space are therefore often vague.

Commonly, we look at successional changes in habitat quality by using spatial samples of different ages as if they were a temporal series, which implies spatiotemporal constancy in successional dynamics. This assumption has served wildlife research well, but in the face of directional climate change and the nearly continuous addition of exotic species, this approach is becoming increasingly untenable. We need to embrace the idea that postdisturbance succession is increasingly unlikely to produce communities similar to those that the disturbance altered: short-term successional patterns are likely to be influenced by the large and dynamic pool of exotic plants and animals and longer-term succession by directional climate change.

Together, these observations indicate a need for studies of disturbance and succession at larger spatiotemporal scales. It is, however, important to assess the feasibility of scaling up studies in time and space. Clearly, studies that expand both the spatial and temporal dimensions of data collection quadratically increase costs; longer time frames do not fit into current competitive funding structures and contain a variety of negatives such as the potential loss of data. Broadscale targeted monitoring, however, can provide a framework allowing acquisition of data at these scales and can be designed to produce both immediate and longer-term results.

Disturbance, Succession, and Questions of Scale

Bormann and Likens (1979) defined disturbance as disruption of the pattern of the ecosystem, principally by external physical forces. This idea, however, assumes that ecosystems function as idealized Newtonian systems, in stasis until external energy is applied. But few ecosystems exist in equilibrium (Sousa 1984). Further, Rykiel (1985) noted that whether disturbance is viewed as changing the state of an ecosystem or being part of that state is entirely a function of scale. At one scale, a tree-fall is a state-altering disturbance; at a larger scale, it is part of a continuous process that creates and maintains the state of an old-forest ecosystem. The same thing is true for larger disturbances; at one scale, fire in western US forests represents a significant disturbance to ecological function, radically altering wildlife habitats. At a larger scale, fire is a part of the ecosystem—many ecosystems are dependent on fire to maintain the presence, patterns, and juxtapositions of plant and animal species (Habeck and Mutch 1973; Covington and Moore 1994; Nowacki and Abrams 2008).

Similar-scale dependencies are associated with the effects of disturbance on wildlife habitat. Habitat is often thought of with a particular scale in mind: perhaps a forest stand or home-range area evaluated across a year or the lifespan of an organism. If successional thinking is applied, habitat may be defined as existing within a specific sere (e.g., a species may be considered to be associated with early seral or late successional forests). However, the habitat requirements for population persistence are, like disturbance, complex and span many scales in both space and time. In space, wildlife habitat spans spatial domains measured in meters (e.g., specific resources for denning), to kilometers (e.g., sufficient resources to support a local viable population), to hundreds of kilometers (e.g., a mosaic of resources and connectivity sufficient to support long-term metapopulation [Levins 1969, 1970] persistence and abundance). Additionally, at all spatial scales, juxtaposition and spatial patterns of habitats are important (e.g., Iverson et al. 1987). Temporally, the definition of *habitat* is even more complex. Again, scales vary from almost instantaneous (e.g., the timing of ice breakup in the Arctic) to millennial (e.g., the processes of erosion and deposition that create soils, caves,

and other habitat features). Temporal habitat patterns are as critical for conservation as are spatial patterns, but much harder to study. In many cases, the events that structure landscapes and define species ranges are rare—often singular. For example, the genetic population structures of many species are strongly associated with glacial vicariance that occurred during the Pleistocene (see Shafer et al. 2010 for a review). To study these rare events, we primarily look to the past to gain insight into their frequency, size, and postevent successional and evolutionary trajectories. Further, for practical reasons, we frequently use habitats of different ages as surrogates for the passage of time, making the tacit assumption that if we were to project one area forward (or backward), it would be similar enough to a surrogate area that we can infer its future or past state by studying that surrogate. I refer to this approach as "trading space for time."

The study of temporal patterns of habitat over time, and specifically the validity of using spatial surrogates to infer temporal patterns, is tightly linked to the concept of succession. Here I use this term in a neo-Clemensian fashion (Clements 1916; Daubenmire 1952): succession assumes a pattern of orderly and predictable changes in species presence and abundance that occur over time after disturbance, leading to a fairly stable terminal state, or climax. As Gleason (1927) noted, the idea of succession is very appealing: if understood, it allows us to predict the future and to see into the past without needing any data other than what we collect in the present. However, successional concepts cannot be accepted naively. The validity of this concept, and its resulting popularity (or lack thereof), is directly related to the complexity of the system studied and the scale at which the system is evaluated. Simple systems have fewer succession pathways and fewer species; hence, the vegetative trajectories are more predictable. For example, in the western forests of the United States, tree communities are often simple and contain large areas of intact natural vegetative communities. In these systems, succession-based classification systems (e.g., habitat types; Pfister and Arno 1980) and concepts (e.g., potential natural vegetation [PNV]; Küchler 1964) are popular. In highly modified landscapes concepts, however, concepts like PNV become abstract as none of the vegetation within a study area may exist in its putative potential condition (Zerbe 1998). For these reasons, looked at objectively, succession is often a problematic concept, but one that has proved useful in many communities and without which our ability to study systems and build predictive models would be severely limited. However, regardless of its historical validity, there is good reason to doubt its relevance as a tool to predict future habitat conditions; the likelihood of rapid directional climate change and the increasing presence of exotic species compromise our ability to use past patterns to predict the future. Another concept, community assembly theory, which views communities as being the result of a continuous process of species invasion and extinction (Lodge 1993), may be more germane.

Disturbance and Its Relationship to Wildlife Communities
Defining Disturbance

The spatial scaling of habitat, including the habitat patch structure generated by disturbance, is ultimately defined by the grain at which specific organisms perceive the world (Kotliar and Wiens 1990) and operationally defined at some arbitrary time scale. Thus, disturbances also are subject to these same scalings: an event that constitutes a significant disturbance for a woodland salamander (e.g., *Ensetina* sp.) may be of little importance to a wolverine (*Gulo gulo*). Thus, when we consider disturbance ecology, we should ideally begin with operational environment of the affected organisms, clearly identify the time frame of interest, and define disturbances accordingly. This, however, is seldom done. When we consider disturbance, we often have a particular scale in mind, generally an intermediate scale. When we label events as disturbances, we envision events such as hurricanes or forest fires and particularly anthropogenic activities such as logging, land clearing, and infrastructure development: scientific articles that claim to study disturbance overwhelmingly study these types of events. However, this scale is not the organism's scale; rather, it is our scale. In part, this is simply a matter of semantics: how we define disturbance rather than what we study. For example, we study the effects of tree-fall and subsequent forest gap dynamics on wildlife; Forsman et al. (2010) found eleven studies of gap effects on forest birds. However, while forest gap formation is an obvious disturbance

process, we discuss and label this process as part of a separate body of literature. Similarly, we are prone to view what we do as being more of a disturbance than what other organisms do. For example, beavers (*Castor canadensis*) instigate a variety of disturbances (tree felling, house and dam construction), oftentimes removing all accessible trees from the area adjacent to their pond (Martell et al. 2006) and thereby creating a shifting pattern of disturbance that can, over many generations, affect large areas (Naiman et al. 1986). However, studies of beaver activity are seldom framed or titled as being disturbance studies.

By choosing to concentrate our studies on mid-scale disturbances, such as fires, and the subsequent changes in habitat, we have, unfortunately, also chosen events that occur at temporal scales that are inconvenient—postdisturbance successional trajectories frequently require many years to evolve. To study these processes, it is therefore necessary either to set up very long-term studies or trade space for time, with its associated assumptions and weaknesses. As noted earlier, trading space for time requires the assumption of a high degree of similarity and transferability across disturbance events and subsequent recoveries. Given these constraints, there are two types of disturbances that occur at scales and with frequencies that allow both study and reasonable transferability: anthropogenic changes such as logging and land conversion, and fires; a great deal of what we identify as the study of disturbance therefore concerns these types of disturbances. Likely for these reasons, and based on practical considerations such as the ease of aging trees and hence inferring past disturbance events, this body of literature has a strong bias toward the study of forested systems.

In addition to often being appropriate in scale, anthropogenic disturbance is highly researchable because we seek to control many aspects of our disturbances. For example, in forestry we apply standardized silvicultural treatments: a clear-cut treatment, in addition to removing all trees, generally will also include removal of residual debris and soil scarification, leaving a fairly homogenous postdisturbance environment that is replicated across multiple treatment areas. Treatment blocks also tend to be of similar size—both economics and policy dictate this. On Forest Service lands in the United States, for example, clear-cut treatments have for many years been limited in size and spacing to ten to twenty ha blocks interspersed with uncut areas (Franklin and Forman 1987). They also occur at relatively fixed rates across time and space because we desire fairly constant flows of products supplied to a fixed array of mills. Thus, a system of clear-cuts is a much more regularized disturbance pattern than would occur naturally, and for scientific study, these regularizing factors allow us to achieve a level of replication that is generally absent in natural events. Of natural disturbances, fires are probably the most ideal for study. In fire-prone areas, fires occur frequently enough to be grouped based on covariates such as intensity, aspect, and pre-fire vegetation. Thus, fires form another major class of disturbance studies. In addition to these two heavily studied classes of disturbance, a third major group of published studies concerns disturbance events that we perceive as being "natural disasters": hurricanes, floods, tsunamis, and so forth.

Wildlife Studies Associated with Disturbance

Formal studies of the effects of disturbances on wildlife are almost exclusively limited to the immediate aftermath of the event. This phenomenon is perhaps most apparent in the study of natural disasters. For example, a Google Scholar search on "Hurricane Hugo" revealed a total of thirteen journal publications on the hurricane's effects on animals, of which ten (77%) were published within the first five years. However, few studies are currently underway; I was not able to locate a single journal article with animal responses to Hurricane Hugo as the primary subject published after 2002. This pattern is not limited to the study of wildlife: of seventy-two identified journal publications on all subjects concerning the effects of Hurricane Hugo, sixty-seven (83%) were published in the first five years, and none were published after 2008. Given that changes in vegetation wrought by Hugo will affect habitat in various ways for hundreds of years, there is no biological rationale for this spate of studies of the exceedingly ephemeral habitat conditions associated with its immediate aftermath coupled with low levels of interest associated with the still rapidly evolving habitat conditions that currently exist. Further, this pattern—frequent studies immediately after the disturbance followed by

disinterest—is the norm not just for studies of topical natural disasters but is pervasive across disturbance types. As noted by Fontaine and Kennedy (2012), in a recent meta-analysis of fire effects on small mammals and birds, the meta-analysis was limited to short-term (four years or less) responses because data "at longer time scales were too sparse to permit quantitative assessment" (Fontaine and Kennedy 2012, abstract).

ANTHROPOGENIC DISTURBANCE

Within the constraints noted by Fontaine and Kennedy (2012; taxa limited to small mammals and birds and effects to the period immediately following disturbance), there are still hundreds of published papers documenting the effects of anthropogenic disturbance on wildlife. Luckily, there has been a series of recent meta-analyses that provide both syntheses of these papers' findings and extensive bibliographic references. Kalies et al. (2010) performed a meta-analysis of twenty-two papers on the effects of thinning, prescribed burns, and some wildfire on birds and rodents in the American Southwest. Not all species or taxa were equally represented; birds were most commonly studied. Spatial scales were small (less than four hundred ha) and temporally close to the treatment (twenty years or less posttreatment). Ground-foraging birds and rodents showed neutral density responses to the treatments, whereas aerial, tree, and bole-foraging birds had positive or neutral responses to both small-diameter removal and burning treatments but negative responses to overstory removal and wildfire. Hartway and Mills (2012) looked at a variety of anthropogenic treatments to increase nest success, including prescribed burns (nineteen studies, fifty-two species). They also evaluated the effects of livestock exclusion, prescribed burning, removal of predators, and removal of cowbirds (*Molothrus ater*). Of the four treatments, prescribed burning was not the most effective but did significantly increase nest success for most species, though there was a large amount of variation between species groups. Specifically, precocial bird species were negatively affected by prescribed burning. Vanderwel et al. (2007) performed a meta-analysis of the effects of partial cutting on forest birds. All forty-two studies examined the effects of cutting ten years or less after harvest. Of the thirty-four bird species studied, fourteen declined with increasing harvest intensity and six increased. Not surprisingly, deep forest birds such as ovenbirds (*Seiurus aurocapilla*) and brown creepers (*Certhia americana*) were the species most sensitive to harvesting.

FIRE

Similar to anthropogenic disturbances, a number of recent meta-analyses document the effects of fire on wildlife. Not all of these meta-analyses separate natural fire from anthropogenic treatments such as thinning or prescribed burns. For example, Zwolak (2009) performed a meta-analysis of fire clear-cutting and partial cutting on small mammal populations. Of the fifty-six studies used in the meta-analysis, most (fifty-four) documented short-term changes. Only eleven documented longer-term (ten to twenty years after the disturbance) effects (Zwolak 2009, supplemental, appendix A). Zwolak found that deer mice (*Peromyscus maniculatus*) increased and red-backed voles (*Myodes gapperi*) decreased after disturbance, with the effect size increasing with rising disturbance intensity. These patterns relaxed as time since disturbance increased. Fontaine and Kennedy (2012) performed a meta-analysis of forty-one papers, documenting the short-term effects of fire and fire surrogates on small mammals and birds. They, like Zwolak (2009), found that postdisturbance effects on the abundance of species increased with increasing fire severity. A total of 119 bird species and 17 small mammal species were analyzed by Fontaine and Kennedy (2012). While generalizations across this broad of a group are difficult, in many ways their results mirror those of Vanderwel et al. (2007) and Zwolak (2009): the largest positive effect was on deer mice in severe burns whereas forest birds such as ovenbirds and hermit thrush (*Catharus guttatus*) were negatively affected. Interestingly, brown creepers were positively affected by moderate intensity disturbances. While the period immediately following disturbance may exhibit the most profound shifts in structure and composition with a relaxation toward the predisturbance state over time (e.g., Zwolak 2009), this is not always the case. Hossack et al. (2013), studying salamander response to fire in Montana, found little change in abundance during the period immediately after wildfire but significant declines seven to twenty-one years after disturbance. Because we lack very many long-term studies, the gen-

erality of the pattern observed by Hossack et al. (2013) is unknown.

General Strengths and Weaknesses of Disturbance-Related Meta-analyses

Throughout these meta-analyses, a general pattern emerges: bird response to disturbance is much more heavily studied than is mammal response—confining their scope to the effects of fire alone, Leidolf and Bissonette (2009) identify 512 studies on birds—and mammal study is almost exclusively limited to small mammals. The general paucity of habitat disturbance studies for mammals other than small mammals is likely due to several factors. Perhaps the most critical factor is that for larger mammals, home range sizes and/or densities are not compatible with abundance estimation within disturbance classes. Another is that they are far more difficult to locate and count than are small mammals and particularly birds. Even for small mammals, there is no approach that provides as efficient an estimate of occurrence and relative abundance as a bird point-count. Thus, specific disturbance studies of many mammalian species of interest such as large carnivores are impeded both by the utilization of large areas that likely exceed areas of disturbance and difficulties in obtaining sufficient sample sizes. The effects of disturbance on other groups such as herptofauna are far less studied.

In addition to being limited in types of species analyzed, meta-analyses are weakened by the need to reduce studies to groups that can be evaluated based on a common metric (effect size) and common covariates such as time since disturbance. While such grouping is necessary, all disturbances are not created equal; postdisturbance abundances may be correlated with abundances within the general landscape (e.g., Brotons et al. 2005), a covariate that is generally missing from disturbance studies. Similarly, the effects of past disturbances may have altered systems leading to different responses to the studied disturbance when viewed across many systems. Fox (2011), for example, notes that dry, disturbance-prone grasslands generically have fewer granivores than other grasslands; these systems would therefore be expected to have different responses to any specific fire because they are already adapted to the occurrence of frequent fires. Assuming that a species can persist through a disturbance and within the immediate postdisturbance landscape, likely the postdisturbance vegetative trajectory is more important than the time since disturbance. Monamy and Fox (2000), for example, found that recolonization trajectories for an old forest specialist differed between two fires based on time since disturbance but that the two trajectories were concordant when vegetative trajectory rather than time was evaluated. This pattern appears logical and is likely general. If this is the case, then standard approaches to successional studies where time since disturbance is used to characterize organism response are likely produce weaker results than studies that measure vegetative conditions directly.

Bird Studies Where Evaluation Close to the Disturbance Event is Essential

Clearly, more common and replicable sorts of disturbance, like fire, will be more likely to engender meaningful adaptations. However, even for common disturbance types, such as fire in dry environments, immediate postfire conditions will be relatively rare on the landscape. To take advantage of these conditions, population responses must be quick and the organisms need either to be common enough to be on or adjacent to the fire when it occurs or to be mobile enough to find these disturbances quickly so as to take advantage of the ephemeral postdisturbance environment. It is not surprising that, for example, deer mice show large increases in postfire environments. They are generalists and therefore present in the area, can reproduce rapidly, and normally favor bare areas over dense grass (Pearson et al. 2001) and should therefore prosper in postfire environments. Similarly, avian species, being highly mobile, can locate desired environments including recent disturbances and can evolve to take advantage of these conditions. For these reasons, the study of avifauna in postfire environments is both active and biologically appropriate. Saab and Powell (2005), in a review of literature on over two hundred species of birds and their associations with fire, showed that aerial, ground, and bark insectivores clearly favored recently burned habitats whereas foliage gleaners preferred unburned habitats. As with other studies, virtually all of the associated papers evaluated effects within

the first five years after the burn. However, unlike other taxa, some bird species are specifically adapted to these immediate postfire conditions, making their study in these environments especially useful. Notable among the bark insectivores is the black-backed woodpecker (*Picoides arcticus*). This species colonizes burned areas within one year after a fire, occupies burned areas for three to five years, and peaks in density around three years after fire (Caton 1996; Saab et al. 2007). Outside of this window, when postfire environments are optimal, black-backed woodpeckers are difficult to locate and are distributed diffusely feeding, opportunistically, on recently killed trees (Tremblay et al. 2010, Dudley et al. 2012).

Succession
Defining Succession

Succession is an idea that emerged in the late 1800s and was championed by Clements (1916). He envisioned the process as an orderly series of plant assemblages from the point of initiation through to the "climax," a stable state. Just as we define disturbance as events that we find disturbing given our viewpoint, similarly, we tend to consider succession in terms of processes that scale reasonably to our lifetimes. For these reasons, the genesis of succession as a concept began with the observations of the dynamics of forests (Clements 1916) and is most commonly still evoked in this context. A great deal of thinking about this concept has occurred since Clements wrote "Plant Succession: An Analysis of the Development of Vegetation" in 1916, but the underlying ideas have remained remarkably stable. Primarily, the concept has been embellished, incorporating ideas such as "secondary" succession, in which initiation contains elements of the predisturbance ecosystem, and various "disclimaxes," where an area is prevented from evolving into its presumed true climax by processes such as frequent disturbance or grazing (see Meeker and Merkel 1984 for a discussion). As mentioned previously, succession has retained popularity in many fields, particularly where ecosystems are simple and primarily composed of native species. Even in these simple systems, however, there is a great deal of variability in these trajectories due to the effects of pathogens and increasingly due to novel weather patterns owing to climate change. For example, Rehfeldt et al. (2009) showed that historically novel weather patterns, but patterns that are predicted to become increasingly common in the future, immediately preceded major aspen (*Populus tremuloides*) die-offs in Colorado.

One critical aspect of successional thinking is that, although the initiation conditions might vary as would, to a certain extent, the trajectory toward climax conditions, the climax conditions are invariant. As Horn (1975) noted, an invariant steady state is a property of stochastic systems with constant transition probabilities (Markov chains). Usher (1979) observed that if succession is described as a Markov process, then transition probabilities are likely not constant and will be affected by many factors such as population processes. Cadenasso et al. (2002), for example, showed that levels of mammal herbivory play a key role in early successional trajectories of old fields. With many exotic species entering ecosystems, the probability of unforeseen biological feedback obviously increases. For example, the successional trajectories of western North American grasslands to disturbance have been radically altered by the introduction of a plethora of exotic plant species. In western Montana, spotted knapweed (*Centaurea maculosa*) has occupied vast areas to the detriment of both native bunch grasses and native forbs (Callaway et al. 1999). In an effort to control knapweed, exotic gall flies were introduced. They failed to control the knapweed but instead became superabundant, leading to a variety of secondary trajectories involving increased densities of deer mice and associated hantavirus (Pearson and Callaway 2006). These grass systems continue to be inundated by new exotics (e.g., leafy spurge [*Euphorbia esula*]) and with climate change, other species, such as yellow star-thistle (*Centaurea solstitialis*), common in adjacent states, may gain foothold.

For a system to move in an orderly manner from disturbance toward a steady-state climax, both the species involved and the climatic background need to remain constant. This strong requirement for constancy was first challenged by Gleason (1927), who argued that a plethora of factors lead to unique postdisturbance trajectories. In the eighty-three years that have elapsed since Gleason's paper was published, the use and acceptance of successional concepts has split along disciplinary lines: disciplines like phylogeography that evaluate

phenomena across broad scales of space and time are necessarily Gleasonian, whereas fields that study small, local phenomena tend to be Clemensian (Hortal et al. 2012). The effects of rapid directional climate change on this debate are to shrink the spatiotemporal domain across which constancy is a reasonable conceit. We are relatively sure that climates will change significantly over the next several centuries and that climate change is already affecting ecosystems, but our understandings of both the details of local climate change and, even more critically, the effects of these changes on habitat and the wildlife that these habitats support are vague (Walther et al. 2002). This uncertainty is obviously increased by the continuous entry of additional exotic species whose interactions with current ecosystems are largely unknown. There is a school of thought that believes that these specifics, when taken together, are important enough to render the entire concept of succession suspect (Ricklefs 2008).

Wildlife Research on the Effects of Succession

The overwhelming number of disturbance papers that measure immediate postdisturbance responses and the associated lack of longer-term studies dictates that the direct study of the site-level effects of succession are sparse to nonexistent. One of the few cases where long-term studies on primary successional dynamics are ongoing is within the area affected by the pyroclastic explosion of Mount Saint Helens (e.g., Spear et al. 2012). Because large-scale population-level studies of the effects of succession on wildlife are difficult and seldom done, most of our understandings of forest succession on wildlife are associated with habitat use data. For example, we know that some species have strong associations with certain habitats and that these habitats are related to disturbance. The Florida sand skink (*Plestiodon reynoldsi*), for example, is obligately associated with Florida scrub habitat (Schrey et al. 2011), a cover type that requires high-intensity fire on a five-to-eighty-year time scale (Laessle 1958). Knowing this habitat association, we can infer the relationship between this species and succession. Given this constraint—that we understand habitat relationships and know the disturbance regime associated with a particular vegetative community—there is a vast literature on the effects of succession on wildlife (see Hunter and Schmiegelow 2011 for a recent review). Interestingly, a great deal of this work focuses on the two ends of the successional spectrum: early successional conditions associated with the immediate aftermath of disturbances and the study of a variety of organisms that are dependent on old forests. This may be because organisms of interest are associated with these two seres. For ungulates, likely the most heavily studied mammal group, the general understanding is that early seral environments are preferred due to both the encouragement of forage and increased forage nutritional content (e.g., Hobbs and Spowart 1984), and a great deal of focus is therefore on the generation and maintenance of these conditions. Similarly, because very old forests are in decline worldwide, dependent species often follow this trend and may become endangered. Species such as the spotted owl (*Strix occidentalis*) and red-cockaded woodpecker (*Picoides borealis*) have become iconic in this regard (Bart and Forsman 1992; Ligon et al. 1986). These older forests are not only in decline but often are associated with higher levels of biological diversity when compared to earlier seres (e.g., Díaz et al. 2005). The emphasis on early seral and old growth forests is also due to a perceived overabundance of mid-aged forests owing to a variety of human activities. Under natural disturbance regimes, forest age structures frequently follow negative exponential or Weibull distributions (Van Wagner 1978), or distributions containing significant areas of both old forest and areas that have recently been disturbed. Conversely, human-caused disturbance generally leads to truncated age distributions and can simultaneously decrease early seral conditions through activities such as fire suppression (e.g., Betts et al. 2010).

Disturbance Viewed in a Landscape Context: Studies of Both Disturbance and Succession

In the case of black-backed woodpeckers, postfire conditions are clearly essential habitat, but for many other species these environments are rare and noncritical. However that may be, based on this review, it is clear that targeted studies of disturbance occur almost exclusively during the immediate aftermath of the event. While it is clear that this has been our approach, it is reasonable to ask whether it is a necessary approach. We could, rather than seek to quantify the specific effects of specific disturbances, endeavor to design studies that would allow comparison and generality between disturbances. In the case of hurricanes on the eastern coast of the United States, for example, rather than studying the effects of Hugo, one could ask how much the immediate postdisturbance conditions related to Hugo resemble those of other recent strong "Cape Verde–type" hurricanes that made landfall on the southeastern coast of the United States (e.g., Fran [1996]; Floyd [1998]; Isabel [2003]). One paper, Rittenhouse et al. (2010), attempted to place the effects of hurricanes in a broader context, including both immediate poststorm effects and overall changes in species composition for birds. Rittenhouse used the North American Breeding Bird Survey (BBS; Link and Sauer 1998) to infer species abundance patterns for the period 1967–2005. This period included a number of hurricane landfalls, notably Hurricane Hugo (September 1989). Study areas were chosen based on the intersection of hurricane tracks and a time series of annual or biennial satellite imagery. The main effect of hurricanes was seen in the year immediately following the hurricane, where modest decreases in abundance and increases in species richness were similar across focal areas. Interestingly, although hurricanes caused a threefold increase in the extent of disturbed forest within the focal areas, this factor was associated with changes in community similarity for only three of the thirteen avian groups examined and was not associated with patterns of avian abundance or species richness. Rittenhouse et al. (2010) attribute the lack of relationships between disturbance levels and population responses to landscape scale heterogeneity; refuges were present in areas where intact forests were the predisturbance land cover. This understanding is similar to that of Brotons et al. (2005), who found that local abundances in areas exterior to disturbances affected postdisturbance abundances within the disturbed area. Thus, to understand postdisturbance trajectories, evaluating the larger landscape may be essential. While Rittenhouse et al. (2010) were able to analyze data at spatial and temporal scales sufficient to infer generalized relationships, studying disturbances at this scale is difficult and was only possible in this case because of the presence of the BBS, which is relatively unique both in its longevity and in its spatially comprehensiveness. However, even though Rittenhouse et al. (2010) were able to apply forty years of BBS data, because hurricanes occurred sequentially within the time period, statistical statements were limited to a period within ten years after disturbance.

A rare example of a meta-analysis of both disturbance and succession is found in Schieck and Song (2006). They performed a meta-analysis of studies on boreal forest birds for a period of 0–125 years after disturbance. Interestingly, they found that while bird communities changed after a disturbance, composition changes were largely associated with numbers of occurrences rather than actual changes in species composition. All species with measurable counts (greater than five) were present in all forest types. The ability of organisms to persist in all forest types and conditions obviates the need to colonize areas except for after the most extreme events. It also suggests that immediate postfire abundance studies may not provide informative data on the landscape-level responses of populations to disturbance and succession.

Species Relationships to Disturbance and Succession: The Case of the Spotted Owl

Given the current approach of conducting largely independent and descriptive studies of postdisturbance organism response, it is reasonable to ask how well these studies composite into generalized understandings for any organism. The spotted owl, being one of the most heavily studied species, provides a good example of the difficulties associated with determining the relationships between an organism and disturbance while applying this methodology. As mentioned previously, the spotted owl has become an iconic organism representing an old forest obligate (Bart and Forsman 1992), and for this reason, the long-term viability of owls in fire-prone forests has been questioned (Spies et al. 2006). However, spotted owls exist in a wide variety of forests, including some that are highly modified (Thome et al. 1999). Short-term relationships to anthropogenic disturbance have been highly studied, with uniformly negative relationships (e.g., Forsman et al. 1984; Franklin et al. 2000), except where the dominant prey are woodrats (*Neotoma fuscipes*; Thome et al. 1999). Fire is much less heavily studied, and the effects are less clear. In some cases, low-severity fire appears to have little effect (Roberts et al. 2011). Clark et al. (2013) associated negative effects with fire, but sites they studied had been burned and salvage logged, and pre-fire timber harvest had occurred. Due to sample size constraints, Clark et al. (2013) could not separate anthropogenic and natural disturbance factors. Further, while Franklin et al. (2000) found short-term negative effects, the heterogeneous fire-generated landscape of northwestern California featured high reproductive and survival rates (Franklin et al. 2000)—the fire mosaic in general appeared to produce excellent owl habitat. In short, the relationship between spotted owls and disturbance is complex and our understandings are still very nebulous. Using derived information to conserve spotted owls into the future is also complicated by the twin factors of climate change and exotic species introduction. Owl demographics are sensitive both to regional climate and local weather (Glenn et al. 2010). Further, fire frequency is anticipated to increase in the Pacific Northwest with climate change (McKenzie et al. 2004). Lastly, the invasion of the barred owl (*Strix varia*) is having a profound negative effect on spotted owl demographics across its range (Dugger et al. 2011), fundamentally changing the expected population responses to specific vegetative patterns. In short, the spotted owl provides a case study in all of the complexities highlighted in this paper. The effects of disturbance are complex, depending both on time since disturbance and the scale at which the question is asked. Effects of any specific type of disturbance are difficult to quantify because, at the scale of spotted owl home ranges, a variety of disturbance types have often occurred: fire, logging, and postfire salvage. Lastly, our understandings concerning the effects of postdisturbance succession on the quality of spotted owl habitat have been fundamentally altered by the invasion of an exotic competitor.

Approaches to the Problem

It is clear that our traditional approaches, involving primarily small-scale, short-duration descriptive studies within disturbed areas (and sometimes including undisturbed control areas), while producing much knowledge of local importance, do not provide a coherent approach to increasing our knowledge of the more general effects of disturbance and succession on wildlife. It is equally clear that the traditional approach of trading space for time to study succession is becoming less and less tenable due to the increasing pool of exotic organisms and the directional nature of climate change. What is less clear is how to move forward in a manner that leads to more rapid increases in our knowledge of these processes while existing in a world that is in a rapid state of biological flux. The simple answer is to expand our studies in both space and time. This would allow the formalization of the context in which disturbance occurs, including the effects of landscape pattern and habitat juxtaposition, and would avoid invoking constancy assumptions to infer future states. There are, however, major problems associated with this approach. The first is cost. Increasing the size and duration of a study quadratically increases the costs. Further, funding large, long-term projects is difficult given the competitive grant paradigm under which a large proportion of science is funded. Competitive grants generally have both spending and time limits that preclude their direct utility for long-term funding; long-term studies are often patched together by acquiring multiple grants, often from different sources. Additionally, there are many ways that these programs can fail and many examples of historical failures due to funding drying up, data loss due to inadequate archiving, failures to maintain quality control across time and space, and poor initial designs leading to the collection of large quantities of relatively useless data. Lastly, there is the strong, but often unacknowledged, power of serendipity associated with many independent studies. The ability to capture serendipity is decreased if scientific resources divert from a large number of small, highly independent studies and commit to a small number of highly directed, long-term projects.

There are, however, approaches that merge the fields of science and monitoring and which may prove useful in documenting landscape-level changes associated with disturbance and subsequent succession. Targeted monitoring (Yoccoz et al. 2001; Nichols and Williams 2006) combines principles of scientific design including clearly articulated goals and specified error levels with the large-scale, long-term requirements for biological monitoring. Increasing the proportion of monitoring that is specifically targeted could allow a reallocation of extant long-term funding into efforts to collect appropriate data to assess biological processes across broader spatiotemporal domains. Because of the high level of scientific design associated with targeted monitoring, intermediate short-term products can both increase the immediate utility of the monitoring and provide feedback on its overall efficacy. Secondly, monitoring efforts should take advantage of new technologies and, specifically, genetic sampling (Schwartz et al. 2007). Not only do these methods often provide the least expensive approach to sampling species (e.g., Kendall and McKelvey 2008), but genetic data are nearly unique in that, if the samples are retained, these samples can be reanalyzed at a later date. Even without the formal retention of genetic data, the past collection of materials that contain genetic data has allowed retrospective studies of populations. For example, Miller and Waits (2003) were able to retrospectively calculate the effective population size of grizzly bears in Yellowstone National Park by analyzing 110 museum specimens collected between 1912 and 1981. Because of this ability to retrospectively analyze data collected in the past, genetically based monitoring is less dependent on current technologies than are most other monitoring approaches. With genetic data, new technologies can be applied to extant samples, allowing a fully modern analysis of all data regardless of collection time. This ability also provides a safeguard against data loss and allows post-hoc testing quality controls: assays can be repeated both to recover data and to ensure that the original analyses were correct.

Conclusions

There has been little to no attempt in the literature to view disturbance in terms of the operational environment of the studied species; we define disturbances at anthropocentric scales. While this is biologically problematic, it is so ubiquitous that it is impossible to write a review article without also adopting this conceit. However, given this constraint, there is a vast body of literature associated with both the effects of disturbance and succession on wildlife. However, due both to practicalities and interest levels, the vast majority of specific studies on the effects of disturbance focus on the period immediately postdisturbance; few studies extend beyond five years postdisturbance. Further, most studies are small in scope and are not designed to test concepts; most generalizations are based on compositing these small-scale primary studies through literature reviews and meta-analyses, of which there are many. However, as the case of the spotted owl demonstrates, it is difficult, even with many largely unrelated small-scale studies of specific areas, to generate coherent understandings of general relationships between an organism and its disturbance environment. Successional studies generally trade space for time, assuming that patterns associated with older habitats generated by past disturbances will indicate future patterns in newly disturbed areas. While this practice is directly related to the paucity of longer-term disturbance-related studies, it is also a necessity as many of the successional processes we care about occur across hundreds of years. This approach has historically proven to be extremely useful, but it contains a variety of hidden assumptions that are unlikely to be met as the rate of biological change increases worldwide.

We are currently seeing forest mortality at levels unprecedented in recent history (Allen et al. 2010). These large-scale disturbances provide the opportunity for ecosystems to evolve to meet climatic exigencies and are the engine of predicted broad-scale biome shifts. Given this, we should not assume that relationships based in large part on disturbance events in the past will necessarily provide adequate guidance when predicting the ecological trajectories associated with current processes.

Rittenhouse et al. (2010) demonstrated that the utility of broad-scale temporally stable monitoring for evaluating the effects of disturbance on wildlife communities and monitoring efforts such as BBS will most likely become increasingly important as climate change and exotic species produce increasingly novel ecosystems (Hobbs et al. 2009); we should support these efforts and seek to increase the number of taxa that are monitored. However, these sorts of generalized surveys will always be weak for evaluating specifics of disturbance and succession. Given the pivotal role disturbance plays in ecological change, more specific focus is probably warranted. To meet this need, we can design targeted monitoring studies specifically designed to assess the effects of disturbance and succession on wildlife populations. Ideally, these would represent a redirection of existing monitoring efforts, allowing us to ascertain whether successional trajectories and associated wildlife communities were following expectations as well as when and where systems were changing in unexpected ways.

LITERATURE CITED

Allen, C. D., A. K. Macalady, H. Chenchouni, D. Bachelet, N. McDowell, M. Vennetier, T. Kitzberger, A. Rigling, D. D. Breshears, E. H. Hogg, P. Gonzalez, R. Fensham, Z. Zhang, J. Castro, N. Demidova, J.-H. Lim, G. Allard, S. W. Running, A. Semerci, and N. Cobb. 2010. "A Global Overview of Drought and Heat-Induced Tree Mortality Reveals Emerging Climate Change Risks for Forests." *Forest Ecology and Management* 259:660–84.

Bart, J., and E. D. Forsman. 1992. "Dependence of Northern Spotted Owls *Strix occidentalis caurina* on Old-Growth Forests in the Western USA." *Biological Conservation* 62:95–100.

Betts, M. G., J. C. Hagar, J. W. Rivers, J. D. Alexander, K. Mcgarigal, and B. C. Mccomb. 2010. "Thresholds in Forest Bird Occurrence as a Function of the Amount of Early-Seral Broadleaf Forest at Landscape Scales." *Ecological Applications* 20:2116–30.

Bormann, F. H., and G. E. Likens. 1979. *Pattern and Process in a Forested Ecosystem: Disturbance, Development and the Steady State*. Springer-Verlag, New York.

Brotons, L., P. Pons, and S. Herrando. 2005. "Colonization of Dynamic Mediterranean Landscapes: Where Do Birds Come from After Fire?" *Journal of Biogeography* 32:789–98.

Cadenasso, M. L., S. T. A. Pickett, and P. J. Morin. 2002. "Experimental Test of the Role of Mammalian Herbivores on Old Field Succession: Community Structure and Seedling Survival." *Journal of the Torrey Botanical Society* 129:228–37.

Callaway, R. M., T. H. Deluca, and W. M. Belliveau. 1999. "Biological-Control Herbivores May Increase Competitive Ability of the Noxious Weed *Centaurea maculosa*." *Ecology* 80:1196–1201.

Caton, E. L. 1996. "Effects of Fire and Salvage Logging on the Cavity Nesting Bird Community in Northwestern Montana." PhD dissertation, University of Montana, Missoula, Montana.

Clark, D. A., R. G. Anthony, and L. S. Andrews. 2013. "Relationship between Wildfire, Salvage Logging, and Occupancy of Nesting Territories by Northern Spotted Owls." *Journal of Wildlife Management*. doi: 10.1002/jwmg.523.

Clements, F. E. 1916. *Plant Succession: An Analysis of the Development of Vegetation*. Carnegie Institute of Washington, Washington D. C.

Covington, W. W., and M. M. Moore. 1994. "Southwestern Ponderosa Forest Structure: Changes Since Euro–American Settlement." *Journal of Forestry* 92:39–47.

Daubenmire, R. 1952. "Forest Vegetation of Northern Idaho and Adjacent Washington, and Its Bearing on Concepts of Vegetation Classification." *Ecological Monographs* 22:301–30.

Díaz, I. A., J. J. Armesto, S. Reid, K. E. Sieving, and M. F. Willson. 2005. "Linking Forest Structure and Composition: Avian Diversity in Successional Forests of Chiloé Island, Chile." *Biological Conservation* 123:91–101.

Dudley, J. G., V. A. Saab, and J. P. Hollenbeck. 2012. "Foraging-Habitat Selection of Black-Backed Woodpeckers in Forest Burns of Southwestern Idaho." *The Condor* 114:348–57.

Dugger, K. M., R. G. Anthony, and L. S. Andrews. 2011. "Transient Dynamics of Invasive Competition: Barred Owls, Spotted Owls, Habitat, and the Demons of Competition Present." *Ecological Applications* 21:2459–68.

Fontaine, J. B., and P. L. Kennedy. 2012. "Meta-analysis of Avian and Small-Mammal Response to Fire Severity and Fire Surrogate Treatments in U.S. Fire-Prone Forests." *Ecological Applications* 22:1547–61.

Forsman, E. D., E. C. Meslow, and H. M. Wight. 1984. "Distribution and Biology of the Spotted Owl in Oregon." *Wildlife Monographs* 87:1–64.

Forsman, J. T., P. Reunanen, J. Jokimäki, and M. Mönkkönen. 2010. "The Effects of Small-Scale Disturbance on Forest Birds: A Meta-analysis." *Canadian Journal of Forest Research* 40:1833–42.

Fox, B. J. 2011. "Review of Small Mammal Trophic Structure in Drylands: Resource Availability, Use, and Disturbance." *Journal of Mammalogy* 92:1179–92.

Franklin, A. B., D. R. Anderson, R. J. Gutiérrez, and K. P. Burnham. 2000. "Climate, Habitat Quality, and Fitness in Northern Spotted Owl Populations in Northwestern California." *Ecological Monographs* 70:539–90.

Franklin, J. F., and R. T. T. Forman. 1987. "Creating Landscape Patterns by Forest Cutting: Ecological Consequences and Principles." *Landscape Ecology* 1:5–18.

Gleason, H. A. 1927. "Further Views on the Succession-Concept." *Ecology* 8:299–326.

Glenn, E. M., R. G. Anthony, and E. D. Forsman. 2010. "Population Trends in Northern Spotted Owls: Associations with Climate in the Pacific Northwest." *Biological Conservation* 143:2543–52.

Iverson, G. C., P. A. Vohs, and T. C. Tacha. 1987. "Habitat Use by Mid-Continent Sandhill Cranes during Spring Migration." *Journal of Wildlife Management* 51:448–58.

Habeck, J. R., and R. W. Mutch. 1973. "Fire-Dependent Forests in the Northern Rocky Mountains." *Quaternary Research* 3:408–24.

Hartway, C., and L. S. Mills. 2012. "A Meta-analysis of the Effects of Common Management Actions on the Nest Success of North American Birds." *Conservation Biology* 26:657–66.

Hobbs, N. T., and R. A. Spowart. 1984. "Effects of Prescribed Fire on Nutrition of Mountain Sheep and Mule Deer during Winter and Spring." *Journal of Wildlife Management* 48:551–60.

Hobbs, R. J., E. Higgs, and J. A. Harris. 2009. "Novel Ecosystems: Implications for Conservation and Restoration." *Trends in Ecology and Evolution* 24:599–605.

Horn, H. S. 1975. "Markovian Properties of Forest Succession." In *Ecology and Evolution of Communities*, edited by M. L. Cody and J. M. Diamond, 196–211. Belknap Press of Harvard University Press, Cambridge, Massachusetts.

Hortal, J., P. De Marco Jr., A. M. C. Santos, and J. A. F. Diniz-Filho. 2012. "Integrating Biogeographical Processes and Local Community Assembly." *Journal of Biogeography* 39:627–28.

Hossack, B. R., W. H. Lowe, and P. S. Corn. 2013. "Rapid Increases and Time-Lagged Declines in Amphibian Occupancy after Wildfire." *Conservation Biology* 27:219–28.

Hunter Jr., L., and F. K. A. Schmiegelow. 2011. *Wildlife, Forests and Forestry: Principles of Managing Forests for Biological Diversity.* 2nd edition. Prentice Hall, Upper Saddle River, New Jersey.

Kalies, E. L., C. L. Chambers, and W. W. Covington 2010. "Wildlife Responses to Thinning and Burning Treatments in Southwestern Conifer Forests: A Meta-analysis." *Forest Ecology and Management* 259:333–42.

Kendall, K. C., and K. S. McKelvey. 2008. "Hair Collection." In *Noninvasive Survey Methods for North American Carnivores*, edited by R. A. Long, P. MacKay, J. C. Ray, and W. J. Zielinski, 141–82. Island Press, Washington D.C.

Kotliar, N. B., and J. A. Wiens. 1990. "Multiple Scales of Patchiness and Patch Structure: A Hierarchical Framework for the Study of Heterogeneity." *Oikos* 59:253–60.

Küchler, A. W. 1964. "Potential Natural Vegetation of the Conterminous United States." *American Geographical Society Special Publication* 36:1–54.

Laessle, A. M. 1958. "The Origin and Successional Relationship of Sandhill Vegetation and Sand-Pine Scrub." *Ecological Monographs* 28:361–87.

Leidolf, A., and J. A. Bissonette. 2009. "The Effects of Fire on Avian Communities: Spatio-Temporal Attributes of the Literature, 1912–2003." *International Journal of Wildland Fire* 18:609–22.

Levins, R. 1969. "The Effects of Random Variation of Different Types on Population Growth." *Proceedings of the National Academy of Sciences* 62:1061–65.

———. 1970. "Extinction." In *Some Mathematical Questions in Biology*, edited by M. Gerstenhaber, 77–107. American Mathematical Society, Providence, Rhode Island.

Ligon, J. D., P. B. Stacey, R. N. Conner, C. E. Bock, and C. S. Adkisson. 1986. "Report of the American Ornithologists' Union Committee for the Conservation of the Red-Cockaded Woodpecker." *The Auk* 103:848–55.

Link W. A., and J. R. Sauer. 1998. "Estimating Population Change from Count Data: Application to the North American Breeding Bird Survey." *Ecological Applications* 8:258–68.

Lodge, D. M. 1993. "Species Invasions and Deletions: Community Effects and Responses to Climate and Habitat Change." In *Biotic Interactions and Global Change*, edited by P. M. Kareiva, J. G. Kingsolver, and R. B. Huey, 367–87. Sinauer Associates, Sunderland, Massachusetts.

Martell, K. A., A. L. Foote, and S. G. Cumming. 2006. "Riparian Disturbance Due to Beavers (*Castor canadensis*) in Alberta's Boreal Mixedwood Forests: Implications for Forest Management." *Ecoscience* 13:164–71.

McKenzie, D., Z. Gedalof, D. L. Peterson, and P. Mote. 2004. "Climatic Change, Wildfire, and Conservation." *Conservation Biology* 18:890–902.

Meeker Jr., D. O., and D. L. Merkel. 1984. "Climax Theories and a Recommendation for Vegetation Classification: A Viewpoint." *Journal of Range Management* 37:427–30.

Miller, C. R., and L. P. Waits. 2003. "The History of Effective Population Size and Genetic Diversity in the Yellowstone Grizzly (*Ursus arctos*): Implications for Conservation." *Proceedings for the National Academy of Science* 100:4334–39.

Monamy, V., and B. J. Fox. 2000. "Small Mammal Succession is Determined by Vegetation Density Rather Than Time Elapsed Since Disturbance." *Austral Ecology* 25:580–87.

Naiman, R. J., J. M. Melillo and J. E. Hobbie. 1986. "Ecosystem Alteration of Boreal Forest Streams by Beaver (*Castor canadensis*)." *Ecology* 67:1254–69.

Nichols, J. D., and B. K. Williams. 2006. "Monitoring for Conservation." *Trends in Ecology and Evolution* 21:668–73.

Nowacki, G. J., and M. D. Abrams. 2008. "The Demise of Fire and 'Mesophication' of Forests in the Eastern United States." *BioScience* 58:123–38.

Pearson, D. E., and R. M. Callaway. 2006. "Biological Control Agents Elevate Hantavirus by Subsidizing Deer Mouse Populations." *Ecology Letters* 9:443–50.

Pearson, D. E., Y. K. Ortega, K. S. McKelvey, and L. F. Ruggiero. 2001. "Small Mammal Community Composition, Relative Abundance, and Habitat Selection in Native Bunchgrass of the Northern Rocky Mountains: Implications for Exotic Plant Invasions." *Northwest Science* 75:107–17.

Pfister, R. D., and S. F. Arno. 1980. "Classifying Forest Habitat Types Based on Potential Climax Vegetation." *Forest Science* 26:52–70.

Rehfeldt, G. E., D. E. Ferguson, and N. L. Crookston. 2009. "Aspen, Climate, and Sudden Decline in Western USA." *Forest Ecology and Management* 258:2353–64.

Ricklefs, R. E. 2008. "Disintegration of the Ecological Community." *The American Naturalist* 172:741–50.

Rittenhouse, C. D., A. M. Pidgeon, T. P. Albright, P. D. Culbert, M. K. Clayton, C. H. Flather, C. Huang, J. G. Masek, and V. C. Radeloff. 2010. "Avifauna Response to Hurricanes: Regional Changes in Community Similarity." *Global Change Biology* 16:905–17.

Roberts, S. L., J. W. van Wagtendonk, A. K. Miles, and D. A. Kelt. 2011. "Effects of Fire on Spotted Owl Site Occupancy in a Late-Successional Forest." *Biological Conservation* 144:610–19.

Russell, R. E., J. A. Royle, V. A. Saab, J. F. Lehmkuhl, W. M. Block, and J. R. Sauer. 2009. "Modeling the Effects of Environmental Disturbance on Wildlife Communities: Avian Responses to Prescribed Fire." *Ecological Applications* 19:1253–63.

Rykiel Jr., E. J. 1985. "Towards a Definition of Ecological Disturbance." *Australian Journal of Ecology* 10:361–65.

Saab, V. A., and H. D. W. Powell. 2005. "Fire and Avian Ecology in North America: Process Influencing Pattern." *Studies in Avian Biology* 30:1–13.

Saab V. A., R. E. Russell, and J. G. Dudley. 2007. "Nest Densities of Cavity-Nesting Birds in Relation to Postfire Salvage Logging and Time Since Wildfire." *The Condor* 109:97–108.

Schieck, J., and S. J. Song. 2006. "Changes in Bird Communities throughout Succession Following Fire and Harvest in Boreal Forests of Western North America: Literature Review and Meta-analyses." *Canadian Journal of Forest Research* 36:1299–1318.

Schrey, A. W., A. M. Fox, H. R. Mushinsky, and E. D. Mccoy. 2011. "Fire Increases Variance in Genetic Characteristics of Florida Sand Skink (*Plestiodon reynoldsi*) Local Populations." *Molecular Ecology* 20:56–66.

Schwartz, M. K., G. Luikart, and R. S. Waples. 2007. "Genetic Monitoring as a Promising Tool for Conservation and Management." *Trends in Ecology and Evolution* 22:25–33.

Shafer, A. B. A., C. I. Cullingham, S. D. Cote, and D. W. Coltman. 2010. "Of Glaciers and Refugia: A Decade of Study Sheds New Light on the Phylogeography of Northwestern North America." *Molecular Ecology* 19:4589–4621.

Sherry, T. W., and R. T. Holmes. 1988. "Habitat Selection by Breeding American Redstarts in Response to a Dominant Competitor, the Least Flycatcher." *The Auk* 105:350–64.

Sousa, W. P. 1984. "The Role of Disturbance in Natural Communities." *Annual Review of Ecology and Systematics* 15:353–91.

Spear, S. F., C. M. Crisafull, and A. Storfer. 2012. "Genetic Structure among Coastal Tailed Frog Populations at Mount St. Helens is Moderated by Post-disturbance Management." *Ecological Applications* 22:856–69.

Spies, T. A., M. A. Hemstrom, A. Youngblood, and S. Hummel. 2006. "Conserving Old-Growth Forest Diversity in Disturbance-Prone Landscapes." *Conservation Biology* 20:351–62.

Thome, D. M., C. J. Zabel, and L. V. Diller. 1999. "Forest Stand Characteristics and Reproduction of Northern Spotted Owls in Managed North-Coastal California Forests." *Journal of Wildlife Management* 63:44–59.

Tremblay, J. A., J. Ibarzabal, and J.-P. L. Savard. 2010. "Foraging Ecology of Black-Backed Woodpeckers (*Picoides arcticus*) in Unburned Eastern Boreal Forest Stands." *Canadian Journal of Forest Research* 40:991–99.

Usher, M. B. 1979. "Markovian Approaches to Ecological Succession." *Journal of Animal Ecology* 48:413–26.

Vanderwel, M. C., J. R. Malcolm, and S. C. Mills. 2007. "A Meta-analysis of Bird Responses to Uniform Partial Harvesting across North America." *Conservation Biology* 21:1230–40.

Van Wagner, C. E. 1978. "Age-Class Distribution and the Forest Fire Cycle." *Canadian Journal of Forest Research* 8:220–27.

Walther, G.-R., E. Post, P. Convey, A. Menzel, C. Parmesan,

T. J. C. Beebee, J.-M. Fromentin, O. Hoegh-Guldberg, and F. Bairlein. 2002. "Ecological Responses to Recent Climate Change." *Nature* 416:389–95.

Wiens, J. A. 1989. "Spatial Scaling in Ecology." *Functional Ecology* 3:385–97.

Yoccoz, N. G., J. D. Nichols, and T. Boulinier. 2001. "Monitoring of Biological Diversity in Space and Time." *Trends in Ecology and Evolution* 16:446–53.

Zerbe, S. 1998. "Potential Natural Vegetation: Validity and Applicability in Landscape Planning and Nature Conservation." *Applied Vegetation Science* 1:165–72.

Zwolak, R. 2009. "A Meta-analysis of the Effects of Wildfire, Clearcutting, and Partial Harvest on the Abundance of North American Small Mammals." *Forest Ecology and Management* 258:539–45.

12 — Wildlife Habitat Restoration

Kathi L. Borgmann and
Courtney J. Conway

As the preceding chapters point out, many wildlife species and the habitat they depend on are in peril. However, opportunities exist to restore habitat for many imperiled wildlife species. But what is wildlife habitat restoration? We begin this chapter by defining habitat restoration and then provide recommendations on how to maximize success of future habitat restoration efforts for wildlife. Finally, we evaluate whether we have been successful in restoring wildlife habitat and supply recommendations to advance habitat restoration. Successful restoration requires clear and explicit goals that are based on our best understanding of what the habitat was like prior to the disturbing event. Ideally, a restoration project would include: (1) a summary of prerestoration conditions that define the existing status of wildlife populations and their habitat; (2) a description of habitat features required by the focal or indicator species for persistence; (3) an a priori description of measurable, quantitative metrics that define restoration goals and measures of success; (4) a monitoring plan; (5) postrestoration comparisons of habitat features and wildlife populations with adjacent unmodified areas that are similar to the restoration site; and (6) expert review of the entire restoration plan (i.e., the five aforementioned components).

Introduction

More than 50% of the land in six out of fourteen of the world's biomes has been converted to anthropogenic land uses (primarily agriculture) since the 1900s (Millennium Ecosystem Assessment 2005). The rate of decline in species diversity is greater than at any other time in human history, and the decline is due primarily to loss or degradation of habitat (Millennium Ecosystem Assessment 2005). As previous chapters point out, a substantial portion of some species' geographic range has been degraded due to one or more anthropogenic changes (Francis, this volume; Wiens and Van Horne, this volume). Hence, the need to restore wildlife habitat within degraded areas to help prevent further losses in biodiversity is now a worldwide focus (Suding 2011). In fact, numerous restoration projects have been initiated all over the world, including large-scale salt-marsh restoration in the San Francisco Bay (EDAW et al. 2007), seabird colony restoration worldwide (Jones and Kress 2012), and restoration of the Everglades in south Florida (US Army Corps of Engineers and South Florida Water Management District 2006). *Restoration* has, in fact, become a familiar term to many of us, but what is restoration? Restoration is "the process of assisting the recovery of an ecosystem that has been degraded, damaged, or destroyed" (Society for Ecological Restoration International Science and Policy Working Group 2004). Restoration can encompass many different aspects; we will focus more narrowly on wildlife habitat restoration. Wildlife habitat restoration describes efforts that focus specifically on improving the resources necessary to promote occupancy or improve habitat suitability for one or more wildlife species (Hall et al. 1997; Morrison et al. 1998). We assume that *habitat* restoration efforts are a subset of ecological restoration efforts because they are directed explicitly at improving habitat conditions for one or more focal species. However, because the term *habitat* is inherently species-specific, restoration efforts that seek to improve habitat conditions to promote occupancy for one species may not necessarily provide suitable habitat for other species. Some have argued that efforts to restore areas to explicitly benefit a single (or even a few) species are shortsighted, and that management actions should instead target entire ecological communities (Hobbs and Norton 1996; Palmer et al. 1997; Menninger and Palmer 2006). Although conservation of entire communities is often the goal of management effort, many habitat restoration efforts will necessarily focus on one or a few species due to legislative mandates or political necessity. In cases where wildlife habitat restoration is targeted at one or more focal species, restoration plans should also be designed to not adversely affect other endemic species in the community.

Our goal in this chapter is not to provide in-depth summaries of the numerous books and papers that have been written about habitat restoration (Hobbs and Norton 1996; Block et al. 2001; Morrison 2001, 2002; Falk et al. 2006). Instead, our goals are to highlight some common problems shared by many past wildlife habitat restoration efforts, describe key components of a successful habitat restoration project, and discuss some steps practitioners can take to improve the efficiency and maximize the effectiveness of future habitat restoration efforts. To help identify areas where past restoration efforts could be improved, we first conducted a literature review to assess the effectiveness of past wildlife habitat restoration projects. We searched ISI Web of Knowledge (www.isiwebofknowldege.com) for "habitat restoration" OR "habitat enhancement" and also included an additional search for "restoration AND habitat AND influence, OR impact*, OR response* AND bird*, OR mammal*." Our search undoubtedly overlooked many published (and even more unpublished) restoration efforts, but we assume that the results of the search are representative of habitat restoration projects. This search resulted in 1,458 references, but only 58 of these references contained information suitable for inclusion in our review (i.e., summarized restoration activities that were conducted to benefit terrestrial wildlife species). The other 1,400 papers merely mentioned habitat restoration in the abstract or provided a summary of habitat restoration

projects but did not provide information on any specific restoration action(s).

While reviewing these restoration studies, we noted problems that were common to many wildlife habitat restoration projects. Several habitat restoration efforts (1) failed to consider the habitat requirements of species (Morrison 1998; Block et al. 2001; Miller and Hobbs 2007; Maslo et al. 2011), (2) lacked quantitative criteria to assess the degree of success of the restoration effort (i.e., pre- or postrestoration monitoring) (Block et al. 2001), and (3) failed to adequately share the outcome with practitioners or the public (Block et al. 2001; Bernhardt et al. 2005; Miller and Hobbs 2007; Suding 2011). Well-planned habitat restoration efforts can improve habitat suitability for wildlife species, but many habitat restoration efforts were conducted with a very general understanding of the habitat requirements of wildlife species. For example, beach stabilization plans failed to consider nesting requirements of the federally endangered piping plover (*Charadrius melodus*; Maslo et al. 2011). Failing to consider habitat requirements of the focal species or suite of species during restoration plans can result in failure to achieve restoration goals at the expense of the species. Common among many restoration studies was the failure to adequately evaluate the successes and failures of restoration efforts. Monitoring wildlife before and after restoration is required to determine if restoration has been successful. Failing to measure wildlife response to restoration stymies our ability to improve future restoration projects and leaves many practitioners with little guidance regarding how to learn from past habitat restoration efforts. In the following discussion, we describe key components of a successful habitat restoration project.

Developing a Restoration Plan

A common criticism of many restoration projects is the lack of an explicit "desired state" or "desired outcome." Oftentimes, the desired outcome of restoration is to restore an area to pre-European conditions (Thorpe and Stanley 2011). However, sufficient details about the conditions that existed prior to European settlement are typically not available for most areas (Thorpe and Stanley 2011). Hence, restoration efforts that seek to return areas to pre-European conditions have led some authors to criticize restoration ecology as being past oriented and static (Hobbs and Norton 1996; Davis 2000; Choi 2007; Hobbs et al. 2011). Instead of focusing on one point in time (i.e., pre-European settlement) as the desired condition for all restoration efforts, we would likely be better served by an approach that restores the natural disturbance regime or removes the anthropogenic factor(s) that resulted in suboptimal conditions. The restoration point then becomes the predisturbance condition (i.e., the likely condition prior to anthropogenic changes that created the need for restoration) and not some predetermined point in time. Careful consideration of predisturbance data, where available, combined with conditions at suitable reference sites can often provide a meaningful benchmark for restoration. If predisturbance conditions are not known, restoration can be targeted so that the restored areas are near the middle of the full range of environmental variation that existed historically at the site (Hobbs and Norton 1996). In either case, restoration efforts should focus less on returning an area to some static state that existed prior to disturbance in the future but instead on a longer-term goal of achieving a range of conditions that will eventually create a self-sustaining system without the need for continual human intervention.

We suggest that the first step in developing a restoration plan is to summarize what is known about past conditions at the site and evaluate how those differ from existing conditions at the proposed site. This will help answer the question: "Why does the site currently need restoration?" Other useful questions to ask at the outset include: For which suite of species has habitat suitability or abundance declined? What ecosystem functions did the site provide prior to anthropogenic changes (that it no longer provides)? Documenting the range of historic environmental conditions and the natural disturbance processes that likely existed prior to anthropogenic changes will help determine if particular species have likely been lost from the site and will help practitioners develop the goals of the restoration project. Predisturbance accounts of the site or conditions at nearby reference sites that support the suite of endemic species can be used to help guide the selection of the desired postrestoration condition (Morrison 2002; Society for Ecological Restoration International Science and Policy Working Group 2004). In some situations, practitioners will not have an accurate list of the species that once occurred in a particular

location or the relative abundance of species prior to human-induced changes, but we can use data collected within the region (historic or current) to suggest what species are likely absent from the proposed restoration site but present in nearby undisturbed sites. By comparing historic conditions (conditions that were likely present prior to anthropogenic changes) with existing conditions at the proposed restoration site, we can start to develop a detailed, quantitative list of desired postrestoration conditions (i.e., goals). For example, if reference sites or information based on conditions present prior to anthropogenic changes suggest that a bird species that relies on willows for breeding declined in abundance, then a restoration action might be to re-establish willow densities within a range that existed prior to anthropogenic changes or that exists currently at reference sites that support healthy populations of the species. Data sources that might help define conditions prior to human-induced changes include the Breeding Bird Survey, Christmas Bird Counts, naturalists' records, and museum specimens, all of which may help to identify the suite of species formerly present at the proposed restoration site.

In addition to examining predisturbance data, practitioners should also sample the existing conditions at the proposed restoration site. Descriptions of existing conditions are useful for (1) identifying species or habitat components that have been lost, (2) determining the environmental attributes that are most in need of restoration, (3) setting goals and benchmarks, and (4) examining postrestoration changes. For wildlife habitat restoration efforts, existing conditions should include surveys of the plant and animal community within the proposed restoration site. Comparisons to former survey data (if such data are available) or with nearby reference communities will help determine the environmental attributes that are most in need of restoration. For example, if no willow flycatchers (*Empidonax traillii*) are present at a proposed restoration site, but predisturbance data from the same area suggest that willow flycatchers were formerly abundant prior to anthropogenic changes, then some important habitat feature(s) are probably lacking at the proposed site. An important point to consider is why willow flycatchers no longer occur at the proposed site. Comparisons among reference sites, predisturbance conditions, and existing conditions will provide an indication of what habitat components are suboptimal. Existing condition or prerestoration data also provides the foundation from which to compare restoration success after restoration efforts have been implemented. With existing condition data, practitioners can document the "effect size" achieved by their restoration efforts (i.e., the extent to which restoration actions changed key environmental conditions).

In addition to examining predisturbance records, the niche requirements and limitations (e.g., dispersal distance) of a suite of focal species (or species of highest concern) should also be considered (Zedler 2000b; George and Zack 2001; Scott et al. 2001; Morrison 2002; Miller and Hobbs 2007). Failure to consider site-specific and species-specific habitat requirements could result in failure to meet the project's goals. For example, the project will likely not meet its goals if the proposed restoration site is not big enough to support a viable population of one of the focal species. Thus, an understanding of the niche requirements will be helpful to create an effective restoration plan (Morrison 2002; Rodewald, this volume). Additional factors to consider include landscape context and spatial factors that can affect occupancy (George and Zack 2001; Scott et al. 2001). For example, species that are sensitive to fragmentation may not occupy a restored site if the site exists in a heavily fragmented landscape. Dispersal capabilities and metapopulation structure should also be considered when designing a restoration plan (Scott et al. 2001; Morrison 2002; Menninger and Palmer 2006). Restoration of physical habitat attributes may not result in the desired outcome (colonization and persistence of one or more desired species) if dispersal barriers exit or if regional abundance is insufficient to allow for colonization of the newly restored area (Schrott et al. 2005; Menninger and Palmer 2006). For example, burning and mowing failed to increase plant species diversity in prairie remnants in Iowa because the patches were too small to support many of the focal species, and past land-use practices significantly reduced the local seed bank, limiting recruitment (Van Dyke et al. 2007). Although all factors that may influence populations cannot be assessed in every restoration project, identifying key ecological attributes and associated niche requirements will result in a more comprehensive plan that has a higher probability of enhancing habitat conditions for the suite of focal species.

The key to successful restoration is to bring together as much information as is known about the focal species and the community to create a more informed restoration plan. Although habitat restoration efforts often originate due to legislative or political impetus that focus on one focal species, we caution against focusing exclusively on a single species. Habitat restoration efforts that focus solely on individual species may have unintended consequences for the rest of the community, hence practitioners should consider other species in the community when designing a restoration plan.

Setting Goals

One of the most important components of a restoration project is the identification of well-defined, quantitative goals (Hobbs and Norton 1996; Hobbs and Harris 2001). Well-defined goals allow practitioners to communicate the importance of the restoration project to policy makers and the public as well as judge the success of restoration actions. Vague goals (e.g., "return the site to historic conditions") are subjective and provide no means by which to judge the success of restoration actions. Because "success" in restoration ecology has often been seen as a vague and subjective term (Zedler 2007), we need to include more explicit, quantitative measures with which to judge the outcome of habitat restoration efforts. Determining the extent of change for each metric that is considered biologically meaningful to the suite of focal species, and doing so prior to restoration, is one way we can reduce the subjective nature of determining the success of restoration efforts. The Society of Ecological Restoration suggests that an ecosystem is restored when the ecosystem is capable of producing a self-sustaining population of species (Society for Ecological Restoration International Science and Policy Working Group 2004).

Despite the recognized need to include metrics of success in restoration efforts, only two (Klein et al. 2007; Maslo et al. 2011) of the fifty-eight projects we reviewed provided information on the criteria used to measure the success of restoration efforts. Because few projects provided criteria to assess the success of restoration efforts, determining whether the actions improved habitat for any wildlife species was difficult. Of the fifty-eight published papers that summarized habitat restoration efforts, 62% of the studies claimed that restoration actions were successful or partially successful, although no criteria were provided by which to judge the success. We could not determine from the paper if the restoration actions were successful in 22% of the projects because not enough information was included to determine if the project met its stated goals.

Many past restoration projects, however, failed to establish goals at all. Of the fifty-eight studies we reviewed in the published literature, only 14% described an explicit goal of the restoration actions. Instead, published restoration efforts typically addressed whether restoration actions (often vaguely defined) affected wildlife species as a post-hoc assessment. Of the studies we reviewed, 50% examined the effects of restoration actions on wildlife species several years after the actions took place. Failure to consider the impacts of restoration on wildlife species at the outset can limit our ability to infer success of habitat restoration efforts. While a restored area may appear visually improved, if wildlife species that were once common (prior to disturbance) do not reestablish in the restored area, habitat restoration efforts may not have been successful. Hence, endemic wildlife species provide a benchmark from which to assess habitat restoration goals. Previous reviews of restoration projects have also highlighted the failure of many restoration efforts to identify explicit goals; 20% of river restoration projects failed to establish restoration goals (Bernhardt et al. 2005). Moreover, coordinators of restoration/restoration projects could rarely provide information about project objectives in a manner that described the desired conditions following restoration (Follstad Shah et al. 2007).

Restoration goals should be set in conjunction with the development of the desired conditions described previously. Goals should be realistic, reviewed by outside experts, and guided by the needs of the focal species (Hobbs and Norton 1996; Morrison 2002). When setting goals, practitioners should take into account the ecological constraints of the system as well as social and financial constraints that are likely to affect the project (Ehrenfeld 2000; Morrison 2002; Miller and Hobbs 2007). If, for example, the proposed restoration site is too small to support the desired species, and the area surrounding the site is nonhabitat or cannot be restored, the ecological constraints of the system will likely prevent the goals from being attained. In most cases, project goals should equate to (or at least include) the desired conditions. Clearly articulated goals based on detailed analysis of the full-range of predisturbance conditions (to the extent available), reference conditions, species-specific needs, and societal values will bolster support from policy makers and the public and will make restoration projects more likely to succeed (Hobbs 2007; Miller and Hobbs 2007).

Monitoring and Evaluating

The field of restoration ecology has primarily focused on community ecology and plant succession (Morrison 1998; Young 2000) or on restoring ecosystem function while taking an implicit approach to wildlife that can best be described as "if you build it they will come" (Palmer et al. 1997; Zedler 2000b). However, restoring an area to a former vegetative community may not always result in the restoration of habitat for endemic wildlife species (Zedler 1996, 2000a; Maslo et al. 2011). For example, prescribed fire and/or mechanical thinning treatments have been implemented on a large portion of western coniferous forests to reduce fire severity and restore forests to presettlement conditions (Moore et al. 1999; Dickson et al. 2009; Kalies et al. 2012). However, we cannot assume that these restoration projects improved habitat conditions for the suite of endemic species without an assessment of the species present. Monitoring endemic wildlife can provide an effective means to measure the extent to which a restoration project has returned the site to its natural state. If the goal of a restoration project is to return a site to its natural condition, we would expect that the suite of species associated with pine woodlands, for example, would be present, and populations would be self-sustaining. If one of the goals of a restoration effort is to improve habitat suitability for endemic wildlife, postrestoration monitoring efforts must be included as an integral part of the restoration project to document whether project goals are met. Surprisingly, restoration projects designed to benefit wildlife often fail to include pre- or postrestoration wildlife monitoring (Block et al. 2001; Bernhardt et al. 2005). For example, past reviews of restoration projects found that only 10% of river restoration projects included an assessment of any plant or wildlife species (Bernhardt et al. 2005). Although many of these river restoration projects were not designed as wildlife habitat restoration projects, we suggest that monitoring wildlife species should be incorporated into the project as another method to judge project success. Lack of pre- and postrestoration monitoring efforts greatly reduces our ability to infer whether restoration actions have been successful. Failure to incorporate a monitoring plan in conjunction with the restoration plan could also lead to inadequate or inappropriate monitoring. For example, if the restoration site is small, attempting to estimate abundance of multiple wildlife species and species richness will likely not provide enough information to assess the effectiveness of the restoration (Block et al. 2001). Thus, practitioners should determine the most appropriate monitoring metrics while developing a monitoring plan. The parameters chosen as the focus of a monitoring effort would relate directly back to a project goal. Morrison (2002) provides details regarding what to monitor and how to monitor wildlife habitat restoration projects. Most restoration projects that monitor wildlife species often record abundance of the focal species after restoration. However, practitioners may also want to consider measuring productivity as abundance can be a misleading indicator of habitat quality (Van Horne 1983). Of the fifty-eight published projects we reviewed, 81% did not incorporate measures of productivity into their monitoring plan. Measuring productivity can be costly but doing so helps to ensure that habitat restoration efforts do not merely create a habitat sink for focal species.

Many restoration efforts focus on a single site (without any spatial replication), which limits the inferences that can be made regarding the effectiveness of the restoration actions elsewhere. Hence, developing an appropriate sampling design that adequately assesses the wildlife responses to restoration is critical to improve our ability to make inferences beyond the boundaries of any one restoration site. However, some past studies have used suboptimal sampling designs to monitor responses to restoration. Of the fifty-eight publications we reviewed, 78% did not include any measurements of site conditions prior to restoration, which limits the conclusions researchers can make regarding changes in wildlife populations as a result of restoration. Often these studies measured wildlife response at a site that was enhanced at some point in the past and measured the same metrics at a control or reference site to determine if the enhanced site exhibited the same or different conditions as the control site. However, the effects of restoration actions cannot be accurately ascertained if prerestoration data are not collected and no replication exists. Other restoration projects measured conditions before and after restoration but did not measure the same conditions at control or reference sites. Lack of control sites does not allow one to account for more widespread (i.e., regional) changes that may be responsible for changes over time at the restoration site. For example, a region-wide drought may cause the abundance of the focal species to decrease at the restoration site and hence hinder one's ability to assess the effectiveness of the restoration actions if data from reference sites are not available. Although 90% of the fifty-eight publications we reviewed monitored a control site, only 19% of projects monitored the restoration and control sites both before and after restoration. Other reviews have also identified a failure to include monitoring of control sites in most restoration efforts. Less than one-third of river restoration projects in the southwestern United States included monitoring at reference or control sites (Follstad Shah et al. 2007). In another recent review of 240 studies, only 20% included prerestoration monitoring and only 58% compared changes at the restoration site to those at control sites (Jones and Schmitz 2009).

Before-after control-impact (BACI) designs are the preferred alternative for evaluating the effectiveness of restoration efforts. BACI designs, as the name implies, compare metrics before and after restoration at treated sites and control sites to determine the impact of a treatment (i.e., restoration actions). A BACI design is optimal because this sampling design allows researchers to assess preexisting conditions (useful for developing desired conditions) and to distinguish natural variation from effects caused by restoration actions. A BACI study without replication (i.e., only one restoration site) cannot make inferences beyond the specific study site but is still a great improvement over restoration efforts that do not include any prerestoration monitoring or a control site. Monitoring of restoration effectiveness is viewed by many restoration practitioners as too cost prohibitive, but monitoring provides the only way to determine whether the thousands of dollars spent on restoration actions were worthwhile. Thus, it is critical to develop a well-thought-out restoration and monitoring plan that carefully balances costs against monitoring needs. In our experiences, restoration projects either seek to monitor too many metrics that are not aligned with project goals or the metrics monitored are not sufficient. Both scenarios tell us little about the effects of restoration actions. Developing a detailed, quantitative restoration plan that undergoes external peer review by experts is key to the successful implementation of restoration activities that will likely have a positive effect on wildlife species (Morrison 2002).

Restoration projects must also consider how long to continue postrestoration monitoring efforts. Of the fifty-eight studies we reviewed, 48% monitored wildlife for two or fewer years after restoration efforts were implemented. However, the full effectiveness of many restoration actions may not be realized until years after the restoration actions are implemented (Jones and Schmitz 2009). The duration of monitoring should be long enough to account for the time it will likely require the plant community to respond to restoration actions or to produce optimal habitat conditions for the focal species. For example, if vegetation height is an important habitat component for one of the focal species, monitoring abundance of the focal species before the vegetation has had a chance to respond to restoration actions will not allow for assessment of the effectiveness of the restoration actions. A more appropriate monitoring plan would be to monitor the abiotic conditions that affect vegetation height in the first couple of years following restoration to ensure that vegetation growth is on schedule and then monitor the suite of focal wildlife species in subsequent years. Critical thresholds or triggers should also be considered when determining how long to plan postrestoration monitoring activities. For example, waiting five years after restoration actions to sample salt-marsh vegetation may be sufficient if the vegetation is not expected to reach desired condition until year five. One caveat, however, is that monitoring at only one point in time could prove costly if vegetation does not respond to restoration activities as expected and a small change soon after restoration could have fixed the problem. Thus, detailed monitoring could be based on the time interval established for specific triggers, but more qualitative assessment of habitat conditions could occur at shorter intervals to ensure that the restoration project is on target to meet the project's goals. Whatever the duration of postrestoration monitoring chosen, suc-

cess will improve if practitioners estimate the timing of anticipated responses of both vegetation and the suite of focal wildlife species to restoration actions and use those estimates to determine the duration of postrestoration monitoring efforts.

Restoration projects will be successful if they include an a priori monitoring plan with detailed descriptions of what, how, when, and how long to monitor the project site. Monitoring plans must also be adaptive. An adaptive design is often a key component to a successful monitoring plan (Block et al. 2001; Lindenmayer and Likens 2009; Block 2013). Adaptive monitoring plans include benchmarks or thresholds that trigger additional actions to improve the chances of meeting the project goals. Establishing thresholds or benchmarks will help determine whether postrestoration monitoring efforts need to be extended and may help avert a crisis if effective changes are implemented when those thresholds are not met. When developing benchmarks, an important aspect to consider is the natural range of variability within a community for each metric. Determining where each metric lies in relation to the natural variability of the metric across the area sampled can help determine whether restoration efforts are working or whether modifications are needed (Hobbs and Norton 1996).

Are We Achieving?

In 1937 the Pittman-Robertson Act made funds available for wildlife restoration. Since that time, the Pittman-Robertson Act has helped improve 41,962,211 acres of wildlife habitat, provided improvements to 604,659 acres of waterfowl impoundments, and funded thousands of research projects (US Fish and Wildlife Service 2012). But the real question is: have these improvements benefited wildlife? Has habitat restoration been successful?

In our review of habitat restoration for mammals and birds, 62% reported that restoration was at least partially successful. Several other studies have also conducted reviews of restoration projects to determine how successful we have been in restoring habitats and ecosystems. Jones and Schmitz (2009) examined 240 ecosystem recovery studies published from 1910 to 2008 and found that restoration was completely successful (all restoration goals were met) in 35% of ecosystem recovery studies examined and was partially successful (some, but not all, restoration goals were met) in another 38% of studies. In another review of restoration projects conducted in a variety of ecological communities across the globe, 44% of eighty-nine restoration projects resulted in increased measures of biodiversity (abundance, species richness, or diversity) compared to prerestoration levels but lower relative to reference sites (Rey Benayas et al. 2009). In a third review of 621 wetland restoration projects, biological structure remained 26% lower compared to reference sites even ten years postrestoration (Moreno-Mateos et al. 2012). The proportion of restoration projects that were deemed successful varied among the three reviews of restoration projects; however, determining project success is challenging and depends on the metrics used to measure success of the restoration project and the stated goals of the restoration project. Of the fifty-eight restoration projects we reviewed, 62% indicated that restoration actions improved conditions, but this number likely overestimates the percentage of successful restoration projects because unsuccessful restoration efforts are less likely to be published (Zedler 2007) and practitioners likely have a strong impetus to deem their project successful to appease their funding sources. Objective measures of project success is also difficult because few restoration efforts developed metrics to measure their successes or failures (Hobbs and Norton 1996) and few determined a priori what constitutes project success (Suding 2011). Only 8% of river restoration projects based project success on pre-determined criteria (Follstad Shah et al. 2007). Moreover, reporting and dissemination of postrestoration monitoring results is rare (Suding 2011), and many restoration projects are not published in academic journals and are therefore difficult to summarize effectively. In fact, 75% of river restoration projects did not disseminate the results of their projects (Follstad Shah et al. 2007).

Are habitat restoration projects effectively enhancing habitat suitability enough to increase the probability of occupancy for the suite of focal species? The answer to this question is project specific, but collectively the answer remains largely unknown because comprehensive assessments of restoration success and failures are uncommon (Suding 2011), and existing datasets lack the information necessary from which to judge project success (Follstad Shah et al. 2007).

Advancing Habitat Restoration

The field of restoration has been criticized for lacking coherent general principles that allow for the transfer of knowledge among practitioners (Hobbs and Norton 1996; Suding 2011). One factor that contributes to poor knowledge transfer and limits our ability to use past knowledge to create more effective restoration plans is the lack of publication of successes or failures (Zedler 2000b; Follstad Shah et al. 2007; Suding 2011). In fact, results of many restoration studies are frequently not even reported or are only available in gray literature, further limiting knowledge transfer (Follstad Shah et al. 2007; Suding 2011; Maslo et al. 2012). For example, only 25% of river restoration projects in the southwestern United States shared results with agencies, funders, or program managers (Follstad Shah et al. 2007). While increased publication of restoration efforts could help remedy this problem, a more effective solution would be to track restoration projects in a publically accessible database as many restoration projects may not be suitable for publication in peer-reviewed journals (or no one person has the time or responsibility to do so), but they do contain valuable information that can help improve future restoration efforts. A database of restoration practices that includes goals, monitoring metrics, progress, and outcomes for every restoration project will aid in tracking restoration progress as well as inform restoration practices. In an attempt to remedy the lack of accessible data on river restoration projects, Bernhardt et al. (2005) synthesized river restoration projects into the National River Restoration Science Synthesis (NRRSS) database. The NRRSS database is exactly what restoration ecology needs to move forward. This database contained information on more than 30,000 river restoration projects across the United States, including project costs, goals, monitoring metrics, and outcomes (Bernhardt et al. 2005). Unfortunately, the National Biological Information Infrastructure (NBII), which housed the database, was terminated, and the data are currently not publically accessible. A similar project called the Global Restoration Network (http://www.globalrestorationnetwork

.org/) was started by the Society for Restoration Ecology, in which users input information on their restoration projects. The database contains valuable information on monitoring, evaluation, and restoration actions. However, entering information into the database is voluntary and no entries have been added since 2009.

A database that houses information on restoration projects and is used by the restoration community is needed (Suding 2011). A restoration database (similar to the global restoration network or the NRRSS) could provide practitioners with a readily accessible source of information regarding the location of other similar restoration projects, the restoration actions that were most successful at improving habitat suitability for one's suite of focal wildlife species, and the restoration actions that had little success. A restoration database could hold information about specific restoration actions that are often not reported in peer-reviewed publications. In the fifty-eight published projects we reviewed, 21% of the projects did not report the size of the treated area, and 57% did not provide detailed information on the restoration activities performed. The database would also help keep track of the cumulative area of different vegetation communities that have been part of restoration efforts and help identify areas or communities that are restoration priorities. Focusing restoration projects in strategic locations is one mechanism that may help maximize the probability of success. The database could also include restoration projects in the planning phase, which might promote more cross-disciplinary collaboration and external peer review (and hence improvements). Databases such as this can also provide information on restoration impacts at a larger scale as many individual restoration projects are often conducted at a small scale with little spatial replication (Follstad Shah et al. 2007).

Another common problem associated with habitat restoration is the length of time required to see the results of the restoration actions. Thus, the dissemination of project information is often delayed and the scientific community has no information on how well the project is progressing. In an attempt to remedy this problem in 2009, the journal *Restoration Ecology* started including a section called "Setbacks and Sur-

prises" to begin learning from unforeseen challenges in restoration (Hobbs 2009). Restoration projects could also be updated annually in the database, to help track pitfalls and successes. A database of habitat restoration projects could provide numerous benefits; however, building a database that would be used by practitioners and keeping it updated has its challenges and will likely only be successful if practitioners are required to enter information into the database as a mandatory component of funding or mitigation contracts for restoration projects.

Wildlife habitat restoration has an increasingly important role to play in our endeavor to conserve species diversity, and the collective success of our future restoration efforts will be maximized if (1) both restoration plans and monitoring plans are well designed and summarize what is known about habitat requirements for the suite of focal wildlife species; (2) goals and objectives are clearly articulated, quantitative (i.e., easily measured), and linked directly to desired outcomes with performance criteria developed by which to judge project success; (3) restoration and associated monitoring activities are planned as part of a BACI design with both spatial and temporal controls; and (4) details and outcomes of restoration actions are readily accessible in a database to ensure that practitioners learn from past mistakes and build on the successes of past efforts. The mentality of "if we build it, they will come" gives a false sense of confidence in the outcome and needs to be replaced with carefully planned restoration projects that consider the needs of endemic wildlife species to help maintain healthy and diverse ecosystems in our ever-changing environment.

Note: Any mention of trade, firm, or product names in this chapter is for descriptive purposes only and does not imply endorsement by the U.S. government.

LITERATURE CITED

Bernhardt, E. S., M. A. Palmer, J. D. Allan, G. Alexander, K. Barnas, S. Brooks, J. Carr, S. Clayton, C. Dahm, J. Follstad-Shah, D. Galat, S. Gloss, P. Goodwin, D. Hart, B. Hassett, R. Jenkinson, S. Katz, G. M. Kondolf, P. S. Lake, R. Lave, J. L. Meyer, T. K. O'Donnell, L. Pagano, B. Powell, and E. Sudduth. 2005. "Synthesizing U.S. River Restoration Efforts." *Science* 308:636–37.

Block, W. A., A. B. Franklin, J. P. Ward, J. L. Ganey, and G. C. White. 2001. "Design and Implementation of Monitoring

Studies to Evaluate the Success of Ecological Restoration on Wildlife." *Restoration Ecology* 9:293–303.

Choi, Y. D. 2007. "Restoration Ecology to the Future: A Call for New Paradigm." *Restoration Ecology* 15:351–53.

Davis, M. A. 2000. "'Restoration'—A Misnomer?" *Science* 287:1203.

Dickson, B. G., B. R. Noon, C. H. Flather, S. Jentsch, and W. M. Block. 2009. "Quantifying the Multi-scale Response of Avifauna to Prescribed Fire Experiments in the Southwest United States." *Ecological Applications* 19:608–21.

EDAW, Philip Williams and Associates, H. T. Harvey and Associates, Brown and Caldwell, and Geomatrix. 2007. "South Bay Salt Pond Restoration Project: Final Environmental Impact Statement/Report." www.southbayrestoration.org.

Ehrenfeld, J. G. 2000. "Defining the Limits of Restoration: The Need for Realistic Goals." *Restoration Ecology* 8:2–9.

Falk, D. A., M. A. Palmer, and J. B. Zedler, eds. 2006. *Foundations of Restoration Ecology*. Island Press, Washington, D.C.

Follstad Shah, J. J., C. N. Dahm, S. P. Gloss, and E. S. Bernhardt. 2007. "River and Riparian Restoration in the Southwest: Results of the National River Restoration Science Synthesis Project." *Restoration Ecology* 15:550–62.

George, T. L., and S. Zack. 2001. "Spatial and Temporal Consideration in Restoring Habitat for Wildlife." *Restoration Ecology* 9:272–79.

Hall, L. S., P. R. Krausman, and M. L. Morrison. 1997. "The Habitat Concept and a Plea for Standard Terminology." *Wildlife Society Bulletin* 25:173–82.

Hobbs, R. J. 2007. "Restoration Ecology: Are We Making an Impact?" *Restoration Ecology* 15:597–600.

———. 2009. "Looking for the Silver Lining: Making the Most of Failure." *Restoration Ecology* 17:1–3.

Hobbs, R. J., L. M. Hallett, P. R. Ehrlich, and H. A. Mooney. 2011. "Intervention Ecology: Applying Ecological Science in the Twenty-first Century." *BioScience* 61:442–50.

Hobbs, R. J., and J. A. Harris. 2001. "Restoration Ecology: Repairing the Earth's Ecosystems in the New Millennium." *Restoration Ecology* 9:239–46.

Hobbs, R. J., and D. A. Norton. 1996. "Towards a Conceptual Framework for Restoration Ecology." *Restoration Ecology* 4:93–110.

Jones, H. P., and S. W. Kress. 2012. "A Review of the World's Active Seabird Restoration Projects." *Journal of Wildlife Management* 76:2–9.

Jones, H. P., and O. J. Schmitz. 2009. "Rapid Recovery of Damaged Ecosystems." *Plos One* 4(5): e5653.

Kalies, E. L., B. G. Dickson, C. L. Hambers, and W. W. Covington. 2012. "Community Occupancy Responses of Small Mammals to Restoration Treatments in Ponderosa Pine Forests, Northern Arizona, USA." *Ecological Applications* 22:204–17.

Klein, L. R., S. R. Clayton, J. R. Alldredge, and P. Goodwin. 2007. "Long-Term Monitoring and Evaluation of the Lower Red River Meadow Restoration Project, Idaho, USA." *Restoration Ecology* 15:223–39.

Lindenmayer, D. B., and G. E. Likens. 2009. "Adaptive Monitoring: A New Paradigm for Long-Term Research and Monitoring." *Trends in Ecology & Evolution* 24:482–86.

Maslo, B., J. Burger, and S. N. Handel. 2012. "Modeling Foraging Behavior of Piping Plovers to Evaluate Habitat Restoration Success." *Journal of Wildlife Management* 76:181–88.

Maslo, B., S. N. Handel, and T. Pover. 2011. "Restoring Beaches for Atlantic Coast Piping Plovers (*Charadrius melodus*): A Classification and Regression Tree Analysis of Nest-Site Selection." *Restoration Ecology* 19:194–203.

Menninger, H. L., and M. A. Palmer. 2006. "Restoring Ecological Communities: From Theory to Practice." In *Foundations of Restoration Ecology*, edited by D. A. Falk, M. A. Palmer, and J. B. Zedler. Island Press, Washington, D.C.

Millennium Ecosystem Assessment. 2005. *Ecosystems and Human Well-being: Synthesis*. Island Press, Washington, D.C.

Miller, J. R., and R. J. Hobbs. 2007. "Habitat Restoration— Do We Know What We're Doing?" *Restoration Ecology* 15:382–90.

Moore, M. M., W. W. Covington, and P. Z. Fule. 1999. "Reference Conditions and Ecological Restoration: A Southwestern Ponderosa Pine Perspective." *Ecological Applications* 9:1266–77.

Moreno-Mateos, D., M. E. Power, F. A. Comin, and R. Yockteng. 2012. "Structural and Functional Loss in Restored Wetland Ecosystems." *Plos Biology* 10(1): e1001247.

Morrison, M. L. 1998. "Letter to the Editor." *Restoration Ecology* 6:133.

———. 2001. "Introduction: Concepts of Wildlife and Wildlife Habitat for Ecological Restoration." *Restoration Ecology* 9:251–52.

———. 2002. *Wildlife Restoration: Techniques for Habitat Analysis and Animal Monitoring*. Island Press, Washington D.C.

Morrison, M. L., B. G. Marcot, and R. W. Mannan. 1998. *Wildlife-Habitat Relationships*. University of Wisconsin Press, Madison.

Palmer, M. A., R. F. Ambrose, and N. L. Poff. 1997. "Ecological Theory and Community Restoration Ecology." *Restoration Ecology* 5:291–300.

Rey Benayas, J. M., A. C. Newton, A. Diaz, and J. M. Bullock. 2009. "Enhancement of Biodiversity and Ecosystem Services by Ecological Restoration: A Meta-analysis." *Science* 325:1121–24.

Schrott, G. R., K. A. With, and A. W. King. 2005. "Demographic Limitations of the Ability of Habitat Restoration to Rescue Declining Populations." *Conservation Biology* 19:1181–93.

Scott, T. A., W. Wehtje, and M. Wehtje. 2001. "The Need for Strategic Planning in Passive Restoration of Wildlife Populations." *Restoration Ecology* 9:262–71.

Society for Ecological Restoration International Science and Policy Working Group. 2004. "The SER International Primer on Ecological Restoration." www.ser.org.

Suding, K. N. 2011. "Toward an Era of Restoration in Ecology: Successes, Failures, and Opportunities Ahead." *Annual Review of Ecology, Evolution, and Systematics* 42:465–87.

Thorpe, A. S., and A. G. Stanley. 2011. "Determining Appropriate Goals for Restoration of Imperilled Communities and Species." *Journal of Applied Ecology* 48:275–79.

US Army Corps of Engineers and South Florida Water Management District. 2006. "Comprehensive Everglades Restoration Plan." www.evergladesplan.org.

US Fish and Wildlife Service. 2012. "Wildlife Restoration Program—Accomplishments." http://wsfrprograms.fws.gov /subpages/grantprograms/WR/WR_Accomplishments.htm.

Van Dyke, F., J. D. Schmeling, S. Starkenburg, S. H. Yoo, and P. W. Stewart. 2007. "Responses of Plant and Bird Communities to Prescribed Burning in Tallgrass Prairies." *Biodiversity and Conservation* 16:827–39.

Van Horne, B. 1983. "Density as a Misleading Indicator of Habitat Quality." *Journal of Wildlife Management* 47:893–901.

Young, T. P. 2000. "Restoration Ecology and Conservation Biology." *Biological Conservation* 92:73–83.

Zedler, J. B. 1996. "Coastal Mitigation in Southern California: The Need for a Regional Restoration Strategy." *Ecological Applications* 6:84–93.

———, ed. 2000a. *Handbook for Restoring Tidal Wetlands*. CRC Press, Boca Raton, Florida.

———. 2000b. "Progress in Wetland Restoration Ecology." *Trends in Ecology & Evolution* 15:402–7.

———. 2007. "Success: An Unclear, Subjective Descriptor of Restoration Outcomes." *Ecological Restoration* 25:162–68.

— CONCLUSION —

MICHAEL L. MORRISON AND
HEATHER A. MATHEWSON

Synthesis for Advancing Useful Knowledge of Habitat

Unifying Themes or Many Directions?

As we noted in the preface, the purpose of this book is to deliver to a broad audience an understanding of current thoughts on the multitude of impacts influencing wildlife and wildlife habitats and provide recommendations for a path forward that will advance management and conservation. *Habitat* is certainly one of the most frequently used terms in animal and plant ecology (Mathewson and Morrison, this volume). Elementary school students are introduced to the term, and there is unlikely to be a hunter or nature explorer who does not read and use the term frequently. This widespread use of habitat is both a blessing and a curse. It is a blessing because a broad swath of people have some notion of what it is, which opens up ample opportunity for education and hence conservation. Unfortunately, it is also a curse because the scientific community has simply refused to come to any meaningful conclusions on what habitat is and how we should study, manipulate, and, ultimately, conserve it.

Hence, we come back to our aim of writing a book accessible to a "broad audience." In the preface, we defined this audience as being composed of advanced undergraduate and graduate students in natural resource management and conservation (e.g., wildlife, range, conservation biology, recreation and parks); resource managers at local, state, and federal levels (e.g., state wildlife and parks departments, USFWS, BLM, Army Corps); and private land managers. Although this sounds like a difficult goal to achieve, we approached the task knowing that once the scientific community settled on at least a basic set of principles concerning habitat, this "broad audience" would at least be hearing a somewhat similar message (we say "somewhat" because we know scientists just prefer to keep their options open to disagree at any time). Although habitat is an ecological concept and thus not within a strict framework directly amenable for hypothesis testing per se, having scientists wandering around using *habitat* in substantially different contexts does not in any way advance ecological understanding and, ultimately, translation into meaningful management options.

Thus, for this closing chapter we went through each chapter and looked for common messages on habitat. That is, are these authors converging on a common understanding of habitat and how it should be thought about and studied? Was there a commonality in the priority research needs expressed? What is the priority list of research needs? Here, we first briefly review with the intent of relating common sections within each chapter. We then finish by discussing the implications the authors of this book presented for advancing our understanding of animal ecology.

Messages

In the subsections that follow, we briefly highlight common messages we gleaned from across the chapters. We are not attempting to review all topics that were raised but rather focus on topics that were raised as important issues in multiple chapters and thus in different contexts.

Definitions

Guthery and Strickland clearly identified a problem with use of the "habitat term" when they summarized the many uses of the term in the scientific literature. It seems that perhaps in the majority of cases, even the author appears to be unsure of the meaning of the term they used. As has been recommended for decades, Guthery and Strickland urged authors to clearly define their intended meaning in a paper and then use that meaning throughout the paper (we will come back to the issue of definitions in the following discussion).

A theme throughout many of the chapters, raised early on by Mathewson and Morrison, and Guthery and Strickland, is the need to clearly elucidate the theoretical relationship being assumed between what is being described or predicted (dependent variable) and the parameters chosen for measurement; that is, the habitat variables. Making this connection between animal and environment should force the author to explain both the parameters and the spatial and temporal scales relevant to the question(s) being asked.

In the "classical" view of habitat, Guthery and Strickland seek to explain where an organism lives, which includes biotic and abiotic requisites and time. They continue on to note that while this classical version of habitat is logically a dead-end concept, except for descriptive science, defining a minimal set of factors that would describe an animal's habitat in the classical sense is not a trivial activity. Thus, Guthery and Strickland go on to explain, as has been done for decades (Johnson 1980), that a hierarchical approach can be adopted where a largely arbitrary area (level 1; e.g., study area) with (demographically, behaviorally) meaningful subdivisions (level 2) can be subdivided (level 3) and perhaps further subdivided (level 4). Guthery and Strickland explain that the hierarchical view of habitat represents sequential parsing; namely, the study of an entity composed of parts that are in turn composed of parts and so on. They concluded that this approach provides a natural means of studying how animals partition time and occurrence in the third-level units that we call habitats (or as many call "habitat types").

Populations and Quality

Rodewald discussed the general failure to explicitly link considerations of habitat with individual and population-level measures of habitat quality, which can then obscure identification of the highest-quality habitat. She also noted that habitat studies have a heavy bias toward a single life stage or season, usually the breeding season.

Likewise, Guthery and Strickland noted that studies have documented a relationship between recruitment and adult survival, which provides at least some support for spatial or temporal differences in habitat (subcomponent) quality (for a defined area and time). However, they also note that many studies ignore the dilemma of density dependence, where in sustaining populations, increases in productivity entail decreases in survival and vice versa. Van Horne and Wiens carried a similar message, clearly identifying that time lags, legacy effects, or carryover effects also erode the match between habitat occupancy and habitat quality.

The discussions by Mathewson and Morrison, Guthery and Strickland, Rodewald, and others in this book identify a clear theme where our studies of habitat quality, while informative and an advancement over classical descriptive habitat studies, provide us with an incomplete picture at best—and an incorrect picture in many cases—of what factors drive population persistence through space and time. These issues bring us back to the topic of the arbitrary nature of most study areas (see also McKelvey chapter) and the uncertainty associated with the portion of the biological population we chose to study. The issue of identifying the biological population of study has been thoroughly discussed (e.g., see Morrison 2012 and review therein) but has yet to translate into general application in field studies. The chapters in this book go a step further in emphasizing both the spatial and temporal components of habitat quality determination, as well as the related and critical topic of potential density dependence. Van Horne and Wiens continue this theme, noting that it has long been known that the density of a species in an area can influence the range of habitats occupied. When habitat occupancy varies with density, the characteristics of the area at a particular time will be sensitive to the species density at that time, which in turn might not say much about the relationships with habitat that influence individual fitness.

Rodewald's emphasis on separating individual- versus population-level concepts of habitat quality is critical and relates directly back to the initial designation of the biological population being studied and the linked spatial scale of study that such work should entail. As reviewed by Morrison (2012), most studies pick a convenient study area (e.g., a park, ranch, or portions thereof) without any stated regard for the biological population being sampled (this is in contrast to stating the statistical population, which usually has little or no meaningful relationship to the biological population). Clearly, individuals make up the biological population. As noted by Morrison (2012), it is how we sample from within that biological population that holds the key to making our studies meaningful in a broad ecological and, ultimately, management-conservation sense.

Heterogeneity

The issue of habitat fragmentation is woven throughout many chapters (e.g., Smallwood, Van Horne and Wiens, Waits, Francis, Lockwood and Burkhalter) and certainly qualifies as a major threat to the persistence of many plant and animal species. However, conceptualizations of habitat fragmentation and corridors vary widely in the literature, and the science supporting these concepts needs to improve (see Smallwood chapter). The issues of spatial and temporal scales are closely linked with how we view fragmentation. Smallwood concluded that quantitative tests of fragmentation and corridor effectiveness should be conducted that measure habitat variables rather than using overly simplistic maps of vegetation cover types; the effectiveness of corridors is highly species specific.

As detailed by Smallwood, definitions of habitat fragmentation and corridors need to be more explicit. Scientists need to more effectively convey their concepts of habitat fragmentation and corridors to biologists in natural resource agencies and environmental consulting firms, as well as to attorneys, political decision makers, and the public. These concepts really matter where decisions are being made about whether and how to further fragment habitat, and how to mitigate the effects of this fragmentation. These decisions are being made now and involve very large areas of wildlife habitat.

A frequent theme throughout this book is the potential and real impacts of fragmentation on animal populations, ranging from genetic changes to outright extinction. Smallwood focused specifically on fragmentation and related topics, while most of the other chapter authors intertwined fragmentation into the topic they were emphasizing. For example, biological invasions by both plants and animals are substantially promoted by fragmentation (Lockwood and Burkhalter); fragmentation can result in a decrease in effective population size, increases in genetic drift and loss of genetic diversity, while increased isolation of fragments leads to decreases in gene flow and loss of genetic diversity (Waits and Epps); reduction in patch size is a frequent outcome of human-related development activities (Francis); and, of course, fragmentation is a major issue in the development of habitat restoration plans (Borgmann and Conway).

Smallwood reviewed the fundamental problem we have in often not knowing why the animals we counted or observed were located where we found them. The aggregations we encounter can be temporary based on resource availability, based on social interactions, constrained because of patch size, and so forth. Clearly, these issues substantially impact (i.e., bias) how we view habitat use.

Applications

Schwartz et al. explained why one must take care in understanding the underlying biological distribution of the entity (species, group of species, an index) of interest when designing a monitoring plan. It is not sufficient to simply "monitor," because the pattern that results over time might not reflect the actual changes in the biological entity of interest.

Borgmann and Conway (see also Rodewald chapter) invoked the "if you build it, they will come" view of restoration and management when discussing the ways that population dynamics can be influenced by individual behaviors related to habitat selection, because such a view fails to recognize that a diverse suite of environmental, biological, and social cues are used by animals to select habitats (e.g., conspecific attraction). Thus, in cases where cues for settlement are lacking, locations that otherwise would be suitable for settlement would not be occupied. This theme runs through many chapters. For example, Van Horne and Wiens noted that it is frequently assumed that places where a species occurs contain good habitat and places where it is absent do not. It has long been known, however, that the density of a species in an area can influence the range of habitats occupied. Issues of source-sink, ecological traps, and related concepts are additional considerations.

Themes

We did not begin this project with an a priori expectation of what the authors would say, and we certainly did not expect that some grand answer would emerge from the chapters (or that we could even recognize it). We, like everyone reading this book, understand to varying degrees of specificity that ecology is typified by interactions, and any understanding thereof requires a synthesis of a lot of information. We are not sure if a grand answer did emerge, but we are confident that a grand direction did emerge. Although all of the authors were provided drafts of each chapter, this was done so they could cross-reference related topics as they deemed appropriate; there was no attempt for them to all get their stories straight. In writing this closing chapter, we went through each chapter and looked for major conclusions in hopes of finding a number of themes based on a similarity in conclusions; the result was synthesized and summarized earlier. It became apparent very quickly that there was, indeed, a central theme that ran across all of the chapters. As would be expected in the field of ecology, the theme is multiparted, with each subpart inexorably linked; part of the theme is based on biology, while the other part is based on study design. We can summarize the primary theme as:

Advancing understanding of animal ecology (and thus habitat) requires that we identify and study a biological population in appropriate space and time.

This primary theme might seem obvious, but after reading the chapters in this book as well as other related literature (much of which was cited across the chapters), we are confident that most of our studies pay little or (usually) no attention to the biological population when establishing the study. Only after the study is done and one must justify publication is an attempt made to justify extrapolation of results outside the study area.

This theme also directly links our conceptual understanding of how animals select a place to settle (i.e., habitat) with study design and thus study area identification. Every chapter in this book discussed spatial (and temporal) scale in one or more contexts, emphasizing that the process of habitat selection is widely viewed as a hierarchical process wherein individuals use a series of cues to decide ultimately where to be and when to be there, what to eat, where to nest, and so forth; Guthery and Strickland explicitly detailed this process. Although this model of habitat selection is widely held, it is seldom used as a basis for selection of a sampling area; naturally this issue relates directly to our previous discussion of identification of the biological population. Hence, this issue of the process of habitat selection is a subpart of our discussion of identification of the biological population.

A secondary but related theme deals directly with habitat quality. Much attention across the chapters was devoted to explaining how measures of quality, be they recruitment or individual survival, must be conducted on appropriate spatial (i.e., usually larger than we study) and temporal (i.e., usually longer than we study) scales. The topic of habitat quality was discussed throughout the chapters, with issues raised concerning what metrics to measure (e.g., recruitment, survival), the interacting issue of density dependence and its influence on quality conclusions, and the interacting issues of study area size and duration. The issue of spatial scale relates directly to the aforementioned major theme, while the issue of temporal scale also becomes a study design conundrum. Ultimately, we want to know how populations persist in space and time. We can summarize the embedded, secondary theme as:

Advancing understanding of the biological population requires quantification of those metrics that explain population persistence.

Terminology

Guthery and Strickland recommend that because habitat is a concept and thus intrinsically ambiguous, authors should explicitly define their intended meaning and then unambiguously apply that meaning throughout the paper. They thought this process would achieve the goal of authors and readers having a common understanding of meaning. Morrison and Hall (2002) also recommended such an approach for defining habitat on a paper-by-paper basis. We think, however, that such a procedure—while certainly a useful fallback strategy—is not the optimal way to advance how we view and study habitat. Defining terms does allow readers to know what is meant, but it does not mean that anything about the study (and thus paper) was done in a manner that actually adds to our knowledge about how and why animals occur where they do.

Thus, we think recommending definitions, while necessary, is far from sufficient. Rather, we think that the necessary procedure is for each author to explicitly develop and justify how they identified the biological population, how study design decisions were made based on that population, and what ramifications on results these decisions were likely to express. Although many will complain that we are setting the bar too high, we are simply asking that authors discuss the biological population to the best of their ability. Journal editors frequently reject manuscripts because of few or small study areas; such manuscripts are often then published in an appropriate regional journal. All we are asking is that authors be transparent. Morrison (2012) discussed these issues in detail, including considerations of subspecies and especially ecotypes as a basis for study design. In this book, Rodewald succinctly summarizes the steps we need to take to move beyond traditional approaches to studying wildlife-habitat relationships. Fortunately, we now have a number of genetic tools (see Waits chapter) that allow us to more easily explore issues of population structure and limits. There remains a role for what Guthery and Strickland termed "classical" habitat studies. However, we think that failure to recognize and adopt the issues (themes) raised herein will doom us to slow and halting progress in understanding population persistence.

Implications

The ultimate conservation implications of these themes are clear and directly related to how we design and manage land with the aim of population persistence. Specification of a biologically relevant study area relates directly to planning for reserves and other management-conservation areas. As Van Horne and Wiens and McKelvey explained, because of natural succession, climate change, or processes or threats moving across the reserve boundary, conditions within a management area with small and fixed boundaries will likely change in ways that cannot be managed for the benefit of the species. Because most studies are small in scope, the generalizations they produce are based on compositing these small-scale primary studies through literature reviews and meta-analyses. It is difficult, however, even with many largely unrelated small-scale studies of specific areas, to generate coherent understandings of general relationships between an organism and its environment. McKelvey clearly showed that because the condition of any patch of land, regardless of size, changes substantially though time, our usual strategy of creating conservation areas that are fixed in location is usually doomed to failure; or as Van Horne and Wiens pointed out, incredibly expensive to maintain in a desired state.

LITERATURE CITED

Johnson, D. H. 1980. "The Comparison of Usage and Availability Measurements for Evaluating Resource Preference." *Ecology* 61:65–71.

Morrison, M. L. 2012. "The Habitat Sampling and Analysis Paradigm has Limited Value in Animal Conservation: A Prequel." *Journal of Wildlife Management* 76:438–50.

Morrison, M. L., and L. S. Hall. 2002. "Standard Terminology: Toward a Common Language to Advance Ecological Understanding and Application." In *Predicting Species Occurrences: Issues of Accuracy and Scale*, edited by J. M. Scott, P. J. Heglund, M. L. Morrison, J. B. Haufler, M. G. Raphael, W. A. Wall, and F. B. Samson, 43–52. Island Press, Washington, D.C.

Index